T0344896

Advances and Applications in Electroceramics

Advances and Applications in Electroceramics

Ceramic Transactions, Volume 226

Edited by
K. M. Nair
Quanxi Jia
Shashank Priya

A John Wiley & Sons, Inc., Publication

Copyright © 2011 by The American Ceramic Society. All rights reserved.

Published by John Wiley & Sons, Inc., Hoboken, New Jersey.
Published simultaneously in Canada.

No part of this publication may be reproduced, stored in a retrieval system, or transmitted in any form or by any means, electronic, mechanical, photocopying, recording, scanning, or otherwise, except as permitted under Section 107 or 108 of the 1976 United States Copyright Act, without either the prior written permission of the Publisher, or authorization through payment of the appropriate per-copy fee to the Copyright Clearance Center, Inc., 222 Rosewood Drive, Danvers, MA 01923, (978) 750-8400, fax (978) 750-4470, or on the web at www.copyright.com. Requests to the Publisher for permission should be addressed to the Permissions Department, John Wiley & Sons, Inc., 111 River Street, Hoboken, NJ 07030, (201) 748-6011, fax (201) 748-6008, or online at http://www.wiley.com/go/permission.

Limit of Liability/Disclaimer of Warranty: While the publisher and author have used their best efforts in preparing this book, they make no representations or warranties with respect to the accuracy or completeness of the contents of this book and specifically disclaim any implied warranties of merchantability or fitness for a particular purpose. No warranty may be created or extended by sales representatives or written sales materials. The advice and strategies contained herein may not be suitable for your situation. You should consult with a professional where appropriate. Neither the publisher nor author shall be liable for any loss of profit or any other commercial damages, including but not limited to special, incidental, consequential, or other damages.

For general information on our other products and services or for technical support, please contact our Customer Care Department within the United States at (800) 762-2974, outside the United States at (317) 572-3993 or fax (317) 572-4002.

Wiley also publishes its books in a variety of electronic formats. Some content that appears in print may not be available in electronic formats. For more information about Wiley products, visit our web site at www.wiley.com.

Library of Congress Cataloging-in-Publication Data is available.

ISBN: 978-1-118-05999-9
ISSN: 1042-1122

oBook ISBN: 978-1-118-14448-0
ePDF ISBN: 978-1-118-14445-9

Printed in the United States of America.

10 9 8 7 6 5 4 3 2 1

Contents

Preface

New areas of materials technology development and product innovation have been extraordinary during the last few decades. Our understanding of science and technology of the electronic materials played a major role in meeting the social needs by developing innovative devices for automotive, telecommunications, military and medical applications. There is continued growth in this area and electronic technology development has enormous potential for evolving societal needs. Rising demands will lead to development of novel ceramics materials which will further the market for consumer applications. Miniaturization of electronic devices and improved system properties will continue during this decade to satisfy the requirements in the area of medical implant devices, telecommunications and automotive markets. Cost-effective manufacturing should be the new area of interest due to the high growth of market in countries like China and India. Scientific societies should play a major role for development of new manufacturing technology by working together with international counterparts.

The Materials societies understand their social responsibility. For the last many years, The American Ceramic Society has organized several international symposia covering many aspects of the advanced electronic material systems by bringing together leading researchers and practitioners of electronics industry, university and national laboratories. Further, The American Ceramic Society has been aggressive in knowledge dissemination by publishing the proceedings of the conferences in the Ceramic Transactions series, a leading up-to-date materials publication, and posting news releases on its website.

This volume contains a collection of 26 papers from four symposia that were held during the 2010 Material Science and Technology Conference (MS&T'10) held at the George R. Brown Convention Center, Houston, Texas, USA, October 17-21, 2010. These symposia include: "Advanced Dielectric Materials and Electronic Devices," "Magnetoelectric Multiferroic Thin Films and Multilayers," "Recent Developments in High Temperature Superconductivity," and "Multifunctional

Oxides." MS&T'10 was jointing sponsored by The American Ceramic Society (ACerS), the Association of Iron & Steel Technology (AIST), ASM International, the Minerals, Metals & Materials Society (TMS) and the National Association of Corrosion Engineers (NACE).

We, the editors, acknowledge and appreciate the contributions of the speakers, co-organizers of the four symposia, conference session chairs, manuscript reviewers and Society officials for making this endeavor a successful one.

K. M. NAIR, *E.I. duPont de Nemours & Co., Inc., USA*
QUANXI JIA, *The Center for Integrated Nanotechnologies, Los Alamos National Laboratory, USA*
SHASHANK PRIYA, *Virginia Technical Institute & State University, USA*

SYMPOSIA ORGANIZERS

Dielectric Materials & Electronic Devices
K.M. Nair, E.I.DuPont de Nemours & Co., Inc
Danilo Suvorov, Jozef Stefan Institute
Ruyan Guo, University Texas at San Antonio
Rick Ubic, Boise State University
Amar S. Bhalla, University of Texas

Ferroelectrics and Multiferroics
Shashank Priya, Virginia Technical Institute & State University
Paul Clem, Sandia National Laboratories
Chonglin Chen, University of Texas at San Antonio
Dwight Viehland, Virgina Technical Institute & State University
Armen Khachaturyan, Rutgers University

High Temperature Superconductors
Haiyan Wang, Texas A&M University
Quanxi Jia, Los Alamos National Laboratory
Siu-Wai Chan, Columbia University
Timothy J. Haugan, US Air Force Research Lab
M. Parans Paranthaman, Oak Ridge National Laboratory
Venkat Selvamanickam, University of Houston

Multifunctional Oxides
Quanxi Jia, Los Alamos National Laboratory
Pamir Alpay, University of Connecticut
Chonglin Chen, University of Texas at San Antonio
Amit Goyal, Oak Ridge National Laboratory
Xiaoqing Pan, University of Michigan

Dielectric Materials and Electronic Devices

NUMERICAL SIMULATIONS OF A BACK GRINDING PROCESS FOR SILICON WAFERS

A.H. Abdelnaby[1], G.P. Potirniche[1], F. Barlow[2], B. Poulsen[1], A. Elshabini[2], R. Parker[3], T. Jiang[3]
[1]Department of Mechanical Engineering; University of Idaho
PO Box 440902,
Moscow, ID, USA
[2]Department of Electrical and Computer Engineering; University of Idaho
PO Box 440902,
Moscow, ID, USA
[3]Micron Technologies Inc.
8000 S. Federal Way,
Boise, ID, USA

ABSTRACT

The optimization of grinding parameters for silicon wafers is necessary in order to maximize the reliability of electronic packages. This paper describes the work performed to simulate a back grinding process for silicon wafers using the commercial finite element code ABAQUS. The silicon wafer analyzed had a thickness of 120 μm and was mounted on a backing tape. The wafer was thinned to a thickness of 96 μm, by simulating the grinding with a diamond particle cutting through successive silicon layers. The modeled residual stresses induced in the wafer were compared with experimental data, and they were shown to agree well. A shear band of intense plastic deformation with a certain orientation angle was generated in the specimen, and the value of this angle was compared with experimental data for similar materials. The numerical model developed can be used to better understand the local conditions in wafers during this back grinding process.

INTRODUCTION

The development of electronic packages is based on strict weight and size requirements. Smaller, lighter, and higher capacity devices at low cost are what consumers demand. In order to achieve this goal, electronic packaging plays a major role in this industry. The silicon wafer thickness affects the package size, thus, the thinner the wafer the lower the overall package height. The manufacturing process of wafers faces many challenges. One of these challenges is the thinning process, which is performed during the back grinding of the wafer. Significant research efforts have been devoted to the development and improvement of this process. Most of the studies have been experimental or analytical in nature, with few analyses considering numerical simulations that studied the wafer-wheel behavior[1-4]. In order to better understand the details of the wafer grinding process; a micron scale study is needed to clarify the internal stresses, strains, and deformations that take place into the wafer material while and after the grinding process.

In this paper a numerical study is performed to simulate the grinding process at the micron level in order to understand the stresses and deformations developed in the wafer during this process[5,6]. The model developed showed good correlation with the experimental data measured using Raman spectroscopy for similar silicon wafers

NUMERICAL MODEL

The goal of this study is to simulate the back grinding process of the silicon wafer in order to be able to measure the stress field at different locations in the wafer, and to achieve a good understanding of the grinding process. The numerical simulations involved varying operating parameters affecting this manufacturing process to determine the optimum operating conditions.

To simulate the grinding process the commercial finite element software ABAQUS, Explicit was used. In the literature, there are different finite element grinding models, which can be categorized by the scale of the modeling approach as macro-scale or micro-scale models. Macro-scale models consider the overall wheel–workpiece interaction, which captures the aggregate effects of the abrasive wheel on the workpiece with no attempts to study the effect of the individual abrasive grain on the workpiece[7]. The micro-scale models focus on the individual grain–workpiece interactions, which can examine the actual material removal mechanism. Thus, micro-scale models have the potential to allow the estimation of the grinding forces directly without resorting to measurements or empirical modeling. This model simulates the micro-scale grinding process, which includes the effect of a single diamond crystal (abrasive grain) while it removes layers from the silicon wafer (workpiece) as shown in figure 1.

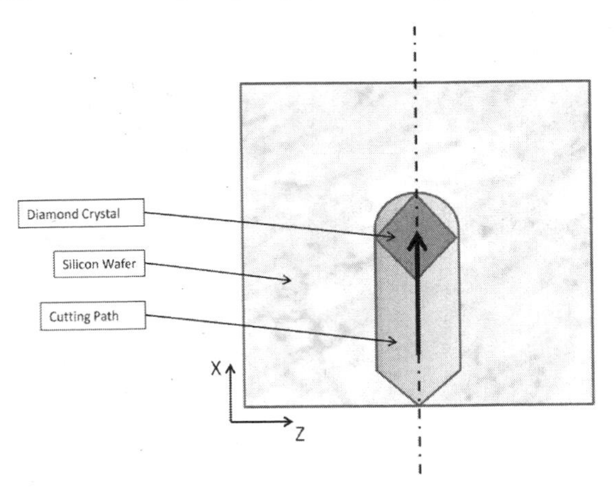

Figure 1. Micro-scale grinding model.

A two-dimensional model was built to simulate a part of the silicon wafer and a small particle of diamond which moves and cuts through successive silicon layers. In the present simulations, the number of the cutting passes was chosen as 12. The cutting depth of each pass was set at 2 μm, where we found a realistic cutting depth for the grinding process. After the completion of the 12 cutting passes the model was allowed a relaxation time. The model considers only a part of the silicon wafer which is completely attached to a plastic backing film Poly Ethylene Terephthalate (PET) as shown in figure 2.

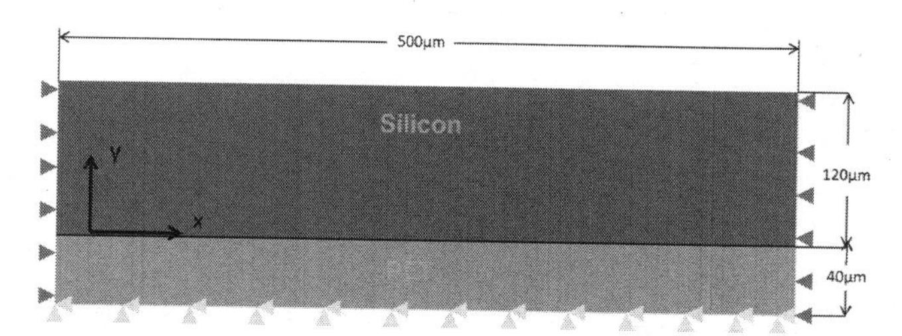

Figure 2. Finite element model used to simulate the micro-scale grinding.

The back grinding process of the silicon wafers is performed in three steps. First is the coarse grinding, the second step is the intermediate grinding, which involves the thinning of the wafer to a thickness of approximately 85 μm. This step is the focus of our study. The third and final step is the fine grinding, when the wafer thickness is reduced to the desired value.

The boundary conditions shown in figure 2 were set as:

1) The bottom of the model is pinned to simulate the attachment to the vacuum chuck.
2) The sides of the model are symmetry boundary conditions along the X-axis to simulate the effect of the remaining parts of the wafer and the adhesive tape on each side which have not been included in the model.
3) The diamond crystal was displaced in the X direction to simulate the cutting action and returning back, and was displaced in the Y direction to simulate the cutting depth.

An initial mesh with a global element size of 5 μm was used for both the silicon and PET material. The mesh was further refined to a global element size of 1 μm. Finally a more refined mesh of 0.5 μm global element size was used for the silicon wafer, and a mesh with a larger element size of 20 μm was used for the PET adhesive tape. This mesh size was used to extract the data presented in this paper.

The model consists of two materials, silicon as the wafer and PET as the backing tape materials. Silicon <100> single crystal at room temperature is a hard and brittle material. As the working temperature is kept at 23 °C by the effect of the coolant during the grinding process, the stress-strain curve for the silicon at 25 °C was used to model the material behavior as elastic-plastic with a damage criterion[8]. A damage criterion and an element deletion scheme have been used to simulate the material removal during grinding.

ELASTIC-PLASTIC AND DAMAGE MODEL

Because of the crystalline nature of the silicon <100>, orthotropic elasticity was chosen to model the elasticity of the material. Linear elasticity in an orthotropic material can be defined by nine independent elastic stiffness parameters. These parameters can be functions of temperature and other predefined fields, such as the strain rate. In our model there is no effect of the temperature due to the cooling process that maintain the wafer at 23°C, and the model is not strain rate dependent. In this case the stress (σ)-strain (ε) relations take the form shown in equation 1, with the constants D^{el} shown in table I[9].

$$\begin{Bmatrix} \sigma_{11} \\ \sigma_{22} \\ \sigma_{33} \\ \sigma_{12} \\ \sigma_{13} \\ \sigma_{23} \end{Bmatrix} = \begin{pmatrix} D_{1111} & D_{1122} & D_{1133} & 0 & 0 & 0 \\ D_{1122} & D_{2222} & D_{2233} & 0 & 0 & 0 \\ D_{1133} & D_{2233} & D_{3333} & 0 & 0 & 0 \\ 0 & 0 & 0 & D_{1212} & 0 & 0 \\ 0 & 0 & 0 & 0 & D_{1313} & 0 \\ 0 & 0 & 0 & 0 & 0 & D_{2323} \end{pmatrix} \begin{Bmatrix} \varepsilon_{11} \\ \varepsilon_{22} \\ \varepsilon_{33} \\ \gamma_{12} \\ \gamma_{13} \\ \gamma_{23} \end{Bmatrix} \quad (1)$$

An isotropic plasticity model with a von-Mises hardening criterion was used to simulate the plasticity of the silicon material. Silicon does not deform significantly in the plastic region before damage onset and fracture. The plasticity of the silicon was modeled by building the effective stress-strain curve using the data from table 2.

Table I. D^{el} matrix constants

D_{1111}	165800 MPa
D_{1122}	63740 MPa
D_{2222}	165800 MPa
D_{1133}	63740 MPa
D_{2233}	63740 MPa
D_{3333}	79620 MPa
D_{1212}	63740 MPa
D_{1313}	63740 MPa
D_{2323}	63740 MPa

A shear damage initiation criterion with an element deletion scheme was used to simulate the material removal that occurs due to the grinding process. The shear damage criterion predicts the onset of damage due to shear band localization, and it is used in conjunction with the von Mises plasticity model[10,11].

Table II. Plasticity constants

Stress MPa	Plastic Strain
1000	0
1100	0.005
1200	0.007

The shear damage criterin model assumes that the equivalent plastic strain at the onset of damage $\bar{\epsilon}_s^{pl}$ is a function of the shear stress ratio θs, and strain rate $\dot{\bar{\epsilon}}^{pl}$.

$$\theta s = \frac{(q + k_s p)}{\tau_{max}} \tag{2}$$

where τ_{max} is the maximum shear stress, k_s is a material parameter, which was set to 0.3, q is the equivalent stress, and p is the pressure stress. The criterion for damage initiation is met when the following condition is satisfied:

$$w_s = \int \frac{d\,\bar{\epsilon}^{pl}}{\bar{\epsilon}_s^{pl}\left(\theta s, \dot{\bar{\epsilon}}^{pl}\right)} = 1 \tag{3}$$

where w_s is a state variable that increases monotonically with the plastic deformation and is proportional to the incremental change in equivalent plastic strain. At each increment during the analysis the incremental increase in w_s is computed as:

$$\Delta w_s = \frac{\Delta\,\bar{\epsilon}^{pl}}{\bar{\epsilon}_s^{pl}\left(\theta s, \dot{\bar{\epsilon}}^{pl}\right)} \geq 0 \tag{4}$$

PET is a hyper-elastic polymeric material used as a thin layer to model the tape on which the wafer is mounted, before both wafer and tape are mounted on the vacuum chuck as one assembly. In order to simulate the behavior of this material, the Moony-Rivlin model for hyperelastic materials was used

$$U = C_{10}\left(\bar{I}_1 - 3\right) + C_{01}\left(\bar{I}_2 - 3\right) + \frac{1}{D_1}\left(J^{el} - 1\right)^2 \tag{5}$$

where U is the strain energy per unit of reference volume C_{10}, C_{01}, and D_1 are temperature-dependent material parameters; $\bar{I}_1$ and $\bar{I}_2$ are the first and second deviatoric strain invariants which are defined as:

$$\bar{I}_1 = \bar{\lambda}_1^2 + \bar{\lambda}_2^2 + \bar{\lambda}_3^2 \tag{6}$$

and

$$\bar{I}_2 = \bar{\lambda}_1^{-2} + \bar{\lambda}_2^{-2} + \bar{\lambda}_3^{-2} \tag{7}$$

where $\bar{\lambda}_1$, $\bar{\lambda}_2$ and $\bar{\lambda}_3$ are the three principal stretches, and J^{el} is the elastic volume ratio, which relates the total volume ratio to the thermal volume ratio. The constant D_1 was set to a large number to eliminate the effect of the thermal expansion because of the effect of the cooling process. The C_{10}, C_{01} constants were chosen 0.3. These values were found from literature by curve fitting experimental data with the Mooney-Rivlin strain energy potential model used by ABAQUS[12].

RESULTS

An intense shear deformation band was observed forming at an angle of 30° with the surface of the specimen. A similar angle, called the shear angle (φ), has been reported in the literature in reference to the machining of metals[13]. Comparing the angle value that was observed in the model and the data from the literature, a similar behavior can be found as in the case of other brittle materials, as shown in figures 3 and 4. Figure 3 also shows surface elements that experienced damage on the surface after machining The damage of the surface elements is due to intense stresses induced by the grinding operation on the surface of the specimen, which forms surface cracks and the usual surface roughness observed in real wafers. Surface cracks and roughness are removed in the fine grinding process. Figure 4 illustrates the stress distribution in the x-direction and the shear stress band in the model.

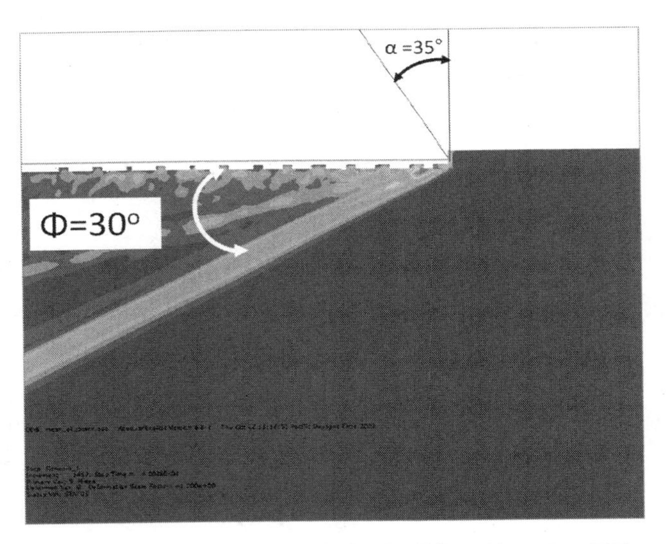

Figure 3. Shear angle φ as observed in the simulation with a value of 30°.

The residual stress induced in the model after grinding was investigated in two ways: first by comparing the stresses at the back of the wafer after grinding with litrature data, and then by comparing the stress distribution through the depth of the wafer with experimental data.

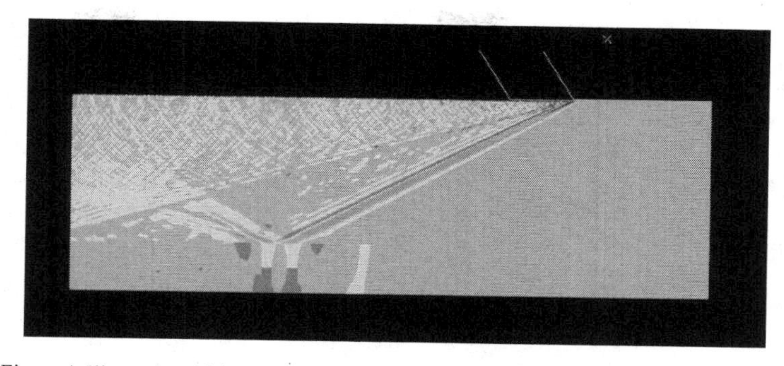

Figure 4. Illustration of the stress distribution in the horizontal direction during the cutting operation and the shear stress band.

The residual stress at the back of the wafer after grinding was found from the literature to be ranging between -100 MPa and -150 MPa [14]. This data was origionaly measured using Raman spectroscopy for grounded wafers. These residual stress values match the simulation results as obseved in figure 5.

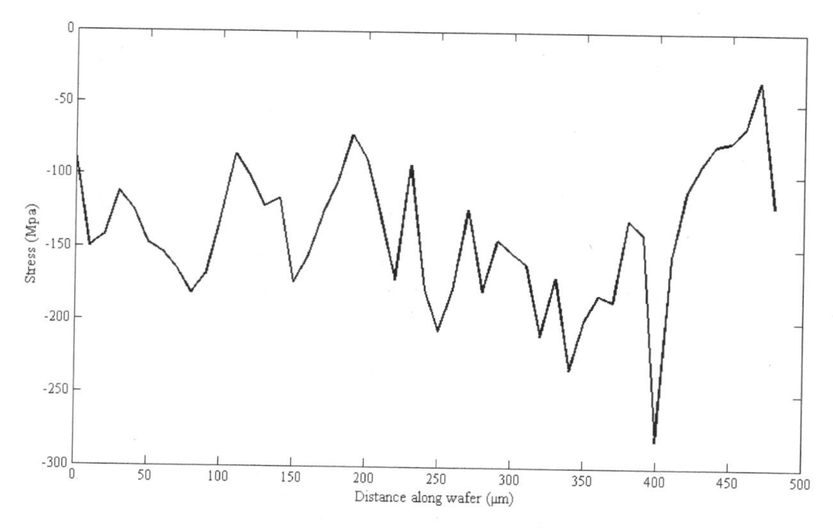

Figure 5. Simulated stress distribution at the back of the wafer after grinding.

The second verification of the residual stresses induced in the model was observed by comparing the simulation results of the stress distribution through the wafer cross section, at a depth of up to 20μm, with Raman spectroscopy data. This data obtained from the literature for the same type of measurement, shows a good agreement between the simulation results and the Raman spectroscopy measurement as shown in figure 6[14].

The graph shows a good agreement between the model results and the measurements at the center of the wafer for a depths grater than or equal to 4 μm. At depths less than 4 μm, there is a small discrepancy between the model and experimental results, which will be investigated in the future.

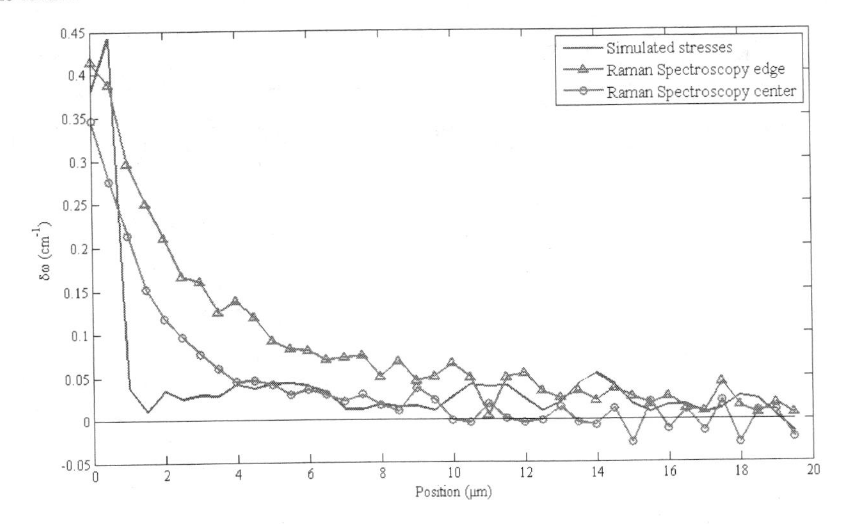

Figure 6. Stress distribution comparisons between model output and Raman spectroscopy

CONCLUSIONS

The continuing development of electronic devices leads to slim, cheap and high capacity packages. The development of such products increases the need of thinner wafers, which are achieved by optimized grinding processes. Experimental studies are expensive and can significantly increase the wafer price. In this paper we built a model that simulates the back grinding processes of a silicon wafer. The accuracy of the model was verified by measuring the shear angle in the cutting operation and comparing it with literature data. Good agreement was observed between the model results and the literature data. The residual stresses induced in the back side of the wafer after the grinding process also agreed with Raman spectroscopy data, after releasing the wafer from the vacuum chuck and the adhesive backing tape. An additional verification showed a good agreement between the stress distributions in the cross-section of the wafer that was measured by Raman spectroscopy with the model output.

The model is able to accurately simulate the back grinding process of silicon wafers, and it can be used to develop a better understanding of the parameters affecting the grinding process.

ACKNOWLEDGMENT

The authors would like to thank the Micron Foundation for their financial support of this work.

REFERENCE

[1]Z. Pei, G.R. Fisher, and J. Liu, "Grinding of silicon wafers: A review from historical perspectives," *International Journal of Machine Tools and Manufacture*, vol. 48, Oct. 2008, pp. 1297-1307.

[2]S. Chidambaram, Z.J. Pei, and S. Kassir, "Fine grinding of silicon wafers: a mathematical model for grinding marks," *International Journal of Machine Tools and Manufacture*, vol. 43, Dec. 2003, pp. 1595-1602.

[3]Y. Lin and S. Lo, "A study of a finite element model for the chemical mechanical polishing process," *The International Journal of Advanced Manufacturing Technology*, vol. 23, 2004, pp. 644-650.

[4]C.A. Chen and L. Hsu, "A process model of wafer thinning by diamond grinding," *Journal of Materials Processing Technology*, vol. 201, May. 2008, pp. 606-611.

[5]X. Chen and W. Brian Rowe, "Analysis and simulation of the grinding process. Part II: Mechanics of grinding," *International Journal of Machine Tools and Manufacture*, vol. 36, Aug. 1996, pp. 883-896.

[6]E. Ng and D.K. Aspinwall, "Modelling of hard part machining," *Journal of Materials Processing Technology*, vol. 127, Sep. 2002, pp. 222-229.

[7]D. Doman, A. Warkentin, and R. Bauer, "Finite element modeling approaches in grinding," *International Journal of Machine Tools and Manufacture*, vol. 49, Feb. 2009, pp. 109-116.

[8]W. N. Sharpe, Jr., K. J. Hemker , R. L. Edwards, "Mechanical properties of MEMS materials," AFRL-IF-RS-TR-2004-76, Final Technical Report, March 2004.

[9]F. Ebrahimi and L. Kalwani, "Fracture anisotropy in silicon single crystal," *Materials Science and Engineering A*, vol. 268, Aug. 1999, pp. 116-126.

[10]Hooputra, H., H. Gese, H. Dell, and H. Werner, "A Comprehensive Failure Model for Crashworthiness Simulation of Aluminium Extrusions, " *International Journal of Crashworthiness*, vol. 9, no.5, pp. 449–464, 2004

[11]Abaqus, Simulia, version 6.8, 2008.

[12]C. F. G. Gerlach, F. P. E. Dunne, "Modeling the influence of filler particles on surface geometry in drawn PET films," Journal of strain analysis, vol. 31, no 1, 1996.

[13]A.G. Atkins, "Modelling metal cutting using modern ductile fracture mechanics: quantitative explanations for some longstanding problems," *International Journal of Mechanical Sciences*, vol. 45, Feb. 2003, pp. 373-396.

[14]J. Chen, I. De Wolf, "Study of Damage and Stress Induced By Backgrinding in Si Wafers," *Semiconductor Science and Technology*, Vol. 18, pp. 261-268, 2003.

SOL-GEL PROCESSING OF SINGLE PHASE $BiFeO_3$ CERAMICS: A STRUCTURAL, MICROSTRUCTURAL, DIELECTRIC, AND FERROELECTRIC STUDY

L. F. Cótica, P. V. Sochodolak, V. F. Freitas, I. A. Santos
Multiferroic and Multifunctional Materials & Devices Group - Department of Physics – Universidade Estadual de Maringá
Maringá-PR. Brazil

D. Garcia, J. A. Eiras
Ferroelectric Ceramics Group – Department of Physics – Universidade Federal de São Carlos
São Carlos – SP. Brazil

ABSTRACT

Multiferroic magnetoelectric are materials that present potential applications where the ferroelectric and magnetic ordered materials have been used. Among these materials, the $BiFeO_3$ is a very promising candidate. Due to the difficult to synthesize this material, nanometric grains are desirables in a powder preparation route. In this sense, the sol-gel synthesis can carry out the requests above mentioned. In this work, a detailed sol-gel process to obtain the $BiFeO_3$ powders and ceramics was studied. A structural and a microstructural study was conducted through X-ray diffraction and scanning electron microscopy. Crystallographic parameters were refined by carrying out a Rietveld analysis. The diffraction peaks were assigned as single $BiFeO_3$ phase. The electron micrographs indicated no compositional fluctuations and that dense ceramics were obtained. The dielectric measurements (impedance spectroscopy) showed a transition near 300 °C. Typical $BiFeO_3$ saturation values and high resistivity shape curves were obtained by ferroelectric hysteresis.

INTRODUCTION

Ferroelectromagnetic materials have attracted much academic and technological attention in the last years.[1] This is because they present potential applications in those areas where (anti)ferroelectric and (anti)ferromagnetic materials are extensively employed.[2,3] In this way, the electric and magnetic order parameter coupling opens the possibility of the integration between the ferroelectromagnetics physical properties through the magnetoelectric effects,[4] and can promote interesting technological advances in many electro-electronic technologies, in spite of the open issues concerning to the origin of the ferroelectromagnetism.[5,6]

Due to their potential applications, the $BiFeO_3$ has reawakened the interest of researchers because it has electrical and magnetic properties simultaneously, i. e., the magnetoelectric properties. In fact, this compound presents ferroelectric ($T_C \sim 830$ °C) and antiferromagnetic orderings (canted weak ferromagnetism – $T_N \sim 370$ °C),[7] which can be explored in many technological applications.

Therefore, many researchers have been developing procedures for the synthesis of the $BiFeO_3$ compound. These procedures involve solid state reaction and chemical routes. The main problem, frequently encountered in the synthesis of $BiFeO_3$, is the presence of unwanted phases such as $Bi_2Fe_4O_9$ and $Bi_{25}FeO_{40}$, which may be due to the time or the atmosphere of calcinations, the purity of precursors, the stoichiometry of the system, among other factors.[8]

According to the phase diagram for the Bi_2O_3-Fe_2O_3 system, proposed by Palai *et al.*,[9] single phase $BiFeO_3$ is only obtained when exactly the same equimolar proportions of Fe and Bi atoms is reached. In this phase diagram there are three possible $BiFeO_3$ states (α, β and γ states). Up to 825 °C the compound has a ferroelectric phase (α), above this temperature the compound undergoes a α - β transition (ferroelectric – paraelectric and rhombohedral – orthorhombic transitions). Above 925 °C, until 933 °C, occurs a β - γ transition corresponding to orthorhombic – cubic transition.

Among the researches that wanted to synthesize $BiFeO_3$ compounds, Fruth *et al.*,[10] discuss the sol-gel formation of $BiFeO_3$ and the influence of precursors and calcination conditions. The obtained material presents several unwanted phases. Palkar *et al.*,[11] using pulsed laser deposition, discussed the synthesis of oxygen controlled stoichiometry $BiFeO_3$ thin films. The single phase was just obtained after washing the powder with dilute nitric acid. This was also an alternative found by Kumar *et al.*[12] They synthesized a powder by solid state reaction and obtained the desired single phase only after washing with nitric acid. However the nitric acid washing of $BiFeO_3$ powders to remove unwanted phases provide coarser powders and the reproducibility of the process is quite poor.

Owing to the unwanted phases, it is difficult to obtain single phase $BiFeO_3$ bulk ceramics. The unwanted phases and low resistivity lies to high leakage current, which make the observation of intrinsic saturated ferroelectric hysteresis loop difficult and is an obstacle to application of $BiFeO_3$ ceramics in practical devices.[12-14] Alternative synthetic routes have been pursued to prepare single phase BiFeO3 ceramics, and some saturated ferroelectric hysteresis loops have been observed in those ceramics.[13,14] High resistivity single phase $BiFeO_3$ ceramics were obtained by hot pressed sinterization, solid state reaction followed by either leaching impurity phases, rapid liquid phase sintering and high heating rate sinterization followed by rapid cooling.[11, 13,15]

An alternate to obtain single-phase $BiFeO_3$ ceramics with high resistivity is to synthesize the ceramics from fine (nanometric) $BiFeO_3$ powders. Sol-gel derived fine powders are usually more homogeneous and reactive than those prepared by conventional solid-state reaction since the mixing of the reagents occurs on atomic-level scale. Thus, in this paper, single phase high resistivity $BiFeO_3$ ceramics were obtained by high heating rate sinterization followed by rapid cooling from sol-gel synthesized nanometric $BiFeO_3$ powders. The structural, microstructural, and electric properties of the obtained powders and ceramics were carefully studied.

EXPERIMENTAL

The $BiFeO_3$ powders were obtained using $Bi(NO_3)_3.9H_2O$ (Aldrich), $Fe(NO_3)_3.9H_2O$ (Aldrich) and NH_4OH (Vetec). Solutions of 0.5 M $Bi(NO_3)_3.9H_2O$ and 0.5 M $Fe(NO_3)_3.9H_2O$ were prepared by dissolving the precursors in distilled water. To precipitate the sol-gel $BiFeO_3$ precursors, the NH_4OH solution was peptized from individual separator funnels into a 1:1 molar ratio Bi:Fe mixture with continuous magnetic stirring to homogenize the system. After precipitation, the precursor was filtered and the cake-like precipitate was washed with distilled water. After processing, the mixture was oven-dried at 80 °C overnight. After that, the dried powders were annealed at 700 °C, in an air atmosphere, during 1h.

Structural, microstructural and chemical studies in annealed powders were conducted through X-ray Diffraction (DRX), Scanning Electron Microscopy (SEM) and Energy Dispersive X-ray Spectroscopy (EDS). The XRD measurements were conducted in a Shimadzu XRD7000 diffractometer, with a Cu Kα radiation. The SEM/EDS micrographs/measurements were obtained using a Shimadzu SuperScan SS-550 microscope. The crystal structure and the crystallographic parameters were refined through Rietveld method. As a refinement tool and reference for consultation, it was used, respectively, the FullProff software[16] and the ICDD-JCPDS database.

The $BiFeO_3$ single phase powders were isostactically cold-pressed (50 MPa) in disc format, and sintered in a high heating rate (30 °C/s) – at 850 °C during 20 minutes in air atmosphere – followed by rapid cooling. The obtained bulk ceramics were characterized by XRD, SEM/EDS, ferroelectric hysteresis loops measurements (performed in a modified Sawyer-Tower circuit), and dielectric measurements (impedance spectroscopy – 4291A Agilent impedance analyzer).

RESULTS AND DISCUSSIONS

Figure 1 shows the Rietveld refined XRD pattern for the 700 oC/1h air heat treated BiFeO3 powder. The refinement assigned the pattern as being single-phase $BiFeO_3$ (R3c space group). The lattice parameters obtained from refinement are $a = 5.5771(3)$ Å and $c = 13{,}871(9)$ Å. In the Rietveld refinement in the sintered ceramics (not shown), a very low level (< 2%) of undesired phases takes place. The lattice parameters of the $BiFeO_3$ compound (R3c space group) are $a = 5.6112(5)$ Å and $c = 13.949$ (5) Å.

The SEM analysis of the 700 °C/1h air heat treated $BiFeO_3$ powder (Figure 2 (a)) showed a very sharp distribution of ~200 nm in diameter spherical particles. The EDS measurements showed no compositional fluctuations, in accordance with the XRD characterization. In the $BiFeO_3$ ceramics (Figure 2 (b)), the SEM micrograph showed an anomalous grain growth with a relatively higher porosity. The relative ceramic density, obtained from the ratio between the theoretical density (ρth,—obtained from calculated unit cell parameters) and the apparent one (ρapp – obtained by Archimedes principium in distilled water), was calculated as being 90 %, which is an acceptable value considering the observed porosity level.

Figure 3 shows the ferroelectric hysteresis loop for $BiFeO_3$ ceramic at 100 Hz and 300 K. The ceramic present typical ferroelectric behavior with remnant polarization of 3.4 $\mu C/cm^2$ and coercive field of 2.09 kV/cm. This behavior can be security attributed to bulk polarization, attesting samples quality indirectly indicating their higher electrical resistivity (1 MΩ).

Focusing the dielectric behavior of the $BiFeO_3$ ceramic, the temperature dependence of the loss factor (tan δ) was collected at 100 kHz. It can be observed a transition near 300 °C besides a loss, which can be probably associated to the higher electrical conductivity, causing higher leakage current. This result corroborates with the SEM ones, where a relatively large porosity, coupled to relatively lower relative density, was observed. Because the low level of the undesired phases have been observed in sintered ceramic (XRD analysis), further experiments and investigations are necessary to reveal the origin of the high loss in the obtained ceramic, and also to synthesize $BiFeO_3$ ceramics appropriated to practical applications.

In summary, the $BiFeO_3$ multiferroic ceramic was synthesized through sol-gel powders, isostactically cold-pressed, followed by a high heating rate and rapid cooling. Even though, further investigations should be conducted to produce ceramics with characteristics interesting for practical applications.

CONCLUSIONS

In this paper were presented the results of detailed structural, microstructural, ferroelectric and dielectric studies in the BiFeO3 powders and ceramics obtained from a sol-gel route with a high heating rate sintering. The samples were investigated by X-ray diffraction, scanning electron microscopy, ferroelectric hysteresis loop measurements, and dielectric measurements (impedance spectroscopy). Analysis of the crystallographic data revealed a single-phase BiFeO3 powder and a low level (< 2%) of unwanted phases in the ceramic body. Typical ferroelectric behaviour was observed in hysteresis loop measurements. The dielectric measurements (impedance spectroscopy) showed a transition near 300 °C.

Further investigations should be conducted to produce ceramics with characteristics interesting for practical applications.

ACKNOWLEDGEMENTS
The authors would like to thank the CAPES (Proc. 082/2007), CNPq (proc. 307102/2007-2 and 302748/2008-3), and Fundação Araucária de Apoio ao Desenvolvimento Científico e Tecnológico do Paraná (Prot. 15727) Brazilian agencies for their financial support. We also gratefully acknowledge the instrumental research facilities provided by COMCAP/UEM.

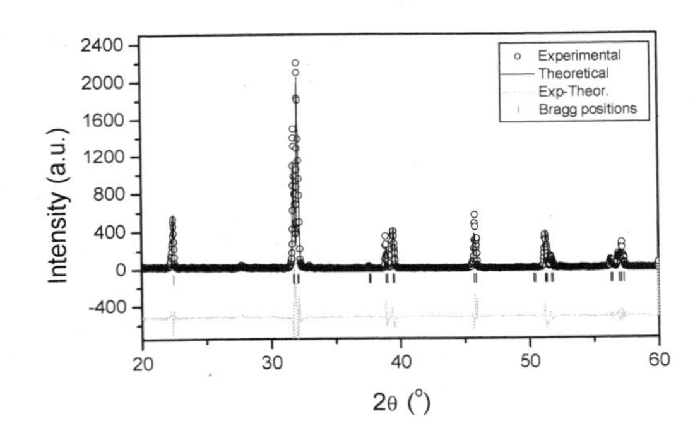

Figure 1. Rietveld refined X-ray diffraction patterns for the 700 oC/1h air heat treated BiFeO3 powder.

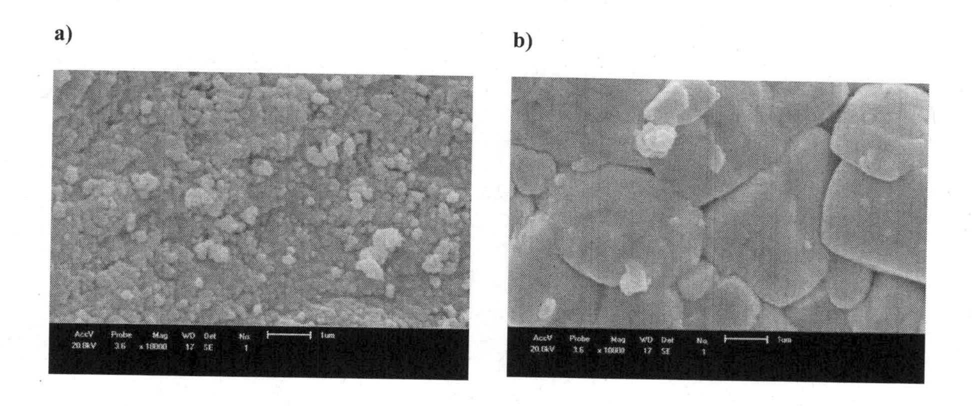

Figure 2 - SEM micrographs for (a) the 700 °C/1h air heat treated $BiFeO_3$ powder and (b) $BiFeO_3$ ceramics.

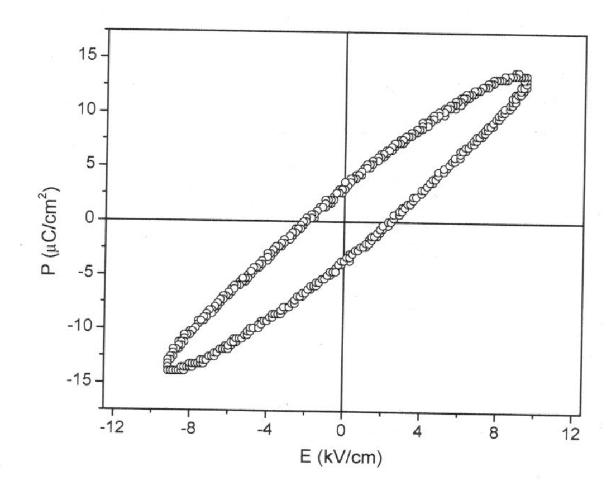

Figure 3 - ferroelectric hysteresis loop for $BiFeO_3$ ceramic

REFERENCES

[1]M. Fiebig, Th. Lottermoser, D. Fr¨ohlich, A. V. Goltsev, and R. V. Pisarev, Observation of coupled magnetic and electric domains, *Nature*, **419**, 818–820 (2002). And references therein.
[2]J.-M. Liu, Q. C. Li, X. S. Gao, Y. Yang, X. H. Zhou, X. Y. Chen, and Z. G. Liu, Order coupling in ferroelectromagnets as simulated by a Monte Carlo method, *Phys. Rev. B*, **66**, 054416-1–11 (2002). And references therein.
[3]N. A. Hill, Why Are there so few magnetic ferroelectrics?, *J. Phys. Chem. B*, **104**, 6694–6709 (2000). And references therein.
[4]K. Ueda, H. Tabata, and T. Kawai, Coexistence of ferroelectricity and ferromagnetism in BiFeO3–BaTiO3 thin films at room temperature, *Appl. Phys. Lett.*, **75**, 555–557 (1999).
[5]M. M. Kumar, A. Srinivas, and S. V. Suryanarayana, Dielectric relaxation in Ba0.96Bi0.04Ti0.96Fe0.04O3, *J. Appl. Phys.*, **84**, 6811–6814 (1998).
[6]M. M. Kumar, A. Srinivas, and S. V. Suryanarayana, Electrical and dielectric properties in double doped BaTiO3 showing relaxor behavior, *J. Appl. Phys.*, **86**, 1634–1637 (1999).
[7]J. De Sitter, C. Dauwe, E. De Grave, A. Govaert, and G. Robbrecht, On the magnetic properties of the basic compounds in the Fe2O3-Bi2O3 system, *Physica B*, **86–88**, 919–920 (1977).
[8]T.P. Gujar, V.R. Shind, C.D. Lokhande, Nanocrystalline and highly resistive bismuth ferric oxide thin films by a simple chemical method, *Mater. Chem. Phys.*, **103**, 142–146(2007).
[9]R. Palai, R. S. Katiyar, H. Schmid, P. Tissot, S. J. Clark, J. Robertson, S. A. T. Redfern, G. Catalan, J. F. Scott, β phase and γ-β metal-insulator transition in multiferroic $BiFeO_3$, *Phys. Rev. B*, **77**, 014110 (2008).
[10]V. Fruth, E. Tenea, M. Gartner, M. Anastasescu, D. Berger, R. Ramer, M. Zaharescu, Preparation of $BiFeO_3$ films by wet chemical method and their characterization, *J. of the Eur. Ceram. Soc.*, **27**, 937 (2007).
[11]V. R. Palkar, J. John, R. Pinto, Observation of saturated polarization and dielectric anomaly in magnetoelectric $BiFeO_3$ thin films, *Appl. Phys. Lett.*, **80**, 1628 (2002).
[12]M. M. Kumar, V. R. Palkar, K. Srinivas, S. V. Suryanarayana, Ferroelectricity in a pure $BiFeO_3$ ceramic, *Appl. Phys. Lett.*, **76**, 2764 (2000).
[13]Y. P. Wang, L. Zhou, M. F. Zhang, X. Y. Chen, J. M. Liu, Z. G. Liu, Room-temperature saturated ferroelectric polarization in $BiFeO_3$ ceramics synthesized by rapid liquid phase sintering, *Appl. Phys. Lett.*, **84**, 1731 (2004).
[14]S. T. Zhang, M. H. Lu, D. Wu, Y. F. Chen, N. B. Ming, Larger polarization and weak ferromagnetism in quenched $BiFeO_3$ ceramics with a distorted rhombohedral crystal structure, *Appl. Phys. Lett.*, **87**, 262907 (2005).
[15]W. N. Su, D. H. Wang, Q. Q. Cao, Z. D. Han, J. Yin, J. R. Zhang, Y. W. Du, Large polarization and enhanced magnetic properties in $BiFeO_3$ ceramic prepared by high-pressure synthesis, *Appl. Phys. Lett.*, **91**, 092905 (2007).
[16]J. Rodriguez-Carvajal, Recent advances in magnetic structure determination by neutron powder diffraction, *Physica B*, **192**, 55 (1993).

ELECTRO CERAMIC PROPERTIES OF POROUS SILICON THIN FILMS ON P-TYPE CRYSTALLINE SILICON

Faruk Fonthal

Advanced Materials for Micro and Nanotechnology Research Group
Facultad de Ingeniería, Universidad Autónoma de Occidente
Calle 25 No. 115 – 85 vía Cali - Jamundi, Valle del Lili, Cali, Colombia
e-mail: ffonthal@uao.edu.co

ABSTRACT

This paper presents the temperature dependence study in the electrical characterization of porous silicon/p-Si heterojunction. Four different Au/porous silicon/Au samples are examined in order to investigate the conduction mechanisms. The Porous Silicon layers were prepared by electrochemical etching in p-type silicon <100> substrates. The resistance-temperature characteristics were described by the thermistor equation. The values obtained for the thermistor constant were in the range of 2553 K – 5570 K at room temperature, which leads to temperature sensitivity of 2.9 – 6.3 %/K. The distribution of tail and deep states in the fabricated samples were qualitatively discussed.

INTRODUCTION

Since the discovery of the porous silicon (PS) photoluminescence at room-temperature [1], the interest of scientists in this material has grown considerably. A number applications based on PS devices have been reported, such as temperature sensor [2], humidity sensor [3], gas sensor [4] and optical sensors [5].

In electronic devices based on PS compatibility with silicon technology, it is important understand the physical mechanisms governing the electrical behaviour in the metal/PS structure. The electrical properties of these devices are strongly dependent not only on the properties of porous silicon, also on the quality of the metal/PS or PS/Si interfaces [6-9].

This paper presents the electrical characterization of four Au/PS/Au devices fabricated under different conditions. Next, it presents the temperature dependence analysis of the resistance in terms of the thermistor parameters. Finally, we estimate the density of states of the PS layer.

EXPERIMENTS

P–type silicon wafers with <100> orientation and resistivities of 4-7 Ω•cm and 7-9 Ω•cm were used as starting material. Aluminium 0.5 μm-thick film was sputter deposited on the backside of the wafer to provide a backside ohmic contact for the electrochemical etching.

The wafer was mounted in an electrochemical cell with the front side in contact with the electrolyte. Then, a positive voltage was applied between the backside aluminium contact and a platinum counter electrode. The PS layers were formed by electrochemical etching in ethanoic HF electrolyte (ethanol:HF 50%, 1:1 volume ratio). The applied etching current density was fixed at 5 mA/cm^2 during 90 s or 180 s. The etching conditions for the four fabricated samples are summarized in Table I.

After removing the samples from the etching cell, thin Au spots were evaporated on the porous silicon layer in order to obtain the top electrical contacts. The diameter of the spots was 1.5 mm and the separation between the centers of the spots was 2.5 mm.
An HP 4145B parameter analyzer was used to measure the DC electrical conductivity in range of ±2 V and temperature range since 300 K to 393 K.

RESULTS AND DISCUSSION

Figure 1a shows the current–voltage (I-V) characteristics of Au/PS/Au structures for the four fabricated samples, in the range of ±2 V at room temperature and in the Figure 1b shows the I-V characteristics of the Au/PS/Au structure for sample B at ±2 V in the range of temperature 300 K – 393 K. The calculated average slopes in the log-log scale are 1.17, which implies that we can assume a quasi ohmic behavior for all samples studied. Only the sample C presents a saturation region at high forward voltage bias.
The higher resistance values were observed for the samples fabricated at lower etching times (samples B and D), i.e. thinner PS layer. When the etching time was doubled (samples A and C), the resistance decreases up to two orders of magnitude which seems to indicate that the measured resistance not only depends on the PS layer but also on the Au/PS interface.
Figure 2(a) shows the temperature dependence of the resistance for the four Au/PS/Au devices, measured from 300 K to 393 K at 0 V. For a temperature T given, the measured resistance R follows the equation:

$$R = R_a \exp\left(- \frac{Ea}{kT}\right) \qquad (1)$$

where R_a is the preexponential factor, k is the Boltzmann constant and Ea is the activation energy. The linear dependence in the Arrhenius plot means that the process is thermally activated. The calculated activation energies are 0.28 eV, 0.31 eV, 0.22 eV and 0.48 eV for samples A, B, C and D, respectively.

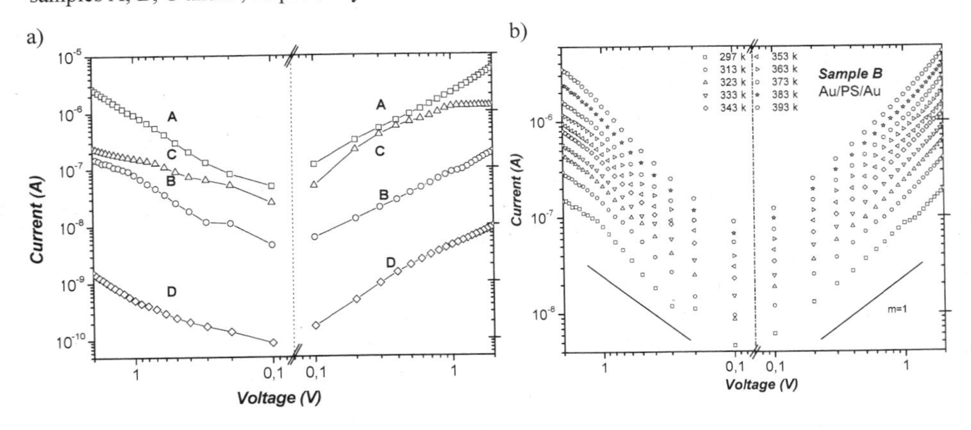

Figure 1. Current-voltage characteristics of the Au/PS/Au devices studied. (a) The four samples at room temperature and (b) Temperature dependence of Sample B, inside shows the schematic structure of the Au/PS contact.

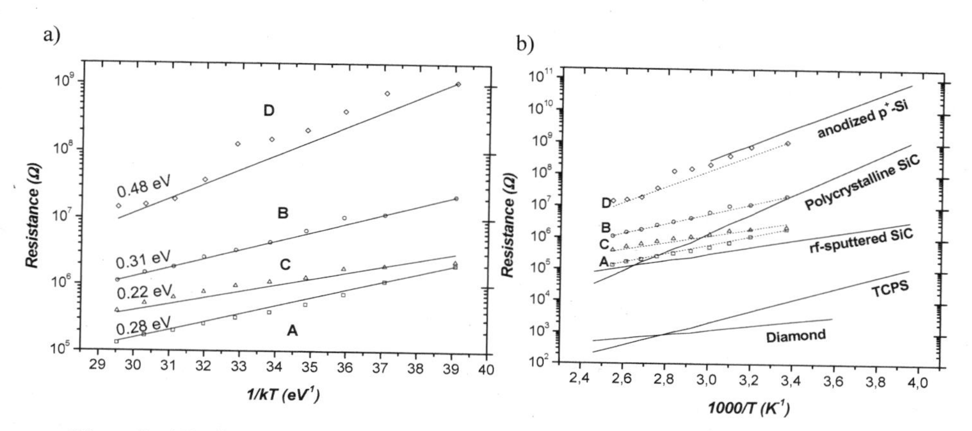

Figure 2. a) Resistance-temperature dependence for the four thermistors studied, at 0 V. Thermistor equation fitted (line) to the measured values (symbol) at room temperature. b) Comparison of the results (symbols) presented in this work with other references (solid lines): thermally carbonized PSi [2], undoped polycrystalline SiC [10], anodised p^+ Si [11], a doped diamond [12] and RF sputtered SiC [13].

Figure 2(b) shows the resistance-temperature characteristic of the four devices as well as the result obtained in other works: Salonen et al [2] for thermally carbonized porous silicon, de Vasconcelos et al [10] for undoped polycrystalline cubic silicon carbide SiC, Ben-Chorin et al [11] for the anodized p^+-Si resistance, Bade et al [12] for doped diamond and Wasa et al for RF-sputtered SiC [13].

The resistance-temperature characteristic of the Au/PS/Au devices can also be described by the thermistor equation:

$$R = R_0 \exp\left[B\left(\frac{1}{T} - \frac{1}{T_0} \right) \right] \tag{2}$$

where R_0 is the measured resistance at T_0=300 K and B is the thermistor constant. In addition it is possible to define the temperature coefficient TC of the resistance as:

$$TC\left(\%/_K \right) = \frac{Ea}{T^2 k \left(1 - \frac{T}{T_0} \right)} \tag{3}$$

The extracted values for these parameters are shown in Table I. The thermistor constant ranges between 2553 K and 5570 K. The temperature coefficient was in the range of 2.9 – 6.3 %/K, in good agreement with [2].

Table I. Fabrication Parameters and thermistor coefficients at 0V and room temperature.

Sample	Resistivity	Etching time (s)	R_0 (MΩ)	B (K)	TC (%/K)
A	4-7 Ω cm	180	2.0	3249	3.7
B	4-7 Ω cm	90	21.6	3597	4.1
C	7-9 Ω cm	180	2.3	2553	2.9
D	7-9 Ω cm	90	1080	5570	6.3

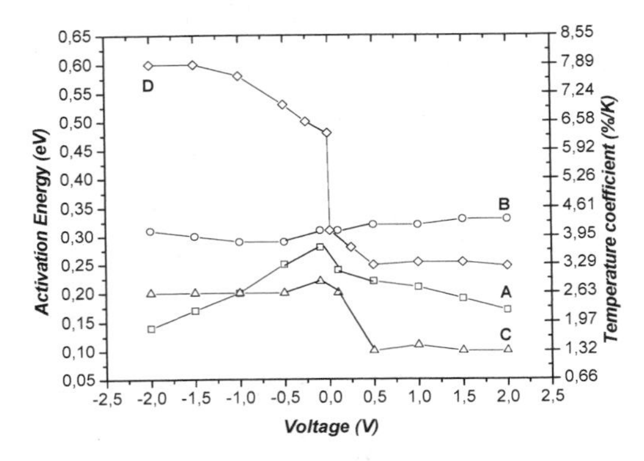

Figure 3. The relationship of Activation energy and Temperature coefficient (%/K) versus voltage characteristics the four Au/PS/Au thermistors studied, at room temperature.

Figure 3 shows the voltage dependence of the activation energy for the four samples. We found that the activation energy for the samples A, B and C does not change significantly with the applied voltage, so we can assume that the Fermi level is pinned around a certain energy level. On the other hand, the activation energy for the sample D increases from 0.25 eV at 2 V to 0.60 eV at -2 V. This voltage dependence suggest that in this sample the Fermi level changes with the applied voltage without limitations.
Because the activation energy changes with the applied voltage, it is possible to calculate the density of states, DOS, for the samples by following the step by step method [14]

$$DOS = \gamma \frac{\Delta V}{\Delta Ea} \tag{4}$$

where γ is a constant that depends on the sample geometry and the PS dielectric properties, $\Delta V=(V_{i+1}-V_i)$ is the difference between two consecutive applied voltages and $\Delta Ea=(Ea_{i+1}-Ea_i)$ is the difference between two consecutive activation energies.
Figure 4 shows the distribution of the density of state versus the activation energy for the four samples, calculated from Eq. 4. This qualitative study shows two different regions. The first region is for low activation energy values (Ea < 0.4 eV), where the DOS exponentially increases

when Ea decreases. This behavior is typical of tail states, probably due to localized states in the gap [15]. Its origin is related to the continuous random network of the non-crystalline materials. The calculated temperature parameters for the exponential distribution for the four samples are in a range from 366 K to 1020 K, in good agreement with other works [16]. The second region is for high activation energy values (Ea > 0.4 eV), where the DOS remains almost constant for different Ea values. This behavior is typical of dangling bonds, probably due to midgap defect density. This trap distribution is usual for polycrystalline and amorphous semiconductors.

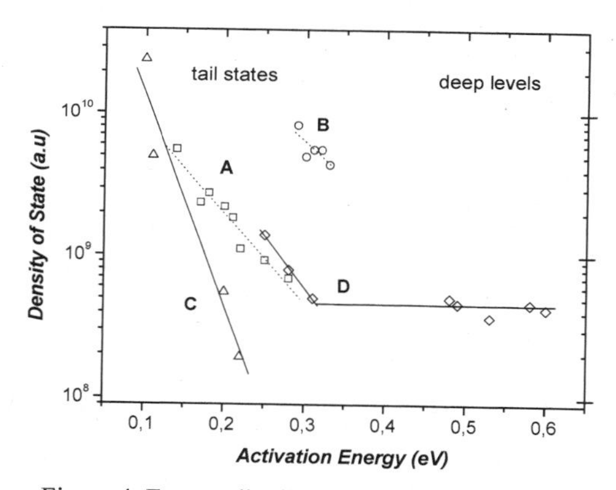

Figure 4. Energy distribution of the localised states

CONCLUSIONS

The DC electrical characterization of Au/PS/Au thermistors based on porous silicon has been presented. A quasi ohmic contact was found for the four fabricated devices. The measured resistance depends more on the etching time than on the wafer resistivity.
The activation energies for the four samples has been determined. The highest activation energy value has been obtained for sample fabricated with the most resistive wafer and the lower etching time (0.48 eV). The same sample presents the highest thermistor constant B (5570 K at 0 V) and the highest temperature coefficient TC (6.3 %/K at 0 V). For the others samples present an electro ceramic behaviour due to the not temperature activation.
The qualitative energy distribution of the density of states has been determined for the four devices. The trap distribution has been explained in terms of exponential (tail states) and uniform (deep levels) profiles, typical of amorphous and polycrystalline materials.

ACKNOWLEDGEMENTS
The authors would like to thank Dr. J. Pallarès (Universitat Rovira i Virgili) for their support in this investigation. This work was supported by the Universidad Autónoma de Occidente (UAO) under Project No. 08INTER-92.

REFERENCES

[1] L.T. Canham, Silicon Quantum Wire Array Fabrication by Electrochemical and Chemical Dissolution of Wafers, *Appl. Phys. Lett.*, **57**, 1046 – 1048 (1990).
[2] J. Salonen, M. Björkqvist, J. Paski, Temperature-Dependent Electrical Conductivity in Thermally Carbonized Porous Silicon, *Sens. and Actuators A*, **116**, 438 – 441 (2004).
[3] J. Tuura, M. Björkqvist, J. Salonen, V.P. Lehto, Electrically Isolated Thermally Carbonized Porous Silicon Layer for Humidity Sensing Purposes, *Sens. and Actuators B*, **131**, 627 – 632 (2008).
[4] H.E. Lazcano, C. Sánchez, A. García, An Optically Integrated NH_3 Sensor Using WO_3 Thin Films as Sensitive Material, *J. Opt. A: Pure Appl. Opt.*, **10,** 104016 (6pp) (2008).
[5] F. Fonthal et al, Electrical and Optical Characterization of Porous Silicon/p-Crystalline Silicon Heterojunction Diodes, *AIP Conf. Proc.*, **992**, 780 – 785 (2008).
[6] F. Fonthal, M. Chavarria, Impedance Spectra Under Forward and Reverse Bias Conditions in Gold/Porous Silicon/p-Si Structures, *Phys. Status Solidi C*, Article first published online: 23 NOV 2010, to be published in print (2011).
[7] M. Chavarria and F. Fonthal, Electrical Characterization and Dielectric Relaxation of Au/Porous Silicon Contacts, *in: Advances in Electroceramic Materials, edited by K.M. Nair, D. Suvorov, R.W. Schwarts, R. Guo, Ceramic Transactions Series*, Wiley, New Jersey, **204,** 113 – 117 (2009).
[8] A.M. Rossi, H.G. Bohn, Photodetectors from Porous Silicon, *Phys. Status Solidi A*, **202**, 1644 – 1647 (2005).
[9] M. Theodoropoulou et al, Transient and AC Electrical Transport Under Forward and Reverse Bias Conditions in Aluminum/Porous Silicon/ p-Si Structures, *J. Appl. Phys.*, **96**, 7637 – 7642 (2004).
[10] E.A. de Vasconcelos, S.A. Khan, W.Y. Zhang, H. Uchida, T. Katsube, Highly Sensitive Thermistors Based on High-Purity Polycrystalline Cubic Silicon Carbide, *Sens. and Actuators A*, **83** 167 – 171 (2000).
[11] M. Ben-Chorin, F. Möller, F. Koch, Nolinear Electrical Transport in Porous Silicon, *Phys. Rev. B.*, **49**, 2981 – 2984 (1994).
[12] J.P. Bade et al, Fabrication of Diamond Thin-Film Thermistors for High-Temperature Applications, *Diamond Relat. Mater.*, **2** 816 – 819 (1993).
[13] K. Wasa, T. Tohda, Y. Kasahara, S. Hatakawa, Highly Reliable Temperature Sensor Using RF-Sputtered SiC Thin Flim, *Rev. Sci. Instrum.*, **50** 1084 – 1088 (1979).
[14] L.A. Balagurov et al, Transport of Carriers in Metal/Porous Silicon/c-Si Device Structures Based on Oxidized Porous Silicon, *J. Appl. Phys.*, **90**, 4543 – 4548 (2001).
[15] M. Hack, M. Shur, Theoretical Modeling of Amorphous Silicon-Based Alloy PIN Solar Cells, *J. Appl. Phys.*, **54**, 5858 – 5863 (1983).
[16] J. Pallarès, L.F. Marsal, X. Correig, J. Calderer, R. Alcubilla, Space Charge Recombination in P-N Junctions with a Discrete and Continuous Trap Distribution, *Solid State Elec.*, **41**, 17 – 23 (1997).

TAPE CAST DIELECTRIC COMPOSITES PRODUCED WITH CAMPHENE AS A FREEZING MEDIUM

E.P. Gorzkowski, M.-J. Pan, and B. A. Bender
Naval Research Laboratory
Code 6351
4555 Overlook Ave., SW
Washington, DC 20375

ABSTRACT

In previous research, freeze casting was used to construct ceramic-polymer composites where the two phases are arranged in an electrically parallel configuration. This resulted in the composites exhibiting a dielectric constant (K) up to two orders of magnitude higher than that of composites with ceramic particles randomly dispersed in a polymer matrix. This technique has been successful with both aqueous and camphene based ceramic slurries that are frozen uni-directionally to form platelets and ceramic aggregates that are aligned in the temperature gradient direction. This paper will provide an update on camphene based samples in order to process the slurry at room temperature. This alleviates the need for liquid nitrogen and a freeze-dryer and opens the possibility for large scale production by creating the composites using a tape casting process. Transitioning the directional freezing technique to use a tape caster will be discussed as well as processing parameters that must be optimized to create high quality composites.

INTRODUCTION

The fabrication of nacre-like laminar ceramic bodies using a novel ice template process was demonstrated by Deville et al. in Science.[1] This technique entails freezing an aqueous ceramic slurry uni-directionally along the longitudinal axis of a cylindrical mold to form ice platelets and ceramic aggregates. Given the proper conditions (including slurry viscosity, water percentage, temperature gradient between the top and bottom of the mold, and starting temperature) the ice platelets are aligned in the temperature gradient direction. The proper starting temperature and temperature gradient must be maintained so that homogeneous freezing occurs and hexagonal ice is formed. This allows the ice front to expel the ceramic particles in such a way to form long range order for both the ceramic and the ice. Upon freeze drying, the ice platelets sublime and leave a laminar ceramic structure with long empty channels in the direction of the temperature gradient. Subsequently, the green ceramic body is sintered to form the final microstructure.

The Deville article focuses on the mechanical properties of the ceramic body, but a ceramic-polymer composite with excellent dielectric properties is also possible by adapting this technique.[2] The adaptation involves 1) using a high K material as the ceramic phase, 2) infiltrating the space between ceramic lamellae with a polymer material, and 3) applying electrodes perpendicular to the ceramic-polymer alignment direction to form an electrically parallel composite dielectric. By utilizing this technique for lead-based dielectric materials, a novel composite was created that exhibits dielectric constant (K) of up to 4000 for PMN-10PT while maintaining low dielectric loss (< 0.05). The finished composites not only exhibit the high dielectric constant of ferroelectric ceramics but maintain the flexibility and ease of post-processing handling of polymer materials.

It is possible to use materials other than water as the freezing medium, such as camphene[3,4,5], tertiary butanol[6], and naphthalene[7]. Using some of these other freezing mediums, such as camphene, enables the freezing temperature to be near room temperature, which eliminates the need for liquid nitrogen and a freeze-dryer, which makes this process suitable for large scale production. This study investigates using a large scale production technique, tape casting, and camphene as a freezing medium to create composites with high dielectric constant that exhibit a graceful dielectric breakdown behavior.

EXPERIMENTAL PROCEDURES

Ceramic slurries were prepared by mixing x wt% Barium Titanate X7R (Ferro Corporation Cleveland, OH) where x is 25, 32.5, 37, and 40 wt.%. and camphene ($C_{10}H_{16}$, Alfa Aesar, Ward Hill, MA) with 0 - 5 wt% Hypermer LP 1(Croda, Edison, NJ), 0 – 1 wt% Hypermer KD 1(Croda, Edison, NJ), 0 - 1.2 wt% Glycerin (JT Baker, Phillipsburg, NJ), 0 – 6 wt% Polystyrene microspheres (Aldrich, St. Louis, MO), 0 – 6 wt% Santicizer 160 (Richard E. Mistler, Inc, Yardley, PA), 0 – 5 wt% Polyalkylene Glycol (Richard E. Mistler, Inc, Yardley, PA), and 0 – 5 wt% Polyethylene glycol 200 (JT Baker, Phillipsburg, NJ). Slurries were warm ball-milled at 50 °C in a high density polyethylene bottle for 12 h with zirconia milling media.

Two methods were used to freeze the slurries. The first method involved pouring the slurries into a Teflon mold (1.5 in. diameter, 0.75 in. tall) and cooling using a custom built freezing setup. The mold is placed between two copper rods that are cooled by liquid nitrogen to – 60 °C at 5 °C/min. There are band heaters attached to the copper rods to control the cooling rate and temperature gradient between the copper rods (10 °C). The second method involved using a freeze-tape-casting apparatus which consists of a standard research tape caster (TTC-1000; Richard Mistler Inc., Yardley, PA) and a chilled plate on the tape caster bed for freezing. Approximately 6 in. x 15 in. tapes were cast using a 6-in. doctor blade assembly and a 30 mil gap. All samples for both freezing methods were dried at room temperature and pressure for 24 h. Tapes were then removed from the mylar carrier in small squares for binder burnout and sintering. This was accomplished by heating the samples at 1.2 °C/min to 300 °C, 0.1 °C/min to 350 °C, 0.6 °C/min to 500 °C, 5 °C /min to 1300 °C for 2 h in order to improve crystallinity. Each square sample was then infiltrated with Epotek 301 Epoxy (Epoxy Technology, Billerica, MA) under vacuum. The square plate capacitor samples were prepared for dielectric testing. This entailed lapping the samples using 400 and 600 grit SiC slurry to create flat parallel faces. Samples were masked and gold coated for capacitance, breakdown and hysteresis measurements. The dielectric constant and loss were measured using an HP 4284A at 0.1, 1, 10, and 100 kHz from 125 down to -60 °C. The breakdown measurements were made using a Hipot tester (QuadTech) at 500 V/s.

Pieces from each of the various samples were mounted onto a stub, carbon coated and masked with conductive tape for Scanning Electron Microscopy (SEM). Images of these surfaces were obtained using a Leo 1550 SEM.

RESULTS AND DISCUSSION

Previously studies utilized water as a freezing medium to create novel polymer / ceramic composites with high dielectric constant and high breakdown strength. Due to the hexagonal structure of the growing ice crystals, the ice forms platelets and thus creates an alternating polymer/ceramic final microstructure. Camphene, on the other hand, forms an interconnected columnar pore structure instead of platelets. Water based samples are ideal for a 2-2 composite, while the camphene based samples are more similar to 3-3 composites. The microstructure of the camphene based samples look like a sponge which can be seen in Figure 1. Figure 1a.) shows the side view and Figure 1b.) shows the top view of the mold processed samples. It can be seen that the pores left behind after the camphene has been sublimed out, align in a columnar fashion making the structure resemble a sponge. The size of the pores can be controlled by altering the freezing rate (for these samples that were cooled at 5 °C/min, it is in the range of 5 μm), the slower the freezing the larger the resultant pores.

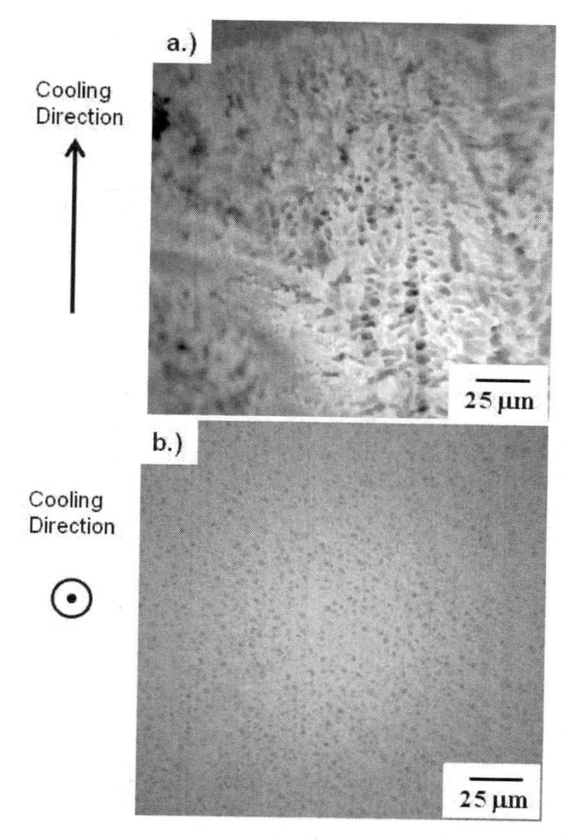

Figure 1 Light Optical micrograph of a.) cross section and b.) top view of a camphene based Barium Titanate X7R sample showing the aligned pore structure.

Once the mold processed samples were filled with Epoxy, the samples dielectric properties were tested. Figure 2 shows the dielectric constant versus temperature data for the mold processed Barium Titanate X7R sample. The room temperature dielectric constant is ~ 800 while the dielectric loss is ~0.06. These values are comparable to the water based samples that were tested in earlier studies. The dielectric constant is slightly lower for equivalent solids loading in the water based slurries, which is most likely due to the extra dimension connectivity of the camphene based composite. In previous studies, it was shown that if side chains connected the platelets in the water processed samples the dielectric constant would go down so it follows that the K would decrease when the ceramic is connected in 3 dimensions instead of 2. In composite theory, this comes from the fact that the side chains in the water processed or the extra dimension of connectivity in the camphene processed add a perpendicular capacitance contribution to the dielectric constant.

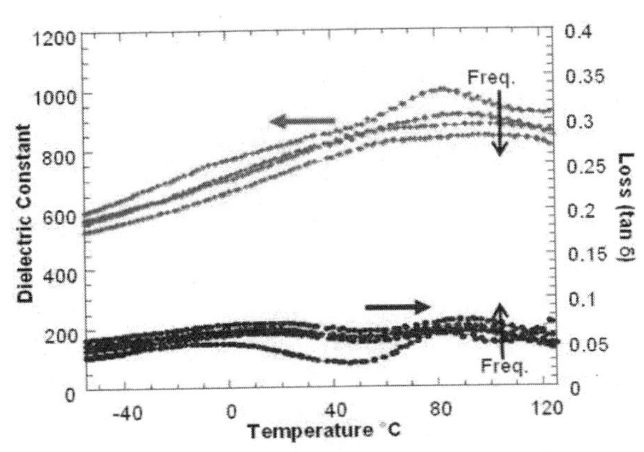

Figure 2 Dielectric constant versus temperature data for a camphene based Barium Titanate X7R sample.

Next, the breakdown strength of the samples was measured. This was tested by increasing the voltage on the sample until a breakdown event occurred. The same sample was then ramped again until another breakdown event occurred. This process was repeated up to 25 times, as shown in Figure 3. No catastrophic failure or fail-short was observed over this testing range. Since the area around the breakdown is "healed" like in most polymer capacitors, voltage can be re-applied. This demonstrates that these composites fail in a graceful manner; however, the mechanism was not studied further. In this study, the breakdown strength increased as the number of breakdown events increased. This is probably due to the established "weakest link" theory, where breakdown occurs at the weakest point of the sample. Since the next weakest spot, the area where the next breakdown occurs is stronger than the first the breakdown voltage goes up. The breakdown strength was ~ 260 kV/cm (in raw numbers), which is closer to the epoxy value ~400 kV/cm than the ceramic value of barium titanate of 80 kV/cm.

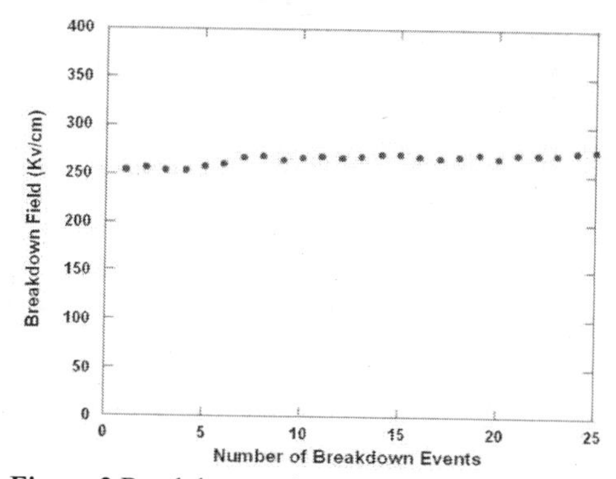

Figure 3 Breakdown voltage results for a camphene based Barium Titanate X7R sample showing graceful failure.

After the mold processed samples were successfully created, the slurries were transitioned to produce tapes of barium titanate X7R. The freeze tape cast setup has a thermally insulating layer between the tape-caster bed and the cooling bed. In this case, the cooling bed is placed in a freezer before tape-casting for 2 hrs to get the aluminum plate cold. The mylar carrier then travels over the freezing bed with the camphene based slurry.

The first processing parameter that needed to be optimized was the amount of binder and dispersant. Figure 4 a.) shows an image of some drying defects (bubbles) that were caused by too many polystyrene spheres. A layer of polystyrene spheres formed at the bottom of the tape, trapping the subliming camphene and creating bubbles. Figure 4 b.) shows a good tape that contains no drying defects with the appropriate amount of all binders and dispersants.

Figure 4 As processed Barium Titanate X7R tapes showing a.) drying defects and b.) no drying defects.

Once good tapes of barium titanate X7R were produced, the effect of solids loading could be determined. A sample of the microstructure of the tape is shown in Figure 5 where a.) is the top view and b.) is the side view of the fired tape. It can be seen that the tape is ~ 160 μm thick, which is typical of all the tape that was made. The microstructure is the same as the mold processed samples in that it appears sponge like. The dielectric constant of each solids loading is 440, 680, and 1180 for 25, 32, and 40 wt% respectively. Again, these values are in line with the mixing rules for electrically parallel composites. Future studies will continue the optimization of the processing parameters along with tape casting directly onto an electroded carrier. This will allow for the creation a prototype capacitor.

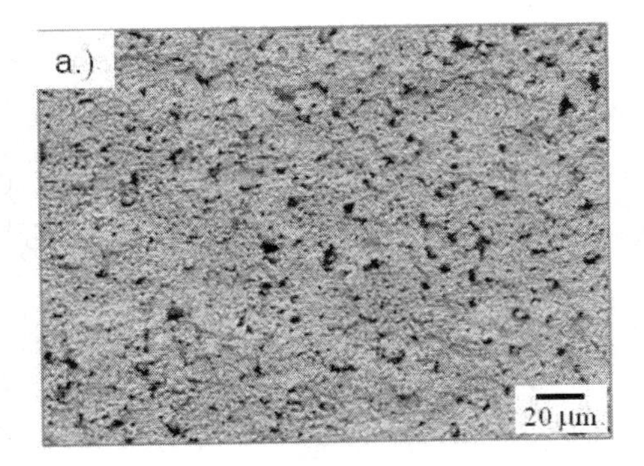

Figure 5 Scanning Electron Micrograph of a.) top view and b.) side view of a Barium Titanate X7R sample fired at 1300 °C for 2hrs.

CONCLUSION

Previously, ceramic-polymer composites in which the two phases are arranged in an electrically parallel configuration were created using water as a freezing medium. The composites had a dielectric constant (K) up to two orders of magnitude higher than that of composites with ceramic particles randomly dispersed in a polymer matrix. This technique was successfully transitioned to camphene based ceramic slurries that were frozen uni-directionally in a mold and on a tape caster. Camphene based samples enable the slurry to be processed at room temperature. This alleviates the need for liquid nitrogen and a freeze-dryer and opens the possibility for large scale production by creating the composites using a tape casting process. The camphene based samples had a dielectric constant as high as 1180 for the highest solids loading sample tested, which is nearly half that of ceramic value while maintaining the graceful failure of polymers. The transition to the tape-caster has been successful and is clearly the way forward to create manufacturing level quantities.

REFERENCES

1. S. Deville, E. Saiz, R. Nalla, and A. Tomsia, "Freezing as a Path to Build Complex Composites," *Science*, **311**, p,515-518 (2006).
2. E. P. Gorzkowski and M. J. Pan, "Ceramic-Polymer Dielectric Composites Produced via Directional Freezing," *Ceramic Transactions* , **204,** pp. 3-10 (2009).
3. K. Araki and J. W. Halloran, "New Freeze-Casting Technique for Ceramics with Sublimable Vehicles," *J. Am. Ceram. Soc.*, **87** [10], pp. 1859–1863 (2004).
4. K. Araki and J. W. Halloran, "Porous Ceramic Bodies with Interconnected Pore Channels by a Novel Freeze Casting Technique," *J. Am. Ceram. Soc.*, **88** [5], pp. 1108–1114 (2005).
5. Y.-M. Soon, K.-H. Shin, Y.-H. Koh, J.-H. Lee, and H.-E. Kim, "Compressive Strength and Processing of Camphene-Based Freeze Cast Calcium Phosphate Scaffolds with Aligned Pores," *Mater. Let.*, **63**, pp. 1548–1550 (2009).

6. R. F. Chen, Y. Huang, C. A. Wang, and J. Q. Qi, "Ceramics with Ultra-low Density Fabricated by Gelcasting: An Unconventional View," *J. Am. Ceram. Soc.*, **90**, pp. 3424 – 3429 (2007).
7. K. Araki and J. W. Halloran, "Room-Temperature Freeze Casting for Ceramics with Nonaqueous Sublimable Vehicles in the Naphthalene-Camphor Eutectic System," *J. Am. Ceram. Soc.*, **87** [11], pp. 2014 – 2019 (2004).

ELECTRONIC TRANSFER BETWEEN LOW-DIMENSIONAL NANOSYSTEMS

Karel Kral
Institute of Physics, Academy of Sciences of Czech Republic, v.v.i.
Prague, Czech Republic

ABSTRACT

The irreversible transfer of electrons or holes between zero-dimensional semiconductor nanostructures is considered theoretically. The mechanism can apply to quantum dot samples, molecular crystals, polymer solids, DNA molecule, etc. The transfer mechanism is studied using a model with inter-molecular electron tunneling mechanism and with the coupling of the charge carrier to the intra-molecular or intra-quantum dot atomic vibrations. The electron-phonon interaction is considered in the self-consistent Born approximation to the nonequilibrium Green's function self-energy. The corresponding kinetic equation shows us a specific diffusion mechanism of an electron along e.g. a molecular solid. This mechanism will be compared to the well-known Marcus theory of electronic transfer. With using the above mechanism we are able to explain earlier experimental data on photoconductivity in charge-transfer crystals, and also rather recent data on the electric conduction of DNA molecules.

INTRODUCTION

In the area of the molecular condensed matter there are at least two kinds of materials, which properties are strongly one-dimensional and in which we observe an electronic transport along the one-dimensional stacks of molecules. The two cases we wish to pay attention to are first of all the DNA molecules [1], in which the carriers of the charge are expected to move from one molecular base to the neighboring one in the DNA molecule. Another example may be served by the Donor-Acceptor (DA) molecular crystal, in which the donor and acceptor molecules form one-dimensional alternating stacks of molecules [2]. In both these kinds of material the one-dimensional stacks remind the stacks of books in shelves. The energies of the single-electron states in the individual molecules (let us call a base of a DNA molecule also simply a molecule) may vary relatively strongly from molecule to molecule in the stack. From this reason we can speak about a rather strong static diagonal disorder influencing seriously the electronic motion along the molecular stacks in these systems. Because of the well known arguments [3] the charge carriers should be localized in such systems, with only elastic scattering of charge carriers, and the electric conductivity should be zero. However, because of experiments, which show effects due to charge carriers moving along the stacks, we ascribe the charge carrier motion to be due to mechanisms which allow the charge carriers to be transferred irreversibly from one molecule of the stack to another one in the same stack, via an inelastic scattering of electrons. We shall treat theoretically this charge carrier motion along the stack. We shall rely on the interaction of the charge carriers with the optical lattice vibrations of the individual molecules in the stack. For simplicity, in what follows we shall not speak about the hole particles as charge carriers, mentioning only the electrons in the conduction band states.

Our theoretical results will be compared with two particular earlier experiments. One experiment deals with the microwave conductivity measurement in DNA molecules [4]. The other experimental paper deals with the photoconductivity of DA molecular crystals [2]. In both experiments we meet the problem of an electronic transfer along the one-dimensional stacks of flat molecules in the presence of a static diagonal disorder. In our theoretical treatment we are not going to calculate directly the electric conduction of the one-dimensional stack, or the DNA molecule. We choose to achieve first a simpler goal motivated by the static diagonal disorder itself. We consider the case when an electron can diffuse from one molecule to the neighboring one using the electron-phonon interaction and dissipating the energy to the environment. In the case of a relatively strong static diagonal disorder there is an important case in which the electron needs to be transferred to the neighboring molecule with the need to overcome a considerable energy barrier caused by the difference in the electronic orbital energies. In contrast to the cases when such an electron is going to be transferred to a molecule with a lower electronic energy, which may be achievable, according to a simple intuition, via releasing energy to the environmental degrees of freedom, the electronic transfer to a higher energy molecule needs a more detailed attention below. We expect that this case of the electron transfer along the stacks represents a certain bottleneck in the whole process of electric conduction and basically influences the conductivity. Let us mention in this respect the paper by Tributsch and Pohlmann [5], in which the authors discuss an effect of a 'positive friction'. In what follows we shall consider only the 'bottleneck part' of the one-dimensional stack of molecules, by considering only a molecular dimer, assuming the situation in which an electron is placed at an orbit at a molecule with low value of the orbital energy. We shall calculate the probability per unit time that this electron is transferred to the other molecule in the molecular dimer with higher electronic orbital energy. We shall calculate the dependence of this transfer rate on the temperature of the molecular lattice vibrations. Such a dependence of the electric conductivity is measured in the DNA experiment [4]. We expect that calculating the same dependence for the bottleneck part of the stack we can compare our theoretical results with the experiments on DNA [4]. At the same time, our calculations will also allow us to give an explanation why the electrons excited in the paper [2] can cause the optical conductivity in the DA crystals measured.

We shall use the microscopic theory based on multiple scattering of the transferred electron on the (intramolecular) optical phonons of both the two neighboring molecules. A similar mechanism has recently been shown to lead to a rapid relaxation of electronic energy in individual quantum dots [6] and to the effect of upconversion of electronic level occupation in quantum dots. We observe that the same effect of the electronic level occupation upconversion can help the electron to overcome the energy barrier of the static diagonal disorder. Because of the previous experience of the authors with the quantum dot theory we approximate the whole system of two molecules by two interacting quantum dots. Using this two quantum dot model is advantageous because a part of the theoretical equipment for the numerical solution has already been developed for individual quantum dots. In the present work we confine ourselves to the above model of 'two quantum dots'. In the development of the calculation of the electron transfer rate we shall therefore speak about two quantum dots and a single electron being transferred irreversibly between them. After achieving the conclusions with using this quantum dot model, we will expect that similar results are obtainable eventually with using a model of interacting molecules, or bases in DNA molecule, etc. Such activities are postponed to a later work. Besides, it is certain that the theoretical model of two interacting quantum dots certainly has important counterparts in electron transfer experiments on samples of quantum dots grown on a surface of insulating substrate.

In this paper, each of the two quantum dots, A and B, will have only one orbitally non-degenerate electronic state. Below we shall specify briefly the Hamiltonian of the system. Then we briefly describe the method of derivation of the formula for the electron transfer rate and show the calculated temperature dependence of this transfer rate. After a brief comparison with experiments about the electric conduction of DNA molecule and about the photocurrent generation in DA molecular crystals, and after making a comment on the relation of the present approach to the well-known Marcus method on the electron transfer [7], we conclude.

HAMILTONIAN OF TWO QUANTUM DOTS

In the present section we specify the Hamiltonian of the system of two interacting quantum dots coupled by the electronic tunneling, for the purpose of considering the process of the irreversible electron transfer from one quantum dot to the other. We shall assume that the two quantum dots are formed of the same material, with the same material constants, differing only by the lateral size of the quantum dot. In order that we can vary the relative positions of the single electron energies of the two molecules we assume that the energies are shifted by a gate potential applied to one of the dots. We shall also choose such magnitudes of the parameters describing the system that it is sufficiently relevant for making a reasonable comparison of the present theoretical conclusions with the experimental properties of the real systems of electronic transfer along DNA molecule [4] and photoconduction carriers generation in the DA crystals [2].

Dealing with only a single electron, we ignore the electronic spin in the present work. The electronic Hamiltonian is $H_e = H_{eA} + H_{eB} + V_t$. Here $H_{eA} = E_A c_A^+ c_A$ and $H_{eB} = E_B c_B^+ c_B$ are the Hamiltonians of an electron on two electronic orbitals, each being localized in a separate quantum dot, A and B, with the respective site energies E_A and B_B. The operator c_A is annihilation operator of electron at the orbital localized at the dot A, the other particle operators having analogical meaning. The operator V_t means the electronic tunneling of the electron between the two sites and has the form $V_t = t(c_A^+ c_B + c_B^+ c_A)$. The parameter t expresses the tunneling efficiency of an electron between the dot sites A and B. The above given Hamiltonian assumes that the Hilbert space of the single electron is determined by two electronic orthonormal orbitals, each is localized within a separate quantum dot.

We shall assume that an electron placed into the system of the two quantum dots interacts with the intra-dot vibrations of the quantum dot atomic lattice. We choose the electron-phonon interaction in the following form: $H_{e-p} = H_{e-p,A} + H_{e-p,B}$. Here $H_{e-p,A}$ is the operator of the electron-phonon interaction in the quantum dot A. In analogy with the earlier theory of the electron energy relaxation in a single quantum dot, and taking into account that there is a single electron orbital per dot, we take this operator in the form which taken over, including notation, from the reference [8,9]: $H_{e-p,A} = \sum_{\mathbf{q}\in\Omega_A} A_q \Phi_A(0,0,\mathbf{q})(b_{\mathbf{q}} - b_{-\mathbf{q}}^+) c_A^+ c_A$. Here A_q is Fröhlich's coupling constant [8,9], given by the material constants of the material of the particular quantum dot. The integration over $\mathbf{q}$ covers the range Ω_A of the phonon wave-vector in the quantum dot A. The sum over $\mathbf{q}$ thus represents the sum over the bulk longitudinal optical (LO) modes of the lattice vibrations of the sample [8,9]. The operator $b_{\mathbf{q}}$ is annihilation operator the LO phonon in the mode with the wave-vector $\mathbf{q}$. The quantity $\Phi_A(0,0,\mathbf{q})$ is the form-factor of the quantum dot A [8,9], taking into account the shape of the quantum dot.

Similarly, for the quantum dot B we have the electron-phonon operator as follows, $H_{e-p,B} = \sum_{\mathbf{q}'\in\Omega_B} A_{q'} \Phi_B(0,0,\mathbf{q}')(b_{\mathbf{q}'} - b_{-\mathbf{q}'}^+) c_B^+ c_B$. Here again $\Phi_B(0,0,\mathbf{q}')$ expresses the form-factor of the quantum dot B at the value of the phonon wave-vector $\mathbf{q}'$. We consider the phonon modes with the wave-vectors $\mathbf{q}\in\Omega_A$ and $\mathbf{q}'\in\Omega_B$ as two different and independent phonon mode systems.

In order to write down the total Hamiltonian, we give also the non-interacting phonon modes Hamiltonian, $H_p = \sum_{\mathbf{q}\in\Omega_A} E_{LO}^{(A)} b_{\mathbf{q}}^+ b_{\mathbf{q}} + \sum_{\mathbf{q}'\in\Omega_B} E_{LO}^{(B)} b_{\mathbf{q}'}^+ b_{\mathbf{q}'}$. This operator consists of two parts, for the two phonon systems A and B. In this work we shall take the optical phonon excitation energies as equal, $E_{LO}^{(A)} = E_{LO}^{(B)} = E_{LO}$. Then the full Hamiltonian operator H of the system is

$$H = H_e + H_p + H_{e-p}. \tag{1}$$

The Hamiltonian (1) is actually one of the forms of the spin-boson Hamiltonian [10].

We choose the magnitude of the difference $E_A - E_B$ and of the tunneling parameter t taking into account the information given by the authors of the references [11-13]. Generally we assume that $|(E_A - E_B)/t| \gg 1$. Assuming that the electron is placed at the molecule B, we wish to learn how large is the rate of the process of the irreversible electronic transfer from the molecule B to the molecule A. Before using the quantum kinetic equations method to obtain this result, we will canonically transform the total Hamiltonian H to a representation in which it is formally similar to the Hamiltonian of the system of an individual quantum dot with the electron-phonon interaction, which we studied earlier [9]. The canonical transformation of the Hamiltonian is described in the next section.

TRANSFORMATION OF THE HAMILTONIAN AND THE RELAXATION RATE

Having a not sufficiently good knowledge about the suitable values of the parameters characterizing the Hamiltonian H as it should be used in, say, DNA molecule, we shall proceed in the following way in our estimate of the electron transfer mechanism. The Hamiltonian of the system is relatively simple from the point of view of the electron-phonon interaction. This interaction could be in principle removed by performing the Lang-Firsov transformation [14]. In the presence of the electronic tunneling interaction in H, proportional to the parameter t, this step would lead to a rather complicated expression for the transformed operator V_t. This way of starting the theoretical analysis would be perhaps suitable in the case, when the depth of the polaron well of the charge carrier in the bulk is large compared to the resonant transfer parameter t and the difference between the unperturbed energies of the charge carrier on the separate noninteracting molecules. We shall not proceed along this way.

In the other case, namely when the resonant transfer t is regarded as large compared to the bulk crystal polaron well width, we start by exactly diagonalizing that part of the Hamiltonian, H_e, which is purely electronic. In the new representation obtained after performing this canonical transformation, the full Hamiltonian H will have the form formally identical with that of the electronic two-level system containing both the transverse and also the longitudinal part of the electron-phonon interaction, which we studied earlier in connection with the electron energy relaxation in individual quantum dots [8,9]. The assumption about the magnitude of the electronic Hamiltonian parameters, $|(E_A - E_B)/t| \gg 1$, allows us to expect that after diagonalizing H_e, one of the new electronic states, say the state with $n=1$, will be very close to the original state A, while the other new electronic state $n=2$ will be very close to the original state B. Basing on this similarity between the electronic states before and after the canonical transformation, we shall calculate the rate $dN_1/d\tau$ of generation of the electronic occupation N_1 in the state with the state index $n=1$, τ is time. We shall regard this rate as a sufficiently exact characterization of the rate of transfer of the electron from the molecule B to the molecule A.

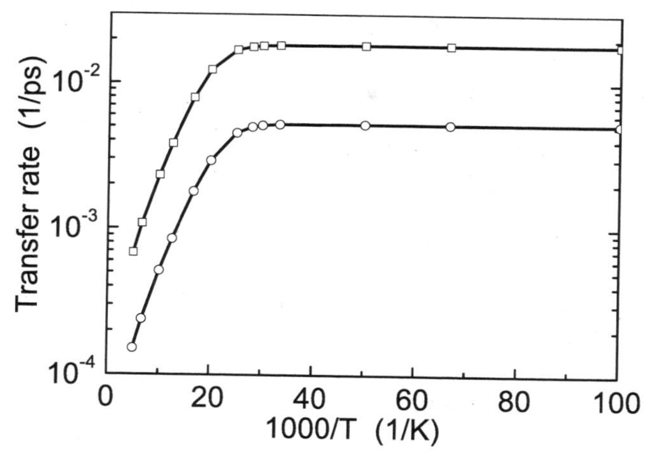

Figure 1. The rate of the irreversible transfer of the electron placed at the electron orbital at the quantum dot B to the quantum dot A, $dN_1/d\tau$. T is the temperature of the lattice. The quantum dot A has the lateral size of $19\,nm$, while the quantum dot B has the lateral size of $21\,nm$. The energy separation $E_A - E_B = 0.2\,eV$. The tunneling parameter t is equal respectively to $0.03\,eV$ (squares) and $0.02\ eV$ (circles). The material parameters of the quantum dot material are those of the bulk GaAs crystal.

The electron-phonon interaction, resulting after performing the canonical transformation, will be included here in the self-consistent Born (SCB) approximation to the electronic self energy [6,8,9]. This means that we shall go beyond the limits of the perturbation calculation, performing summations of the perturbation expansion series up to infinite order. With this approximation the kinetic equation giving the quantity $dN_1/d\tau$ will be determined, without imposing assumptions of an overall thermodynamic equilibrium to the theory. Because of the space limitation, the details of the derivation of the rate $dN_1/d\tau$ will be left to be published elsewhere. The Reader is refered to references [6,8,9] in which the physical meaning of the SCB approximation used here is explained. Let us only briefly remind an intuitive interpretation of the SCB approximation applied to quantum dots: In the small system with two energy levels, like the present system with two molecules is, the electron scatters on the phonon system and then leaves the target. In contrast to bulk systems, in the present small system the electron reflects at the boundaries of the nanostructure and returns back to the target to continue the scattering process. This may theoretically continue up to an infinite order of the multiple scattering process. In our case of the scattering on the optical phonons the resulting state of the phonons can be a coherent multiphonon state of the vibrational system. Such a coherent multiphonon state can be viewed as similar to the coherent light state. From this reason we expect that in contrast to incoherent Fock's state phonons, such coherent states can remind a macroscopic oscillation of the atomic lattice, with which the electron interacts. The macroscopic coherent oscillations of the atomic lattice thus represent a time dependent contribution to the effective Hamiltonian, which influences the motion of the electron. Because of this time dependence, the vibrations can influence the electron nonadiabatically. Also, the concept of the electronic energy becomes not well defined. The resulting formula for the electronic relaxation rate [6,8,9] does not have the very traditional form of the Fermi Golden rule, namely, it does not contain the energy conservation delta function and has only a form of an off-shell formula for the inter-state transition probability. Because of the explicitly time dependent effective Hamiltonian ruling the electronic motion in the small system, the electron and the vibrations do not exchange only heat. Rather, the phonons execute a time dependent force on the electron, and the two subsystems are thus generally not in a thermal (thermodynamic) equilibrium. Therefore the distribution functions of the electron and the distribution of Fock's optical phonons need not be given by the equilibrium distributions with the same temperature. In particular, the electron and the phonons tend to achieve a steady state [6], which is characterized not by the usual detailed balance condition respecting the usually assumed overall thermodynamic equilibrium of the whole system. The steady state of the whole system is rather characterized by another steady state condition, which is in fact contained in the kinetic equation for the quantity $dN_1 / d\tau$ [6]. The electron-phonon motion in our case tends to lead to a steady state, in which the eletronic occupation does not correspond to the temperature of the Fock's phonon excitations. This tendency does not weaken when the temperature of the lattice goes to 0K.

The rate formula is derived using the nonequilibrium Green's functions formalism [15,16]. The rate $dN_1 / d\tau$ is derived under the assumption of the instant collisions approximation and using the common Kadanoff-Baym ansatz.

The magnitudes of the parameters like the electronic level deparation, $E_A - E_B$, and the tunneling parameter t, can be selected using the theoretical works [11-13]. The numerical result of the calculation of the rate of generation of the electron occupation on the molecule A is shown in Figure 1. This temperature dependence of this rate is preliminarily identified with the temperature dependence behavior of the microwave conductivity in DNA [4]. We can see that both the data, theory and experiment, give a temperature dependence which is not thermally activated at the range of the low temperatures. This result seems to support the view that the motion of the charge carriers along DNA chain is facilitated by the multiple phonon scattering mechanism, during which virtual multiphonon states are created in the double dot system. The electron multiply scatters on the phonons of both two molecules and multiply tunnels between these molecules, before it finishes the irreversible scattering act. We can therefore speak about a new special mechanism of electronic diffusion along DNA molecules or along the molecular chains in DA molecular crystals. From the above agreement between theory and experiment it seems to come out that the thermally nonactivated behavior of the electric conduction of DNA molecule can be ascribed to the relative smallnes of the nanostructure units of the DNA bases and to the effect of the multi-phonon excitations in them.

The numerical calculation gives a decrease of the transfer rate with the increasing temperature. In contrast, the experiment based on the microwave measurement gives however an increase of the electric conduction with temperature [4]. The result of our two-electronic level calculation better corresponds to the measurement using the muon labelling [17], in which paper a decrease of the conduction with increasing temperature is reported. The temperature dependence of the DNA conduction at increased temperatures thus remains perhaps not decided.

Referring the Reader to the details of the experiment with the photoconductivity generation in DA molecular crystals, we can explain the observed experimental observation to the same mechanism as that leading to the activationless electric conduction in DA crystals. Without going into details, we can say that the observed generation of the photoconduction in the experiment [2] is due to the dissociation of the charge transfer exciton in the DA crystal, being an implication of the upconversion of the electron energy level occupation, due to the multiphonon scattering of charge carriers in neghboring molecules of a molecular crystal.

Let us remark a certain similarity between the presently used approach to the electron transfer between the quasi-zero dimensional nanostructure elements and the theory of such processes developed by R. Marcus [7]. The theoretical formulation due to R. Marcus, of this irreversible electron transfer process important in various fields of physics and chemistry [5], is a remarkable example of the intuition in physics. As a key concept, the presence of the otherwise non-specified macroscopic fluctuating parameter in the system is assumed, which allows us to circumvent the serious problem of the energy conservation in the course of the reaction of the electron transfer. Another key point of the theory due to Marcus is the use of the electron transfer probability formula at the level of the Fermi Golden rule, taken at the thermodynamic equilibrium with the environment.

The observed agreement of the low temperature behavior, between the presently developed multiphonon theory and the experiment in DNA, allows us to suggest that the virtual multi-phonon states, expected to be present in the pairs of the neighboring molecules, can be tentatively identified with the fluctuating parameter of Marcus. The presence of the virtual LO phonon modes excited to multi-phonon states can be expected to play a role of a factor allowing to circumvent the problem of the energy conservation. Concerning the transition rate rapidity, the present approach to the electron transfer theory allows to consistently derive the transfer rate formula (not repeated here, see refs. [6,8-10]). In addition to this, we can also suggest that the present example of the irreversible electron transfer gives an example of the so called possitive friction effect, which is discussed in the reference [5].

CONCLUSIONS

The agreement is found between the temperature independence of the calculated low-temperature characteristic and that of the electronic conduction in DNA. The present mechanism of the electron transfer can also give an explanation of the mechanism of the photoconduction generation in the Donor-Acceptor molecular crystals. The suggested multiphonon electron transfer mechanism may thus represent a specific mechanism of the electronic diffusion in molecular materials, which depends up to an infinite order on the electron tunneling parameter. The present approach to the electron irreversible transfer between two quasi-zero dimensional nanoelements is suggested to be related to the Marcus theory of this phenomenon.

ACKNOWLEDGMENTS

The support of the projects OC 10007 of MŠMT and AVOZ10100520 is acknowledged.

REFERENCES

[1]S. O. Kelley, and J. K. Barton, *Science*, 283, 375 (1999).
[2]M. Samoç, and D. F. Williams, *J. Chem. Phys.*, 78, 1924 (1983).
[3]D. J. Thouless, *Phys. Rev. Lett.*, 39, 1167 (1977).
[4]P. Tran, B. Alavi, and G. Gruner, *Phys. Rev. Lett.*, 85, 1564 (2000).
[5]H. Tributsch, and L. Pohlmann, *Science*, 279, 1891 (1998).
[6]K. Král, and P. Zdeněk, *Physica E*, 29, 341 (2005).
[7]R. Marcus, *Journal of Electroanalytical Chemistry*, 438, 251 (1997).
[8]K. Král, and Z. Khás, *Phys. Rev. B*, 57, R2061 (1998).
[9]K. Král, P. Zdeněk, and Z. Khás, *Nanotechnology, IEEE Transaction on*, 3, 17 (2004); K. Král, P.
[10]Zdeněk, Z. Khás, Surface Science 566–568, 321–326 (2004).
[11]A. J. Leggett, S. Chakravarty, A. T. Dorsey, Mathew P. A. Fisher, Anupam Garg and W. Zwerger, Rev. Mod. Phys. 59, 1 (1987).
[12]A. Troisi, and G. Orlandi, *Chem. Phys. Lett.*, 344, 509 (2001).
[13]E. M. Conwell, and S. V. Rakhmanova, *Proc. Natl. Acad. Sci.*, 97, 4556 (2000).
[14]E. M. Conwell, *Proc. Natl. Acad. Sci.*, 102, 8795 (2005).
[15]G. D. Mahan, *Many-Particle Physics, 2nd Ed.*, Plenum Press, New York.
[16]E. M. Lifshitz, and L. P. Pitaevskii, *Physical Kinetics*, Butterworth-Heinemann. Reprint edition.
[17]E. Torikai, H. Hori, E. Hirose, K. Nagamine, *Physica B*, 374–375, 441–443 (2006).

COMBINED DILATOMETER-MASS SPECTROMETER ANALYSIS OF THE SINTERING OF BARIUM TITANATE

Murray A. Moss[#] and Stephen J. Lombardo[#,+]
[#]Department of Chemical Engineering
[+]Department of Mechanical and Aerospace Engineering
University of Missouri
Columbia, Missouri, USA

ABSTRACT

A combined dilatometer-mass spectrometer (CDMS) system has been developed and is used to measure the consumption and release of compounds in the gas phase during the sintering of $BaTiO_3$. The CDMS utilizes a dilatometer to measure the length changes of the sample and a quadrupole mass spectrometer to analyze gas phase compounds, thereby allowing real-time data acquisition during sintering. Analysis of these compounds allows for the investigation of chemical reactions involving possible impurities from the production of the ceramic powder or absorbed species being released from the furnace, which may influence the kinetics of sintering. This information can then be used to tailor sintering cycles.

INTRODUCTION

Analytical techniques to identify gas phase species that are evolved from inorganic materials heated to high temperatures have been studied and summarized in detail elsewhere[1-5]. The Knudsen diffusion cell is a method in which the material is heated and the products can be directly measured with a mass spectrometer or contained for later analysis. Another method uses a carrier gas that flows over the sample to transport evolved gas species for analysis. The latter method is more relevant for investigating the relationship between reactions that occur at high temperature and the rate of sintering. Questions that are explored in this work are to what extent, at what concentrations, and at what temperatures do evolved species affect sintering kinetics and material composition. Possible reactions[4-9] that occur at high temperatures include the decomposition of organic material, the evolution of sintering aids, and the removal or modification of surface layers that are different from the bulk. Apart from the reactions that are caused by the sample, the furnace environment may affect sintering by contributing to or even initiating reactions[10].

In this work, high temperature reactions and sample length change were analyzed for a barium titanate sample using a combined dilatometer and mass spectrometer system. Barium titanate was chosen as it is widely used in the electronics industry as capacitors because of its high dielectric constant. Carbon dioxide and sulfur dioxide were assigned to gas species that were evolved, with sulfur dioxide evolving at the onset of sintering. Reference cracking patterns and known isotope abundances were used in identifying the gas phase species.

EXPERIMENTAL

The combined dilatometer-mass spectrometer system is comprised of three parts: the gas flow system, the furnace/dilatometer assembly, and the mass spectrometer. The gas flow system includes gas cylinders and a mixing board that can combine up to three separate streams of gas. The furnace/dilatometer assembly consists of an alumina sample holder and pushrod, and a type B thermocouple that is positioned directly under the sample. A mullite muffle tube is used in the furnace, along with a fused silica transfer capillary that extends through an exterior heated interface and into the mass spectrometer. This transfer capillary allows a stream of gas from the furnace to be introduced into the mass spectrometer.

The interior of the muffle tube is evacuated to 0.6 mbar and backfilled twice with the carrier gas, helium at 99.999% purity, before each experiment. The helium is then set to a flow rate of 130 ml/min at 1 atm total pressure. Although barium titanate is commonly sintered in an oxygen rich or a reducing atmosphere, an inert atmosphere is used here to eliminate reactions between the sample and the gas. To obtain a complete purge of the system and establish flat baselines in the MS signals, the gas flow is maintained for 60-180 minutes. Prior to the start of the heating cycle, selected mass to charge, m/z, ratios are monitored, as well as the sample length change as a function of time and temperature. The thermal cycle includes both heating and cooling at 14°C/min, with a 20 min hold at 1350°C.

Samples are prepared from barium titanate powder (TAM Ceramics, NY) with a particle size of 0.9-1.3 μm and >99.0% purity. One gram of powder is placed in the die and dry pressed to a relative green density of 0.50, based on a theoretical density of 6.02 g/cm^3. No binder or release agent was used during the sample preparation. Calcia stabilized zirconia spacers were used on either side of the sample to prevent it from reacting with the alumina of the sample holder and pushrod.

RESULTS

Figure 1 shows the heating profile, length change of the $BaTiO_3$ sample, and four MS signal intensities: m/z=4, 14, 17, and 18, for a heating rate of 14°C/min in helium gas. The MS baselines are established from times 0-90 min, while the furnace is at room temperature. From 90-170 min (200-1100°C) the sample expands. At 170 min (1100°C) the sample begins to sinter, with a final linear shrinkage of about 17.8%, corresponding to a change in fractional density from 0.50 to 0.93. This is comparable to another study that showed a 0.92 fractional density with a maximum sintering temperature of 1250°C[11].

In Figure 1, the MS intensities of m/z=17 and 18 have similar profiles and the major peaks occur immediately after the start of heating, from 90-140 min (200-700°C). The major intensity of m/z=18 suggests it can be assigned to the evolution of water, with m/z=18 (${}^{1}H_2{}^{16}O$) and m/z=17 (${}^{16}O{}^{1}H$)[12]. The m/z=4 has a sloping baseline throughout the entire cycle and features a consumption or loss of signal from 135-155 min (600-900°C); this signal is assigned to the carrier gas, He. The m/z=14 signal is relatively unchanged and at low intensity throughout the entire cycle, an indication that there is not a significant amount of N_2 in the system.

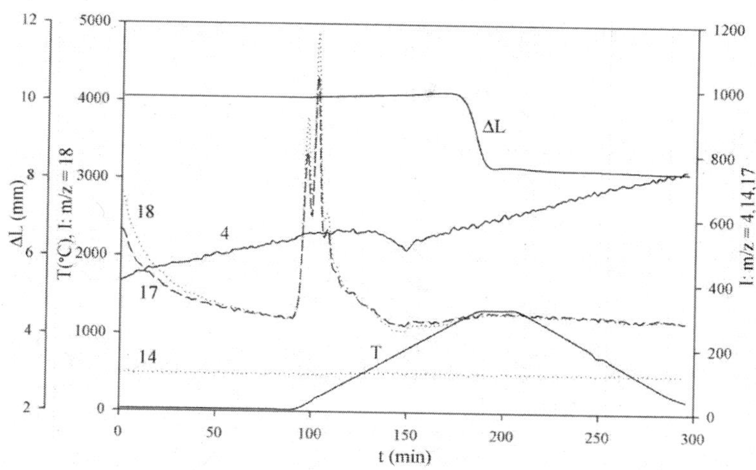

Figure 1. Temperature (T), shrinkage (ΔL), and species intensity (I) versus time (t) for 1 g of $BaTiO_3$ heated at 14°C/min in flowing helium. Species with m/z ratios = 4, 14, 17 and 18 are shown.

Figure 2 shows the heating profile along with five signal intensities that have similar profiles for m/z=12, 16, 28, 44, and 45. These profiles have multiple low temperature peaks from 90-115 min (200-350°C), a low temperature peak from 115-130 min (350-550°C), and a major peak from 130-150 min (550-825°C). The major peak of m/z=44 can be assigned to CO_2 being released, as this temperature range is in agreement with known decomposition temperatures of barium carbonate found as a pure substance and in a mixture with titanium dioxide[14]. The related signals can be attributed to the known cracking pattern of CO_2 (m/z=12 (^{12}C), m/z=16 (^{16}O), and m/z=28 ($^{12}C^{16}O$))[12] and an isotope of CO_2 (m/z=45 ($^{13}C^{16}O_2$))[13]. Even though the m/z=28 and 44 signal intensities are the largest of the observed signals, these peaks occur earlier (before 150 min) than the observed onset of sintering at 170 min (1100°C).

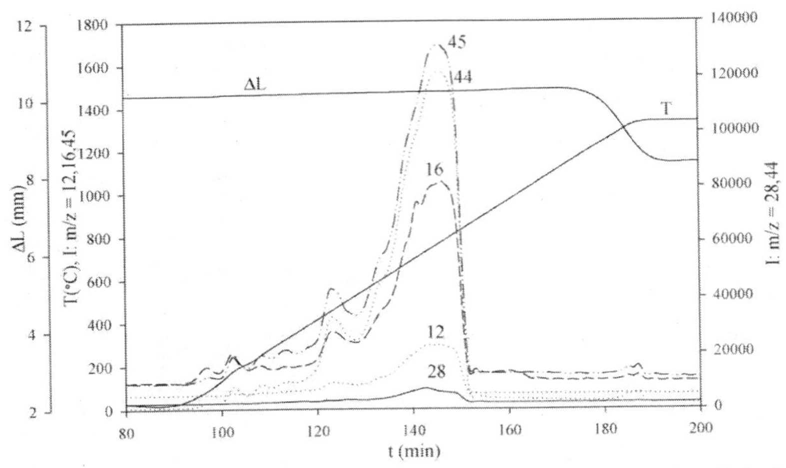

Figure 2. Temperature (T), shrinkage (ΔL), and species intensity (I) versus time (t) for 1 g of $BaTiO_3$ heated at 14°C/min in flowing helium. Species with m/z ratios = 12, 16, 28, 44, and 45 are shown.

Figure 3 shows the heating profile and the intensities of four signals, m/z=16, 32, 48, and 64, with similar shapes at 150-170 min (825-1100°C) and these occur at the onset of sintering. A plausible source for this family of peaks is TiO with peaks at m/z=64 ($^{48}Ti^{16}O$), m/z=48 (^{48}Ti), m/z=32 ($^{16}O_2$), and m/z=16 (^{16}O)[13]. A potential source of the TiO is unreacted TiO_2 powder, which is used to produce $BaTiO_3$. The TiO_2 powder could then decompose[15] into TiO and $1/2O_2$, and this, along with the cracking of TiO in the MS, would result in the observed signals at m/z=64, 48, 32, and 16. A second plausible source for the family of peaks shown in Figure 3 is SO_2 and its cracking pattern: m/z=64 ($^{32}S^{16}O_2$), m/z=48 ($^{32}S^{16}O$), m/z=32 (^{32}S), and m/z=16 (^{16}O)[12]. Sulfur is not commonly listed as a contaminant of $BaTiO_3$ on certificates of analysis; however, barium is mined as $BaSO_4$ and depending on the synthesis technique, could still be present in the sample. From Figure 3, it is also observed that the m/z=32 signal has a net consumption between 100-250 min (175-750°C). The absorption of O_2 before sintering and desorption after sintering is attributed to the mullite muffle tube in the furnace.

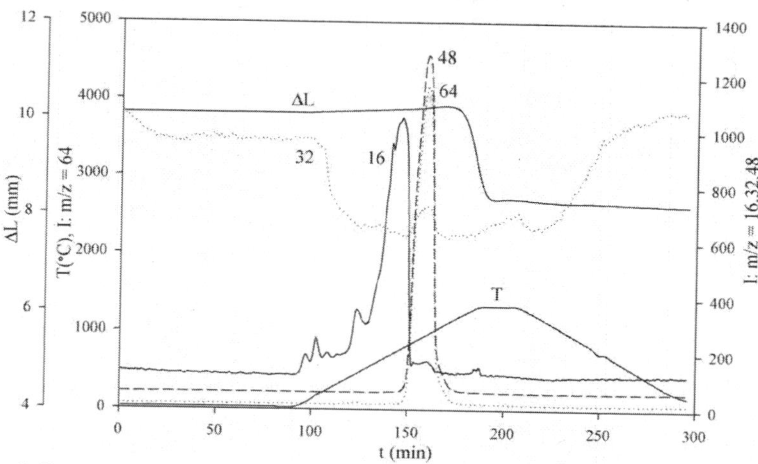

Figure 3. Temperature (T), shrinkage (ΔL), and species intensity (I) versus time (t) for 1 g of $BaTiO_3$ heated at 14°C/min in flowing helium. Species with m/z ratios = 16, 32, 48, and 64 are shown.

In order to determine the origin of the family of signals in Figure 3, the natural isotopes of titanium and sulfur can be used. Figure 4 shows the intensity of five signals with similar shapes over the same time period as the signals in Figure 3 (150-170 min). If the signals are the result of TiO, intensity peaks at m/z=46 (^{46}Ti), m/z=47 (^{47}Ti), m/z=62 ($^{46}Ti^{16}O$), and m/z=63 ($^{47}Ti^{16}O$)[13] would also be observed; however they are not. The intensity peaks that are observed can be attributed to isotopes of sulfur: m/z=34 (^{34}S), m/z=49 ($^{33}S^{16}O$), m/z=50 ($^{34}S^{16}O$), m/z=65 ($^{33}S^{16}O_2$), and m/z=66 ($^{34}S^{16}O$)[13]. Figure 4 also shows intensity peaks of m/z=49, 50, 65, and 66 at time 110-130 min (300-575°C). These peak times coincide with other signal peaks of m/z=38, 39, 41, 51, 55, 58, 78, 91, and 104; however, no definitive assignment has been made to this complicated family of peaks.

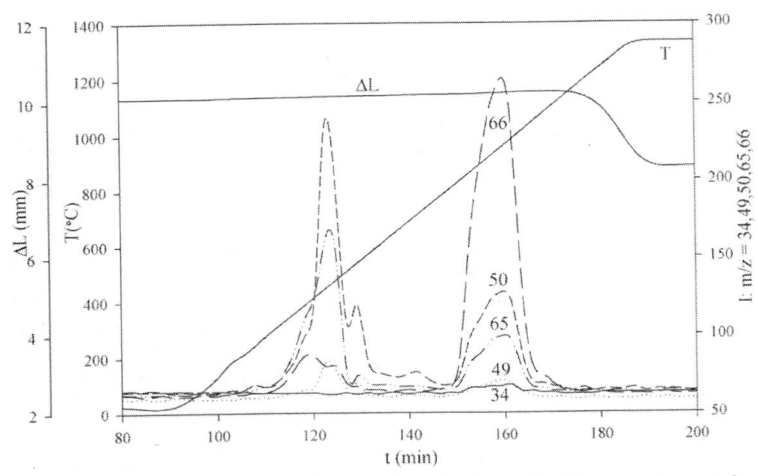

Figure 4. Temperature (T), shrinkage (ΔL), and species intensity (I) versus time (t) for 1 g of $BaTiO_3$ heated at 14°C/min in flowing helium. Species with m/z ratios = 34, 49, 50, 65, and 66 are shown.

DISCUSSION

In this work, high temperature reactions before and during sintering were observed by the use of combined dilatometry and mass spectrometry. The advantages of using mass spectrometry are the rapid detection time and real-time data acquisition. One challenge with the MS is that fragmentation often occurs during ionization, resulting in multiple signals. Another challenge is that the MS only detects mass/charge ratios. Tools that were used in this work to identify and assign gas phase species are reference cracking patterns and isotope abundances. The occurrence of gaseous species during the onset of sintering raises questions about whether or not these species affect the rate of sintering or defect formation in the sample.

CONCLUSIONS

Throughout the sintering cycle of a barium titanate sample, high temperature reactions and length changes were observed using a combined dilatometer and mass spectrometer system. At lower temperature, the evolution of water and some carbon dioxide from the sample were observed. The major evolution of carbon dioxide occurred at high temperature. Immediately following the appearance of carbon dioxide, the species evolved at the onset of sintering was assigned to sulfur dioxide. The identification of the gas phase species was accomplished by using reference cracking patterns and isotope abundances.

REFERENCES

[1]J.L. Margrave, The Characterization of High-Temperature Vapors (John Wiley & Sons, New York, 1967).
[2]J.H. Hastie, High Temperature Vapors: Science and Technology (Academic Press, New York, 1975).
[3]Characterization of High Temperature Vapors and Gases, Vol. 1, Edited by J.H. Hastie (NBS Special Publication 561, U.S. Department of Commerce, 1979).
[4]G.A. Somorjai, Studies of the Sublimation Mechanism of Solids in: Advances in High Temperature Chemistry, Vol. 2, Edited by L. Eyring (Academic Press, New York, 1969).
[5]W.L. Worrel, Dissociation of Gaseous Molecules on Solids at High Temperature in: Advances in High Temperature Chemisty, Vol. 4, Ed. By L. Ering (Academic Press, New York, 1971).
[6]B.V. Hiremath, A.I. Kingon, and J.V. Biggers, Reaction Sequence in the Formation of Lead Zirconate-Titanate Solid Solution: Role of Raw Materials, *J. Am. Ceram. Soc.*, **66**, 790 (1983).
[7]Sin-Shong Lin, Mass Spectrometric Analysis of Vapors in Oxidation of Si_3N_4 Compacts, *J. Am. Ceram. Soc.*, **58**, 160 (1975).
[8]J. Jung, Body-Controlled Sintering and Differential Thermal Analysis up to 2400°C, *Keramische Zeit*, **42**, 830 (1990).
[9]S. Siegel, M. Hermann, and G. Putzky, Effect of Process Atmosphere on Dilatometer Controlled Sintering, *Key Engineer. Mater.*, **89-91**, 237 (1994).
[10]K. Feng and S.J. Lombardo, High-Temperature Reaction Networks in Graphite Furnaces, *J. of Mat. Sci.*, **37**, 2747 (2002).
[11]H. Xu and L. Gao, Hydrothermal synthesis of high-purity $BaTiO_3$ powders: control of powder phase and size, sintering density, and dielectric properties, *Materials Letters*, **58**, 1582 (2004).
[12]S.E. Stein, Mass Spectra in: NIST Chemistry WebBook, NIST Standard Reference Database Number 69, Eds. P.J. Linstrom and W.G. Mallard, National Institute of Standards and Technology, Gaithersburg MD, 20899, http://webbook.nist.gov, (retrieved September 27, 2010).
[13] J.S. Coursey, D.J. Schwab, J.J. Tsai, and R.A. Dragoset, (2010), Atomic Weights and Isotopic Compositions (version 3.0). National Institute of Standards and Technology, Gaithersburg MD, http://physics.nist.gov/Comp. (retrieved September 27, 2010).
[14]M.I. Zaki and M. Abdel-Khalik, Thermal Decomposition and Creation of Reactive Solid Surfaces. I. Characterization of the Decomposition Products of Alkaline Earth Oxalates, *Thermochimica Acta*, **78**, 29 (1984).
[15]A.W. Czanderna and J.M. Honig, Interaction of Oxygen with Titanium Dioxide, *J. Phys. Chem.*, **63**, 620 (1959).

EFFECT OF DC POLING FIELD ON FERROELECTRIC PROPERTIES IN ALKALI BISMUTH TITANATE LEAD-FREE CERAMICS

Toshio Ogawa, Takayuki Nishina, Masahito Furukawa*, and Takeo Tsukada*
Department of Electrical and Electronic Engineering, Shizuoka Institute of Science and Technology, 2200-2 Toyosawa, Fukuroi, Shizuoka 437-8555, Japan
*Materials and Process Development Centre, TDK Corporation, 570-2, Matsugashita Minamihatori, Narita, Chiba 286-8588, Japan

ABSTRACT

The fluctuation of dielectric and piezoelectric properties and the DC poling field dependence of ferroelectric properties were investigated to intend for practical use and to evaluate the mechanism of domain alignment in lead-free ceramics of the forms of $(1\text{-}x)(Na_{0.5}Bi_{0.5})TiO_3\text{-}x(K_{0.5}Bi_{0.5})TiO_3$ (x=0.08-0.28), $0.79(Na_{0.5}Bi_{0.5})TiO_3\text{-}0.20(K_{0.5}Bi_{0.5})TiO_3\text{-}0.01Bi(Fe_{0.5}Ti_{0.5})O_3$ and $(1\text{-}x)(Na_{0.5}Bi_{0.5})TiO_3\text{-}xBaTiO_3$ (x=0.03-0.11). It was found that there were two kinds of fluctuation from the results to evaluate dielectric loss (tanδ); one comes from chemical compositions and another is due to the manufacturing processes such as firing process. For the compositions for improvement of tanδ, it was though that the substitution by $(K_{0.5}Bi_{0.5})TiO_3\text{-}Bi(Fe_{0.5}Ti_{0.5})O_3$ for $(Na_{0.5}Bi_{0.5})TiO_3$, especially $Bi(Fe_{0.5}Ti_{0.5})O_3$, played an important role in reducing tanδ. In addition, tanδ could be improved by $BaTiO_3$ substitution for $(Na_{0.5}Bi_{0.5})TiO_3$ in the case of the small amount of up to x=0.11. For the manufacturing process, since there is a linear relation between tanδ and relative dielectric constant (ε_r), it was found that the firing process affects the fluctuation of tanδ in individual samples of the same firing lot. From DC poling field (E) dependence of ferroelectric properties, asymmetrical shapes in piezoelectric constants vs ±E caused by nonuniform mobility of domain alignment upon the application of positive and negative poling fields were observed in the small amount substitution by $(K_{0.5}Bi_{0.5})TiO_3$ ($x \leqq 0.11$) and $BaTiO_3$ ($x \leqq 0.03$). On the other hand, symmetrical shapes in piezoelectric constants vs ±E were observed in higher piezoelectricity with lower frequency constant accompanied with typical domain clamping. The effects to improve tanδ by $Bi(Fe_{0.5}Ti_{0.5})O_3$ and by $BaTiO_3$ substitutions for $(Na_{0.5}Bi_{0.5})TiO_3$ were also proved by the measurement of P-E hysteresis loops.

INTRODUCTION

Lead-free piezoelectric ceramics have been studied by many researchers, because of replacing $Pb(Zr,Ti)O_3$ (PZT) ceramics. There are three major chemical compositions; alkali niobate[1], alkali bismuth titanate[2] and barium titanate[3]. While relatively high piezoelectricity of 70% in PZT ceramics is realized in alkali niobate [the piezoelectric strain d_{33} constant is 307 pC/N[4] in $0.95(Na,K,Li,Ba)(Nb_{0.9}Ta_{0.1})O_3\text{-}0.05SrZrO_3$ with a small amount of MnO[5]] and barium titanate, low piezoelectricity with low dielectric constant and high electro-mechanical quality factor were obtained in alkali bismuth titanate. In addition, these ceramics have undesirable characteristics in the cases of alkali-content compositions such as large fluctuation in dielectric and piezoelectric properties accompanied with fluctuation in compositions during the manufacturing processes, particularly ball milling (alkali dissolving phenomenon) and firing (alkali vaporization phenomenon).

Improving the fluctuations in the dielectric and piezoelectric properties, investigation for effects of some kinds of substitution to compensate alkali dissolving and alkali vaporization on the decrease of dielectric loss in the ceramics and for mass reproducibility of manufacturing lots was significant from practical points of view in use. Furthermore, improving the piezoelectricity, a study on domain alignment by DC poling field is also important that how to realize higher piezoelectricity in lead-free ceramics. We have already shown the mechanism of domain alignment in PZT[6-10], $PbTiO_3$ (PT)[11], $BaTiO_3$ (BT)[12], alkali niobate ceramics composed of $(Na,K,Li,Ba)(Nb_{0.9}Ta_{0.1})O_3$-$SrZrO_3$[4] and in a relaxor single crystal of $Pb[(Zn_{1/3}Nb_{2/3})_{0.91}Ti_{0.09}]O_3$ (PZNT)[13] by measuring the piezoelectricity vs DC poling field characteristics. Therefore, in this study, the fluctuation of dielectric and piezoelectric properties was investigated in alkaline bismuth titanate ceramics to intend for practical use. Moreover, a direction for new research on lead-free ceramics with higher piezoelectricity is proposed to study the mechanism of domain alignment on the basis of the poling characteristics, especially the DC poling field dependence of dielectric and piezoelectric properties in alkali bismuth titanate ceramics.

EXPERIMENTAL PROCEDURE

The lead-free ceramics evaluated are of the forms of $(1\text{-}x)(Na_{0.5}Bi_{0.5})TiO_3$-$x(K_{0.5}Bi_{0.5})TiO_3$[14] [abbreviate to (1-x)NBT-xKBT] (x=0.08, 0.11, 0.15, 0.18, 0.21, 0.28), 0.79 $(Na_{0.5}Bi_{0.5})TiO_3$-$0.20(K_{0.5}Bi_{0.5})TiO_3$-$0.01Bi(Fe_{0.5}Ti_{0.5})O_3$ [abbreviate to 0.79NBT-0.20KBT-0.01 BFT] and $(1\text{-}x)(Na_{0.5}Bi_{0.5})TiO_3$-$xBaTiO_3$[15] [abbreviate to (1-x)NBT-xBT] (x=0.03, 0.07, 0.11), the compositions of which were chosen through research by trial and error on the chemical compositions necessary to obtain a d_{33} of over 100 pC/N in alkali bismuth titanate. The ceramics were fabricated by a conventional ceramic manufacturing process under firing conditions of 1150-1175 ℃ for 2 h. The DC poling field, poling temperature and poling time were 2.4-5.6 kV/mm, 30-100 ℃ and 15 min, respectively, since the poling conditions depend on the dielectric loss of the ceramic disks (dimensions: 14 mm$^{\phi}$x 0.5 mm^{T}) with a Ag electrode. In the experiment of the poling field dependence of ferroelectric properties, the DC electric field (E) was gradually (0.25 kV/mm or 0.5 kV/mm each) varied from E=0→+4.0→0→-4.0→0 to +4.0 kV/mm under a poling temperature of 70℃ and a poling time of 30 min in (1-x)NBT-xKBT and 0.79NBT-0.20KBT-0.01BFT, and from E=0→+3.0→0→-3.0→0 to +3.0 kV/mm under a poling temperature of 70℃ and a poling time of 30 min in (1-x)NBT-xBT. After each poling, the dielectric and piezoelectric properties were measured at room temperature using an LCR meter (HP4263A), a precision impedance analyzer (Agilent 4294A) and a piezo d_{33} meter (Academia Sinica ZJ-3D), respectively. Hysteresis loops of polarization vs electric field (P-E hysteresis loops) measured at 70℃ were observed using a ferroelectric test system (Radiant RT6000HVS) by applying a bipolar triangle pulse, the period of which was 400 ms.

RESULTS AND DISCUSSION

Fluctuation of dielectric properties before poling

Figure 1 shows the relationships between the substitution amounts of KBT, 0.20KBT-0.01BFT and BT for NBT and dielectric loss (tanδ) and relative dielectric constant (ε_r) measured at 1 kHz in the ceramics before poling. While values of tanδ in the substitution by 0.20KBT-0.10BFT and BT became small, values of tanδ have a tendency to increase with increasing the substitution by KBT. Furthermore, the standard deviation (±σ) of the fluctuation of tanδ became large. However, the ±σ of ε_r became small because of the intrinsic physical property in the ceramics. It was though that the substitution by 0.20KBT-0.01BFT and BT, especially BFT, played an important role in reducing tanδ to compensate alkali vaporization during the firing. As the valence of Fe in $Bi(Fe_{0.5}Ti_{0.5})O_3$ (BFT) is easy to change by its circumstance, it may work to

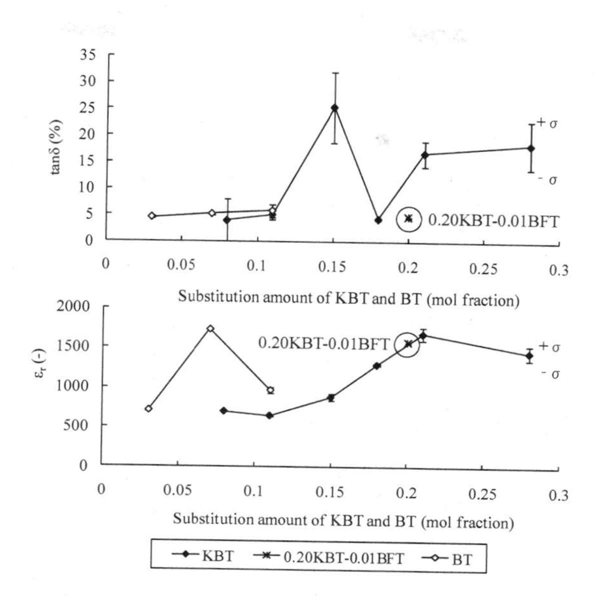

Figure 1 Relationships between dielectric loss (tanδ), relative dielectric constant (ε_r) measured at 1 kHz and substitution amounts of $(K_{0.5}Bi_{0.5})TiO_3$ (abbreviate to KBT), $0.20(K_{0.5}Bi_{0.5})TiO_3$-$0.01Bi(Fe_{0.5}Ti_{0.5})O_3$ (abbreviate to 0.20KBT-0.01BFT) and $BaTiO_3$ (abbreviate to BT). $\pm\sigma$ is standard deviation.

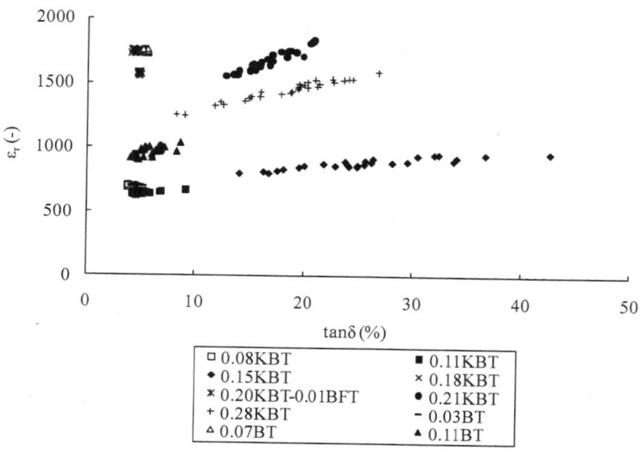

Figure 2 Relationships between ε_r and tanδ (at 1 kHz) of various compositions of (1-x)$(Na_{0.5}Bi_{0.5})TiO_3$-x$(K_{0.5}Bi_{0.5})TiO_3$ (x=0.08-0.28: abbreviate to xKBT), $0.79(Na_{0.5}Bi_{0.5})TiO_3$-$0.20(K_{0.5}Bi_{0.5})TiO_3$-$0.01Bi(Fe_{0.5}Ti_{0.5})O_3$ (0.20KBT-0.01BFT) and (1-x)$(Na_{0.5}Bi_{0.5})TiO_3$-x$BaTiO_3$ (x= 0.03-0.11: abbreviate to xBT).

improve the dielectric properties[16], especially tanδ. Furthermore, in order to clarify the fluctuation in dielectric properties, the relationship between tanδ and ε_r for all the samples before poling was illustrated in Fig. 2. Since there were linear relations between them in the cases of the same compositions with large tanδ over 10%, it was found that the firing process affects the fluctuation in individual samples of the same firing lot. We believe that one of the reasons to appear the line relationships was caused by positions of ceramic green bodies in saggar during firing[17], which correspond to the degree of the alkali vaporization.

Fluctuation of dielectric and piezoelectric properties after poling

Figures 3(a) and 3(b) show the relationships between the substitution amounts of KBT, 0.20KBT-0.01BFT and BT for NBT and tanδ and ε_r measured at 1 kHz in the ceramics after poling, and the relationships between the substitution amounts and planer coupling factor (k_p) and piezoelectric strain d_{33} constant in the cases of different manufacturing lots of 1st lot (1) and 2nd lot (2). Herein, the different manufacturing lot means to produce the ceramics at different date. While the values of tanδ and the ±σ of tanδ, and the ±σ of ε_r were large in the substitution by KBT of the 1st lot, those were improved in the 2nd lot. It depended on manufacturing lots because of fluctuation in alkali dissolving phenomenon in ball milling process and alkali vaporization phenomenon in firing process. On the other hand, the values of tanδ and the ±σ of tanδ, and the ±σ of ε_r were small in the substitution by 0.20KBT-0.01BFT and BT. This means that the reproducibility was preferable to the substitution by 0.20KBT-0.01BFT and BT due to the improvement of tanδ by BFT and the small amount substitution of up to x=0.11 by BT. It was thought that the compositions by KBT substitution may be more active than the compositions by BT substitution regarding alkali dissolving and alkali vaporization in the processes.

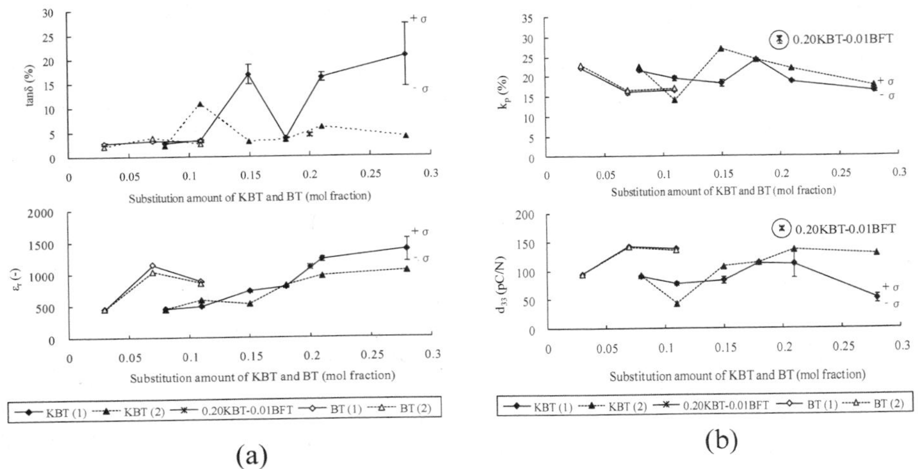

Figure 3 Relationships between (a) tanδ, ε_r (at 1 kHz) and (b) k_p, d_{33} and substitution amounts of $(K_{0.5}Bi_{0.5})TiO_3$ (KBT), $0.20(K_{0.5}Bi_{0.5})TiO_3$-$0.01Bi(Fe_{0.5}Ti_{0.5})O_3$ (0.20KBT-0.01BFT) and $BaTiO_3$ (BT). ± σ is standard deviation; — and --- show (1) 1st lot and (2) 2nd lot, respectively.

The k_p and d_{33} constant dependences of the substitution by KBT, 0.20KBT-0.01BFT and BT show the same tendency in the cases of dielectric properties as smaller tanδ corresponds to higher k_p and d_{33}, because of sufficient poling treatment. The maximum k_p of around 30% and d_{33} of 160 pC/N were obtained in the composition of 0.79NBT-0.20KBT-0.01BFT.

Poling field dependence

Figures 4 and 5 respectively show the effects of a DC poling field (E) on ε_r (at 1 kHz), k_p, fc_p (frequency constant in the k_p mode), and d_{33} at various substitution amounts of KBT, 0.20KBT-0.01BFT and BT for NBT when E was varied from 0 to ±4.0 kV/mm and from 0 to ±3.0 kV/mm. By increasing x from 0.08 to 0.18 in (1-x)NBT-xKBT and 0.79NBT-0.20KBT-0.01BFT (Fig. 4), and x from 0.03 to 0.11 in (1-x)NBT-xBT (Fig. 5), the relationships of ε_r, k_p, fc_p, and d_{33} with E show domain clamping at a specific E. It was considered that the minimum ε_r, k_p, and d_{33} and the maximum fc_p, owing to electrical domain clamping, denoted by ↑↓ [the arrow (↑) means domain alignment], occurred at coercive fields (E_c) corresponding to a specific E, as mentioned earlier. Although the E_c (E at d_{33}=0) decreased with increasing x from x=0.08 to 0.18 in (1-x)NBT-xKBT and 0.79NBT-0.20KBT-0.01BFT (Fig. 4), the E_c decreased from x=0.03 to 0.07, and after that the Ec slightly increased from x=0.07 to 0.11 in (1-x)NBT-xBT (Fig. 5). The reason for the E_c dependence on the composition will be discussed later. The ceramics with the compositions of 0.82NBT-0.18KBT [Fig. 4(b)] and 0.79NBT-0.20KBT-0.01BFT [Fig. 4(c)] show ε_r and fc_p vs E characteristics such as rhombohedral PZT ceramics[8], whereas those of 0.92NBT-0.08KBT [Fig. 4(a)] and 0.97NBT-0.03BT [Fig. 5(a)] show k_p vs E characteristic such as PT ceramics[11]. In addition, asymmetrical shapes in the k_p and fc_p vs +E curves, and in the k_p and fc_p vs -E curves were observed for ceramics of 0.92NBT-0.08KBT [Fig. 4(a)] and 0.97NBT-0.03BT [Fig. 5(a)] because the ceramics showed nonuniform mobility of domain alignment upon the application of positive and negative poling fields. However, the ceramics of 0.82NBT-0.18KBT [Fig. 4(b)], 0.79NBT-0.20KBT-0.01BFT [Fig. 4(c)] and x=0.07 [Fig. 5(b)] and 0.11 [Fig. 5(c)] in (1-x)NBT-xBT exhibited symmetrical shapes in their k_p and fc_p vs ±E curves. These phenomena are due to the substitution amount for NBT and the kinds of the substitution materials such as KBT and BT. Therefore, it was found that BT substitution [x=0.07 in Fig. 5(b)] was preferred by KBT [x=0.08 in Fig. 4(a)] substitution to realize the symmetrical hysteresis shapes in k_p and fc_p vs ±E in the case of the small amount substitution. We believe these compositions with the symmetrical hysteresis shapes correspond to morphotropic phase boundaries (MBP)[18]. Moreover, the maximum k_p (around 25% in 0.82NBT-0.18KBT and 0.79NBT-0.20KBT-0.01BFT and around 15% in 0.93NBT-0.07BT and 0.89NBT-0.11BT) and maximum d_{33} (160 pC/N in 0.79NBT-0.20KBT-0.01BFT and 140 pC/N in 0.93NBT-0.07BT and 0.89NBT-0.11BT) were obtained at the lowest fc_p of 2930 Hz·m and 2950-2980 Hz·m, respectively. Since the domain alignment in the ceramics was accompanied by the deformation of the crystal under a DC poling field, it was clarified that high k_p and high d_{33} in the lead-free ceramics are necessary to realize low fc_p (low Young's modulus) in ceramic compositions, which corresponds to MPB, as shown in Fig. 4 and 5 (fc_p vs E). Moreover, the optimal ceramic composition regarding piezoelectricity shows that the typical domain clamping in the ε_r, k_p, fc_p, and d_{33} vs E curves occurred almost simultaneously at E in 0.82NBT-0.18KBT and 0.79NBT-0.20KBT-0.01BFT [Fig. 6(a) ; see the E to obtain minimum k_p, k_t and d_{33}] and in 0.93NBT-0.07BT [Fig. 6(b)]. The difference between $+E_c$ and $-E_c$ in Figs. 6(a) and 6(b) was thought to be the effect of domain alignment caused by the DC poling of as-fired (virgin) ceramics at E=0→E=+4.0 (+3.0) kV/mm.

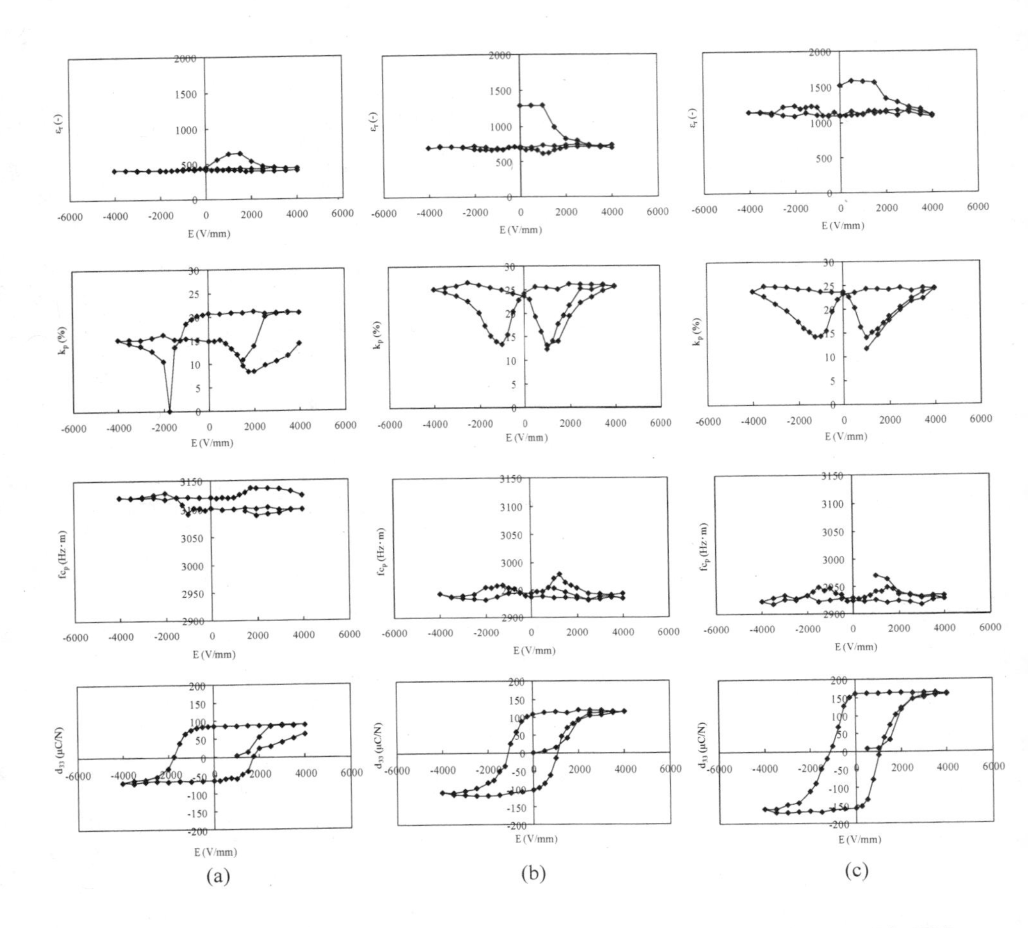

Figure 4 DC poling field dependences of ε_r, k_p, fc_p and d_{33} at (a) $0.92(Na_{0.5}Bi_{0.5})TiO_3$-$0.08(K_{0.5}Bi_{0.5})TiO_3$ (b) $0.82(Na_{0.5}Bi_{0.5})TiO_3$-$0.18(K_{0.5}Bi_{0.5})TiO_3$ and (c) $0.79(Na_{0.5}Bi_{0.5})TiO_3$-$0.20(K_{0.5}Bi_{0.5})TiO_3$-$0.01Bi(Fe_{0.5}Ti_{0.5})O_3$ measured at 70℃.

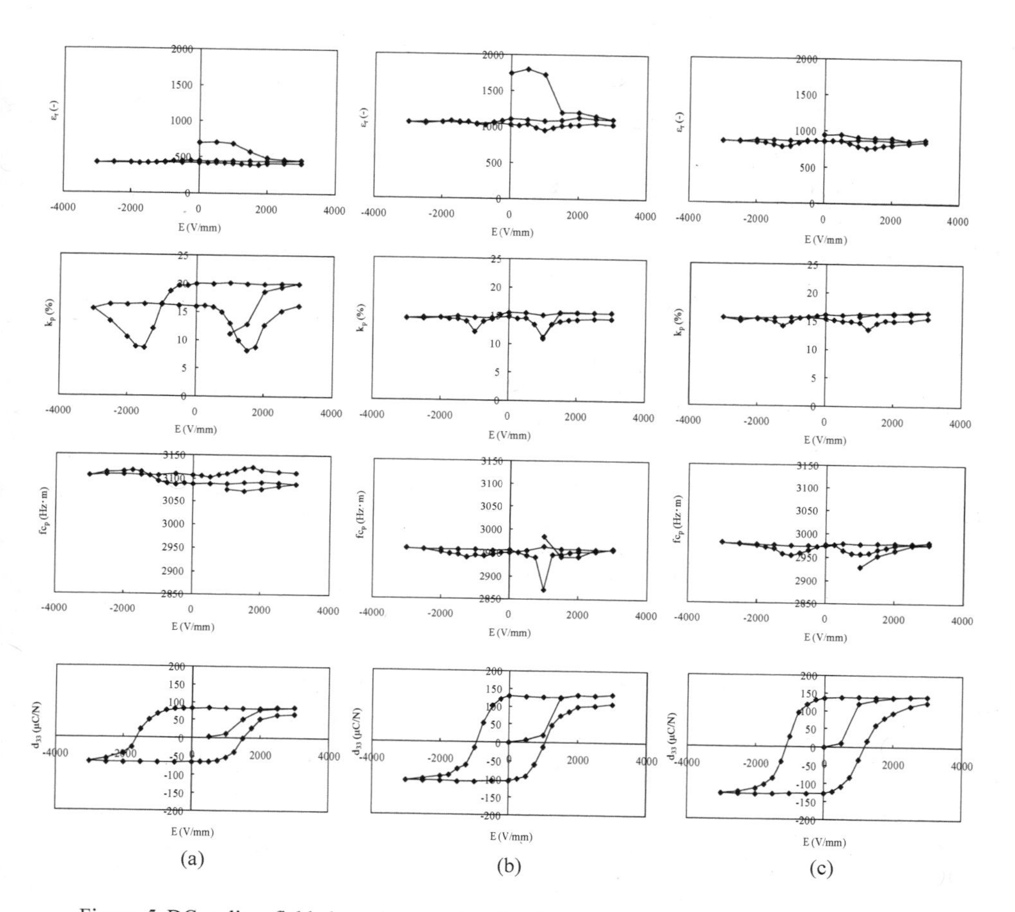

Figure 5 DC poling field dependences of ε_r, k_p, fc_p and d_{33} at (a) $0.97(Na_{0.5}Bi_{0.5})TiO_3$-0.03 $BaTiO_3$, (b) $0.93(Na_{0.5}Bi_{0.5})TiO_3$-$0.07BaTiO_3$ and (c) $0.89(Na_{0.5}Bi_{0.5})TiO_3$-$0.11xBaTiO_3$ measured at 70℃.

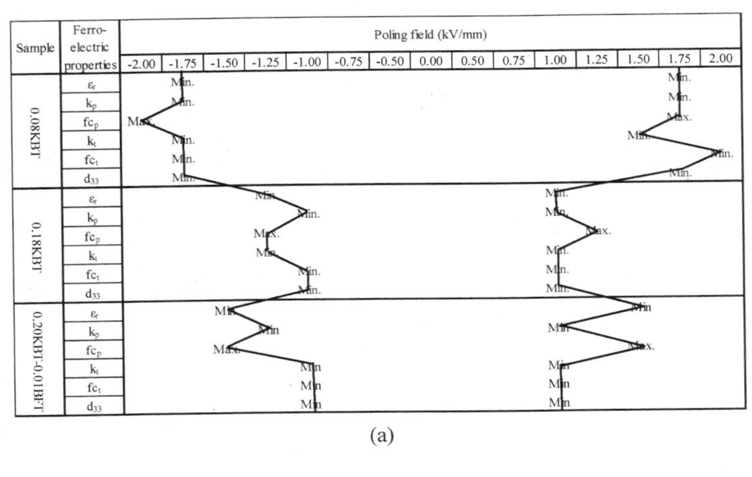

(a)

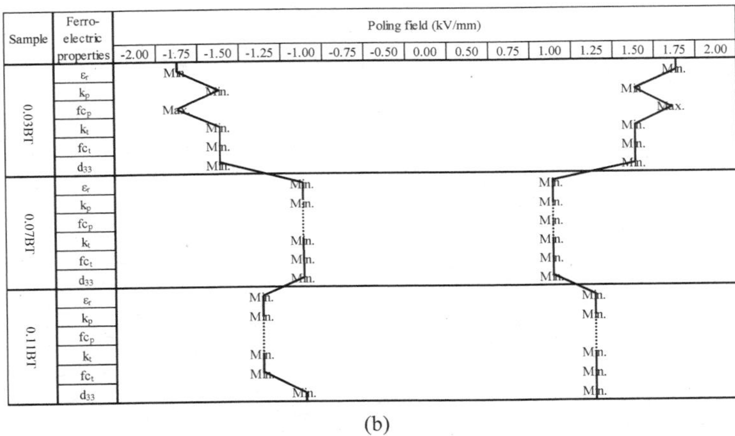

(b)

Figure 6 Relationships between DC poling field (E) required to obtain minimum ε_r, k_p, k_t*, fc_t*, and d_{33} and maximum fc_p, and substitution amounts of xKBT, 0.20KBT-0.01BFT and xBT. Domain clamping appears at the coercive field (E_c) required to realize minimum ε_r, k_p, k_t*, fc_t*, and d_{33} and maximum fc_p. A typical domain clamping state is observed simultaneously at a specific E of ± 1.0 kV/mm by 0.07BT substitution [Fig. 6(b)]. * k_t is the electromechanical coupling factor in the thickness mode of the disk, and fc_t is the frequency constant in the k_t mode.

P-E hysteresis loops

Figures 7 and 8 show hysteresis loops measured at 70 ℃ under various applied fields (E), with a maximum E of ±6.0 kV/mm, at the compositions of (1-x)NBT-xKBT, 0.79NBT-0.20KBT-0.01BFT and (1-x)NBT-xBT, respectively.

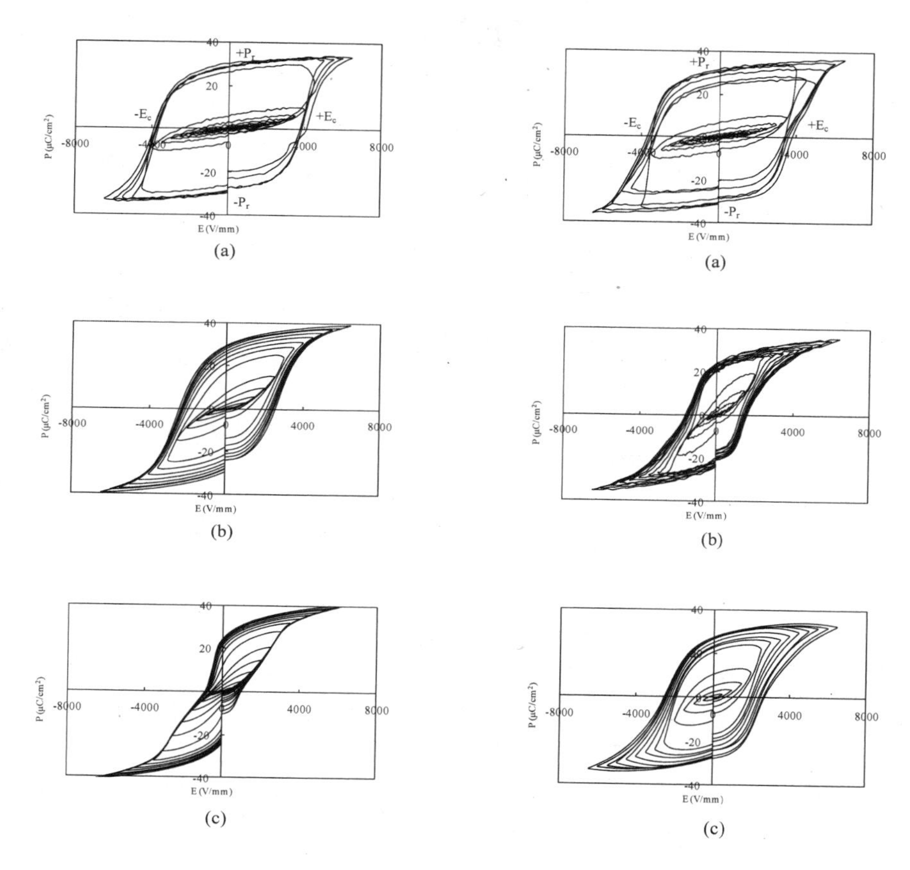

Figure 7 P-E hysteresis loops at (a) $0.92(Na_{0.5}Bi_{0.5})TiO_3$-$0.08(K_{0.5}Bi_{0.5})TiO_3$, (b) $0.82(Na_{0.5}Bi_{0.5})TiO_3$-$0.18(K_{0.5}Bi_{0.5})TiO_3$ and (c) $0.79(Na_{0.5}Bi_{0.5})TiO_3$-$0.20(K_{0.5}Bi_{0.5})TiO_3$-0.01 $Bi(Fe_{0.5}Ti_{0.5})O_3$ measured at 70℃. The remnant polarization ($\pm P_r$) and coercive field ($\pm E_c$) are shown in Fig. 7(a).

Figure 8 P-E hysteresis loops at (a) 0.97 $(Na_{0.5}Bi_{0.5})TiO_3$-$0.03BaTiO_3$, (b) 0.93 $(Na_{0.5}Bi_{0.5})TiO_3$- $0.07BaTiO_3$ and (c) $0.89(Na_{0.5}Bi_{0.5})TiO_3$-0.11x $BaTiO_3$ measured at 70℃. The remnant polarization ($\pm P_r$) and coercive field ($\pm E_c$) are shown in Fig. 8(a).

While almost the same P-E hysteresis loops were observed in 0.92NBT-0.08KBT [Fig. 7(a)] and 0.97NBT-0.03BT [Fig. 8(a)] independent of the substitution amounts (x=0.08 or 0.03), the E_c obviously decreased in 0.93NBT-0.07BT (x=0.07 substitution) [Fig. 8(b)] by comparison with 0.92NBT-0.08KBT (x=0.08 substitution). It was thought that the effect of the substitution materials (KBT and BT) on ferroelectricity was different, especially Curie temperature (T_c), as mentioned in the realization of the asymmetrical and symmetrical hysteresis shapes in k_p and fc_p vs $\pm$E under the poling field dependence. The loop of 0.79NBT-0.20KBT-0.01BFT [Fig. 7(c)] showed a propeller shape and a smaller Ec in comparison with the loop of 0.72NBT-0.18KBT [Fig. 7(b)], because of the introduction of 1 mol% $Bi(Fe_{0.5}Ti_{0.5})O_3$ (BFT). We believe that Fe in BFT generated space charges (the propeller shape hysteresis loop) in the ceramics, which improved tanδ before and after poling [Figs. 1, 2 and 3(a)], and as a result, the piezoelectricity, especially k_p and d_{33} constant were enhanced [Fig. 3(b)]. On the other hand, increasing x from 0.07 to 0.11 in (1-x)NBT-xBT, the $\pm E_c$ in Fig. 8(c) increased as well as the $\pm$Ec obtained from the poling field dependence (Fig. 5). These results corresponded to the change in crystal phase identified by XRD from rhombohedral phase to tetragonal phase while x increased from 0.07 to 0.11. It was thought that BT characteristics such as a low T_c of 130℃ dominated the ferroelectricity regarding the P-E hysteresis loop while the increase of x.

The E dependences of remnant polarization ($\pm P_r$) and coercive field ($\pm E_c$) are illustrated in (1-x)NBT-xKBT, 0.79NBT-0.20KBT-0.01BFT (Fig. 9) and (1-x)NBT-xBT (Fig. 10), respectively.

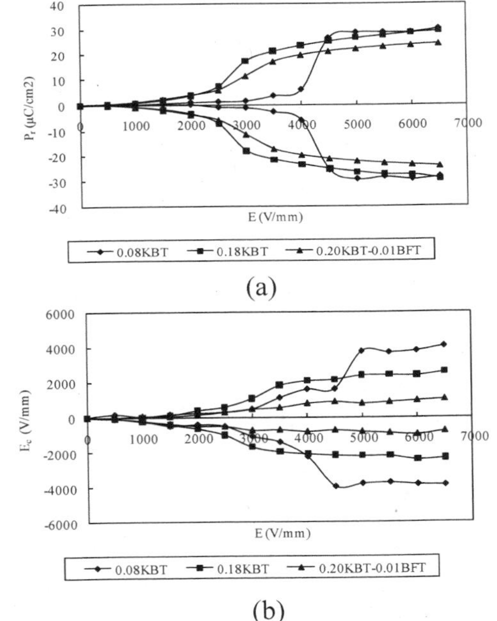

Figure 9 Applied field (E) dependences of (a) remnant polarization ($\pm P_r$) and (b) coercive field ($\pm E_c$) in substitution amounts of 0.08KBT, 0.18KBT and 0.20KBT-0.01BFT measured at 70℃.

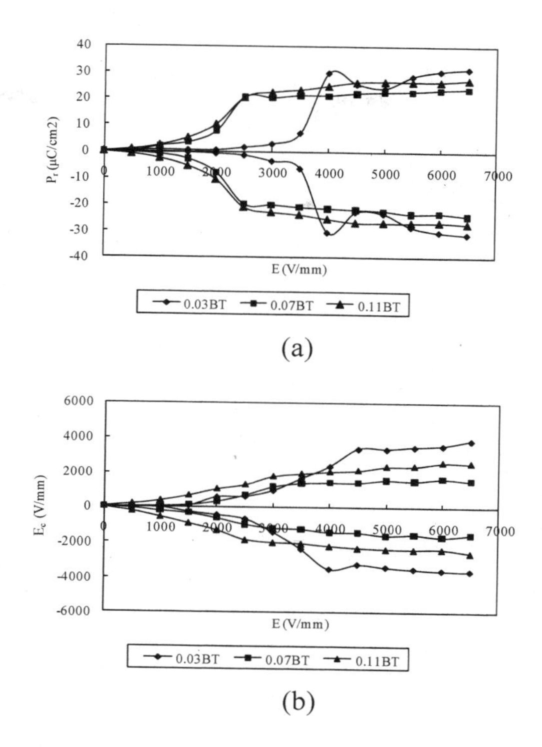

Figure 10 Applied field (E) dependences of (a) remnant polarization ($\pm P_r$) and (b) coercive field ($\pm E_c$) in substitution amounts of 0.03BT, 0.07BT and 0.11BT measured at 70℃.

While the P_r reached to ±25-±30 μC/cm^2 over E=5.0 kV/mm independent of the kinds of the substitution materials and the substitution amount, the Pr in (1-x)NBT-xBT [Fig. 10(a)] became larger than the one in (1-x)NBT-xKBT and 0.79NBT-0.20KBT-0.01BFT [Fig. 9(a)] between the E of 1.0 and 4.0 kV/mm. The smallest Ec dependence on E was confirmed at 0.79NBT-0.20KBT-0.01BFT [Fig. 9(b)] and 0.93NBT-0.07BT [Fig. 10(b)], at the compositions of which higher k_p and d_{33} constant were obtained in the alkali bismuth titanate ceramics. Since the mobility of ferroelectric domains during the application of the polar field was successfully evaluated by the shapes of the P-E loop and the relationships between P_r/E_c and E, it was found that the reason for obtaining higher k_p and d_{33} constant was obtained in the slim P-E loop [Fig. 7(c) and Fig. 8(b)], larger P_r from the rang of lower E (1.0 kV/mm≦E≦4.0 kV/mm) and smaller E_c dependence on E.

CONCLUSIONS

The fluctuation of dielectric and piezoelectric properties and the DC poling field dependences of dielectric and piezoelectric properties were investigated in alkali bismuth titanate lead-free ceramics. The fluctuation of dielectric properties, in particular tanδ, could be suppressed by the substitution by $Bi(Fe_{0.5}Ti_{0.5})O_3$ and $BaTiO_3$ for $(Na_{0.5}Bi_{0.5})TiO_3$. Higher planer coupling factor and piezoelectric strain constant were realized in symmetrical shapes in piezoelectric constants vs DC poling field caused by uniform mobility of domain alignment upon the application of positive and negative poling fields. Furthermore, the typical domain clamping simultaneously occurred at the compositions with low frequency constant. The P-E hysteresis measurement also showed the same results as the poling field dependence.

ACKNOWLEDGEMENTS

This work was partially supported by a Grant-in-Aid for Scientific Research C (No. 21560340) and a Grant of Strategic Research Foundation Grant-aided Project for Private Universities (No. S1001032) from the Ministry of Education, Culture, Sports, Science and Technology and a Research Foundation Grant 2009 jointly sponsored by Academia and Industry of Fukuroi City and the Murata Science Foundation 2009.

REFERENCES

[1]Y. Saito, H. Takao, T. Tani, T. Nonoyama, K. Takatori, T. Homma, T. Nagaya, and M. Nakamura, *Nature*, **432**, 84 (2004).
[2]T. Takenaka and K. Sakata, *Jpn. J. Appl. Phys.*, **19**, 31 (1980).
[3]H. Takahashi, Y. Numamoto, J. Tani, and S. Tsurekawa, *Jpn. J. Appl. Phys.*, **45**, 7405 (2006).
[4]T. Ogawa, M. Furukawa, and T. Tsukada, *Jpn. J. Appl. Phys.*, **48**, 709KD07 (2009).
[5]M. Furukawa T. Tsukada, D. Tanaka, and N. Sakamoto, *Proc. 24th Int. Japan-Korea Semin. Ceramics*, 2007, p. 339.
[6]T. Ogawa, A. Yamada, Y. K. Chung, and D. I. Chun, *J. Korean Phys. Soc.*, **32**, S724 (1998).
[7]T. Ogawa and K. Nakamura, *Jpn. J. Appl. Phys.*, **37**, 5241 (1998).
[8]T. Ogawa and K. Nakamura, *Jpn. J. Appl. Phys.*, **38**, 5465 (1999).
[9]T. Ogawa, *Ferroelectrics*, **240**, 75 (2000).
[10]T. Ogawa, *Ceram. Int.*, **26**, 383 (2000).
[11]T. Ogawa, Jpn. *J. Appl. Phys.*, **39**, 5538 (2000).
[12]T. Ogawa, *Jpn. J. Appl. Phys.*, **40**, 5630 (2001).
[13]T. Ogawa, *Ferroelectrics*, **273**, 371 (2002).
[14]P. Z. Yang, B. Liu, L. L. Wei, and Y. T. Hou, *Mater. Res. Bull.*, **43**, 81 (2008).
[15]Y. J. Dai, J. S. Pan, and X. W. Zhang, *Key Eng. Mater.*, **336-338**, 206 (2007).
[16]C. Zhou, X. Liu, and W. Li, *Mater. Sci. Eng. B*, **153**, 31 (2008).
[17]T. Ogawa, *Ceramic Bulletin*, **70**, 1042 (1991).
[18]W. Zhao, H. P. Zhou, Y. K. Yan, and D. Liu, *Key Eng. Mater.*, **368-372**, 1908 (2008).

MULTIFUNCTIONAL NATURE OF MODIFIED IRON TITANATES AND THEIR POTENTIAL APPLICATIONS

[x]R. K. Pandey[1,2], P. Padmini[2], P. Kale[2], J. Dou[2], C. Lohn[1], R. Schad[2], R. Wilkins[3] and W. Geerts[1]

[1]Texas State University, San Marcos, TX; [2] The University of Alabama, Tuscaloosa, AL

[3] Prairie View A&M University, Prairie View, TX

[x] Corresponding Author: rkpandey@txstate.edu

Key Words: magnetic-semiconductors, radhard electronics, spintronics, magneto-electronics, power amplifiers and tuned varistors.

ABSTRACT

The phenomenal growths of information technology, microelectronics and radhard electronics have warranted the development of new class of materials. Multifunctional oxides, magnetic-semiconductors, multiferroics and smart materials are just a few examples. They are needed for the advancement of technologies such as spintronics, magneto-electronics, radhard electronics, and advanced microelectronics. For these technologies, of particular interest are some solid solutions of ilmenite-hematite (IH) represented by (1-x) $FeTiO_3.xFe_2O_3$ where x varies from 0 to 0.6; Mn-doped ilmenite (Mn-$FeTiO_3$) and pure and Mn-doped pseudobrookite, Fe_2TiO_5 . These multifunctional oxides are ferrimagnetic with the magnetic Curie points well above the room temperature as well as wide band gap semiconductors with band gap $E_g > 2.5$ eV. In this paper, we will discuss: (a) processing of device quality samples for structural, electrical and magnetic characterization, (b) selective magnetic switching of integrated structures, and (c) response of their nonlinear current-voltage characteristics (I-V) when subjected to an external magnetic field as well as when irradiated with high energy radiations. Possibility of fabricating novel magnetic spin valves, magnetically tuned varistor devices, generation of signal amplification are some of the potential applications of these magnetic-semiconductors.

INTRODUCTION

Iron titanates are well known minerals and have been extensively studied by geologists for a very long time. Recent discovery of the existence of magnetic exchange bias of more than 1T in the grains of titanoheamatite ($FeTiO_3$ bearing Fe_2O_3) mineral has renewed interest of geophysicists and magneticians in this material[1]. Its magnetic and semicondcuting properties which varies widely and are dependent upon the inclusions of Fe_2O_3 and some other elements are justification enough to study the fundamental properties of these materials in laboratory processed ilmenite-hematite, $(1-x)FeTiO_3 . xFe_2O_3$ for $0<x<0.6$. The lattice constants of IH varies negligibly with hematite concentration. For IH with x =0.6, the unit cell is hexagonal with the lattice constants of a = 0.505 nm and c = 1.385 nm. This has become even of greater importance because of the surge in processing of new and novel materials showing multifunctional properties. Fe-titanates when doped with magnetic ions such as Fe^{+3} and Mn^{+3} show some very interesting magnetic and semiconducting properties which can be exploited in variety of ways to develop some novel applications in emerging and established technologies such as spintronics, magneto-electronics, radhard electronics, sensors and detectors.

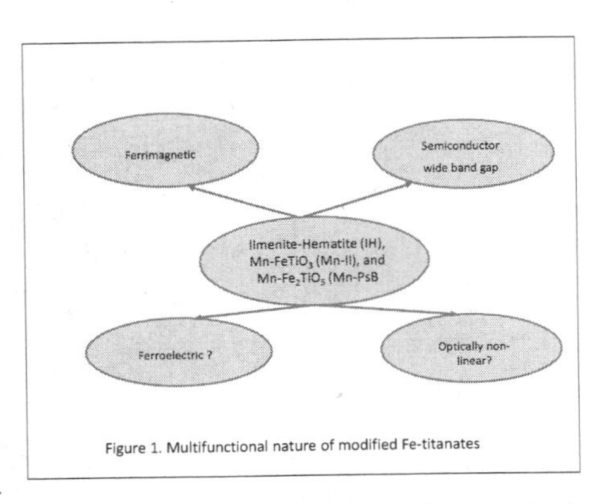

Figure 1. Multifunctional nature of modified Fe-titanates

The race for the discovery of room temperature magnetic semiconductors for the development of MRAM memory (magnetic RAM) using bipolar orientation of magnetic spins (which is at the heart of proposed spintronics technology) began just a decade ago and has continued up to this time. A variety of materials were identified as potentially useful spintronics materials including dilute magnetic systems based on III-V semiconductors, especially GaAs. None of them lived up to the rigorous requirements of a practical spintronics device. A series of ilmenite-hematite solid solution with proper concentration of Fe_2O_3 are well known to exhibit ferrimagnetism with the Curie point well above the room temperature as well as wide band gap semiconductor behavior[2,3]. This led to the motivation for intense research to develop ilmenite-hematite (IH) and similar iron titanates (Fe-titanates) with properties needed for a successful room temperature spintronics device. In spite of the fact that some compositions of $(1-x)FeTiO_3.xFe_2O_3$ (IH) as well as Mn- $FeTiO_3$ (Mn-Il) and Mn-Fe_2TiO_5 (Mn-PsB) have excellent magnetic and semiconductor properties, some of which we will discuss in this paper, we do not believe that they are suitable for the fabrication of a MRAM devices or for spintronics because of their very low mobility at room temperature. Superior materials have been identified to satisfy the requirements of these technologies. Just in October 2010 an integrated structure consisting of GaN substrate and Ga-Mn-nitride film has been identified that shows promise for the success of spintronics technology operating at room temperature[4].

So far as iron titanates are concerned they are very promising materials for other applications some of which we will explore in this paper. Apart from possessing coupled ferrimagnetic and semiconductor properties, modified Fe-titanates also show remarkable immunity to radiations owing to their wide band gap in excess of 2.5 eV. The multifunctional nature of this group of materials is shown in Fig.1.

Because of their multifunctional nature we propose to discuss the properties and device applications based on: (i) ilmenite-hematite solid solutions with chemical formula of (1-x) FeTiO3.xFe2O3 where x =0.33 (IH33) and x=0.45 (IH45), (ii) Mn-doped ilmenite (Mn^{+3}-$FeTiO_3$) with Mn = 0.45 atomic % and (III) Mn-doped pseudobrookite (Mn^{+3}-Fe_2TiO_5) with Mn = 0.45 %. The understanding of their multifunctional nature is crucial for developing some unique devices based on their magnetic, semiconducting and radhard properties.

Magnetic, dielectric and semiconducting properties are well established for the IH system. We find that the dielectric constant of the IH with x = 0.33 is about 9000 at room temperature. Such large values of the permittivity is usually associated with ferroelectric materials. In fact, ferroelectricity was reported for ilmenite, $FeTiO_3$, with the Curie point of $\approx$ 600 K back in 1994[2]. But its confirmation has not been established as yet. IH is the most studied member amongst modified iron titanates primarily because of the discovery of its magnetic and semiconducting properties by geologists long time ago and by the recent emergence of spintronics. It is beyond the scope of this paper to present a complete review of IH . Nevertheless, the interested readers are referred to some of the publications that might allow them a greater familiarity with the subject matter [1-3, 6-20].

SAMPLE PROCESSING

Three principal sample processing techniques were used for processing these materials:

Ceramic Processing: Cold, uniaxial ceramic pressing followed by sintering and annealing steps for using a thoroughly ball milled mixture of constituent chemicals in appropriate amounts to yield the desired compositions. Pellets were pressed using a Carver press at a force of approximately 8-10 kN. The green pellets were sintered at least three times in Ar environments at temperatures ranging from 650 to 900 °C. The duration of sintering was kept constant for 3 hours for each sintering step. The annealing was carried out at 1200 °C in Ar for 12-15 hours. This invariably produced dense and homogenous pellets with uniform properties throughout the bulk. No sample ever showed evidence of microcracking[7,12-14,17-19].

Film Growth: Film growth by pulsed laser ablation method (PLD) using the ceramic targets processed as described above; epitaxial growth achieved by using sapphire (Al_2O_3) single crystal substrate, and highly textured film by using MgO single crystal substrate. A KrF excimer laser of 248 nm wavelength was employed for the film growth. The following processing parameters were maintained: substrate to target distance ~ 5-6 cm, chamber pressure maintained usually around 10^{-7} Torr, laser repetition frequency of 2 Hz with the pulse duration of 20 ns. The fluence was kept around 1.7-1.8 J.cm^{-2}.

Single Crystal Growth: Bulk single crystals of pure PsB and Mn-PsB were grown by the high temperature solution growth method with $PbO.V_2O_5$ as flux. Black crystals, 5-7 mm long x 3-5 mm wide platelets, were grown between 1325 and 850 °C using the cooling rate of approximately1-2 °Ch^{-1}. Platinum-rhodium crucibles were used with sealed lids; and the crystals were harvested by dissolving the flux in mineral acids[15].

IH series were studied processing ceramic samples of a large number of compositions to understand their structural, magnetic and electrical properties[6,7]. The annealing atmospheres used were air, Ar and N_2. The properties such as resistivity, band gap and capacitance were found to have strong dependence on the annealing atmosphere, temperature and time. Based on their magneto-electronic properties it was realized that both IH33 and IH45 offered the best probability of studying the underlying magneto-electronic mechanisms and for device fabrication. Epitaxial films of IH33 and IH45 were grown using Al_2O_3 (sapphire) substrates[8,9]. However, it was also realized that these materials appear to be good candidates for developing magnetically and electrically tuned varistor devices and for radiation immune devices. Both IH33 and IH45 were studied for this purpose both as ceramic samples and textured films. Our results of such studies have been reported earlier[10-16]. Some newer results will be emphasized here.

Magnetic and semiconducting properties of IH system can be summarized as:

- It is ferrimagnetic with the Curie points well above the room temperature when $0<x<0.6$;
- The transition between p- and n-type semiconductor nature occurs for $x\approx0.2$. That is, it is p-type for all compositions with $x=0$ (pure ilmenite) and $x\cong0.2$. For any compositions for $x\geq 0.2$ it is n-type.
- Films grown under different partial pressure of O_2 show a wide variation in the value of its band gap. It can vary from 2.3 to 2.5 eV for high oxygen pressure and from 3.0 to 3 .4 eV for low oxygen pressure[8].
- From the Hall effect measurements done at room temperature it has been confirmed that the magnetic spins are polarized in a film of IH40[20]. In this paper it has been confirmed also that IHF with 40 atomic % hematite is a n-type semiconductor and ferrimagnetic with the magnetic Curie point > 400K.
- Besides widely studied IHC33, IHC45 also exhibits remarkable magnetic and semiconducting properties. It is a n-type material with a band gap of 2,3 eV. Its important magnetic parameters are: Curie point $\approx$ 610 K, coercivity and retentivity, as derived from its hysteresis loop, are $\approx$ 250 Oe and $\approx$3.5 emu/g, respectively, and the saturation magnetization $\approx$ 19.4 emu/g[16].

Many attempts were made to grow bulk single crystals of IH33 and IH45 using the flux growth methods but none of them produced even small crystals. Phase separation during the growth cycle invariably prevented the growth of IH crystals but always yielded well formed small ilmenite crystals. But this method was very successful in growing good quality crystals of pseudobrookite, both pure and Mn-doped, in size ranging between 3 to 7 mm in size[15].

Naturally found ilmenite is a p-type semiconductor and magnetically it can either be feebly ferrimagnetic or paramagnetic depending upon where it is found and what is its impurity contents. Lab processed ilmenite is non-magnetic and also p-type. Its unit cell is hexagonal with the lattice constants of a = 0.504 nm and c=1.374 nm. When doped with Mn it becomes an n-type semiconductor as well as shows strong magnetic moment at room temperature. Mn concentration of about 45 at.% shows the highest magnetic moment as well as the highest magnetic Curie point.

Pseudobrookite (PsB) is a rare mineral found only in a few locations in the world. In nature only tiny single crystals with metallic black luster of PsB is found. It is stable in air and can be processed without the need of special processing environments as is the case with ilmenite. It is a n-type semiconductor and feebly magnetic at room temperature. But when doped with Mn (≈ 40-45 at. %) it becomes strongly magnetic with a very stable hysteresis loop. Its electron mobility $\mu_{eff} = 6.3\ cm^2V^{-1}s^{-1}$. It is important to note that amongst oxides the electron mobility of PsB is relatively high and therefore it is a promising material for the development of electronic devices. Besides ceramic and single crystals its textured films were also grown using the PLD method with home grown targets[15].

PHYSICAL PROPERTIES AND DISCUSSIONS

Ilmenite-Hematite System

The IH system shows remarkable tolerance to neutron, proton and heavy ion radiations. Its non-linear current-voltage characteristics remains virtually unaffected even by heavy exposures to these radiations[13-17]. This is desirable for the design of devices whose performance depends upon their I-V properties such as a varistor, a diode or a transistor. We also find that exposure to proton radiation enhances chemical ordering in IH resulting in higher values of the magnetic moment for all compositions of bulk ceramic samples[17]. The dielectric properties of IHC33 are shown in Figures 2 and 3. In Figure 2 shows the dielectric constant, resistance and reactance as a function of biasing voltage at 1 MHz We see from Figure 2 that its dielectric constant is very high (~9,000 for biasing V =0 volt). Both the permittivity and the reactance

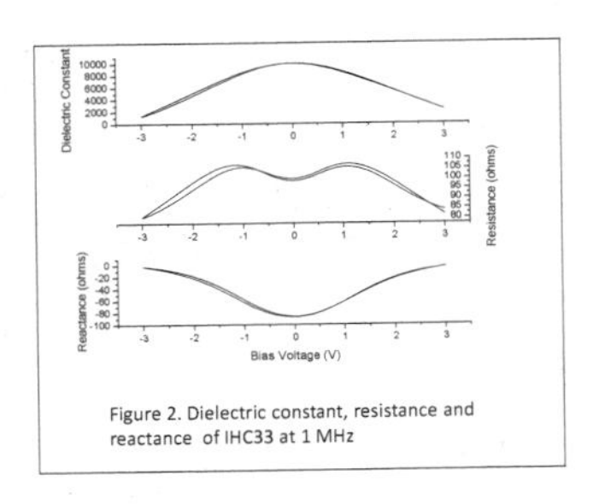

Figure 2. Dielectric constant, resistance and reactance of IHC33 at 1 MHz

show a peak and a trough, respectively corresponding to V=0, whereas the resistance has well pronounced two maxima and one minima which corresponds to V=0. The two maxima are symmetrically located around ±1.5 V. From Figure 3 it is also observed that IHC33 exhibits a pronounced phase shift corresponding to 1 MHz; whereas at 1 kHz it shows no phase shift at all.

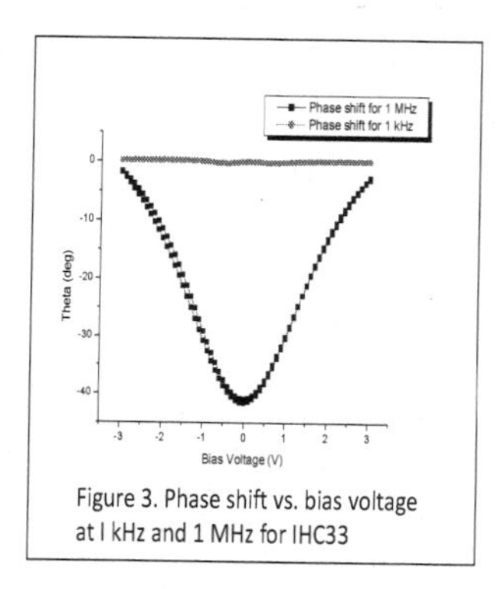

Figure 3. Phase shift vs. bias voltage at l kHz and 1 MHz for IHC33

Pseudobrookite, Fe_2TiO_5 and Ilmenite, $FeTiO_3$

Ilmenite and pseudobrookite are also important members of the Fe-titante family. But because of the absence of strong magnetism in them they were of lesser interest to the geologists. As such they have not been studied as extensively as the IH system. However, ilmenite drew the attention of scientists when it was discovered in one of the earlier moon missions conducted by NASA that ilmenite is abundantly present on the moon and it is also a semiconductor material. When doped with Mn, as we will see a bit later, it becomes both a n-type semiconductor and a good ferrimagnet. Also pseudobrookite (PsB) is a n-type semiconductor with the band gap equal to 2.34 eV as well as weakly magnetic. Mn-doping enhances its magnetic ordering and a doping concentration of Mn ≥ 0.40 leads to a well defined hysteresis loop. PsB

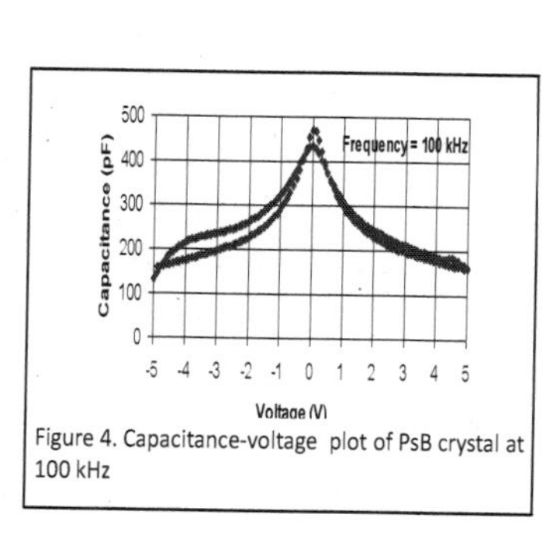

Figure 4. Capacitance-voltage plot of PsB crystal at 100 kHz

crystallizes in orthorhombic structure with the lattice constant of a = 0.979, b = 0.993 and c = 0.372 nm. Also Mn-doped PsB are n-type semiconductors. Like IH PsB shows remarkable tolerance to neutron and heavy ions radiation[15,16].

The C-V plot, as shown in Figure 4, for the PsB crystal generated at 100 kHz is highly symmetrical with respect to the applied voltage with a pronounced peak corresponding to V = 0. Its dielectric constant and dissipation factor (loss tangent) were determined between 100 and 100 kHz and are given in Figures 5a and 5b, respectively. As seen from these plots we notice that the permittivity of PsB is about 1200 for 100 Hz, which is substantially smaller than that of IHC 33. Also, its loss tangent is rather very high which disqualifies it as a desirable capacitor material. Figure 6 a shows the temperature dependence of the resistivity of Mn-PsB crystal confirming its semiconducting nature. The value of its room temperature resistivity is found to be approximately 0.03 Ω.cm and the activation energy of about 0.012 eV. The temperature dependence of the magnetic moment is plotted in Figure 6b. As the temperature increases the magnetic ordering decreases reaching its minimum value corresponding to the magnetic Curie point which is found to be ~550 K for Mn-PsB. The temperature dependence of the magnetization for Mn-PsB sample is similar to the ones found for 3d-ferromagnetic materials like Fe, Ni and Co. Figure 6 c

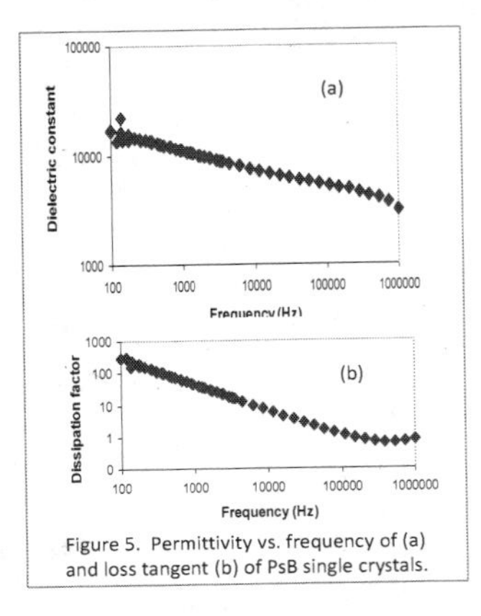

Figure 5. Permittivity vs. frequency of (a) and loss tangent (b) of PsB single crystals.

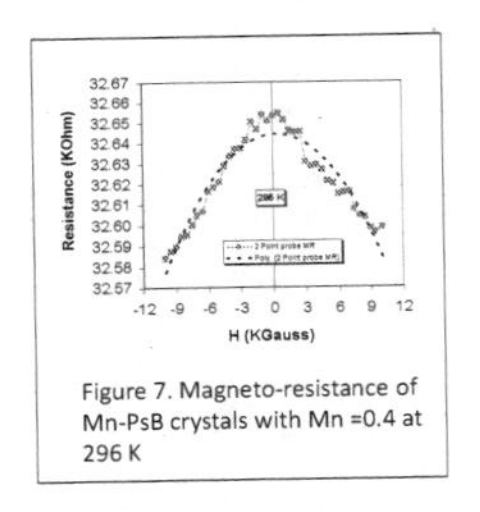

Figure 7. Magneto-resistance of Mn-PsB crystals with Mn =0.4 at 296 K

is the C-V plot of Mn-PsB crystal at frequencies varying from 1 kHz to 1 MHz As expected the capacitance, C, decreases with the increase of frequency and shows symmetry with equal and opposite values for biasing voltage, V. Its permittivity at room temperature is approximately 7000 which is comparable to that of IHC. Mn-PsB with Mn = O.4 also shows strong magneto-resistive property, as seen from Figure 7, with a well defined peak at the zero magnetic field. It is highly symmetrical for the positive and negative values of the magnetic field, H.

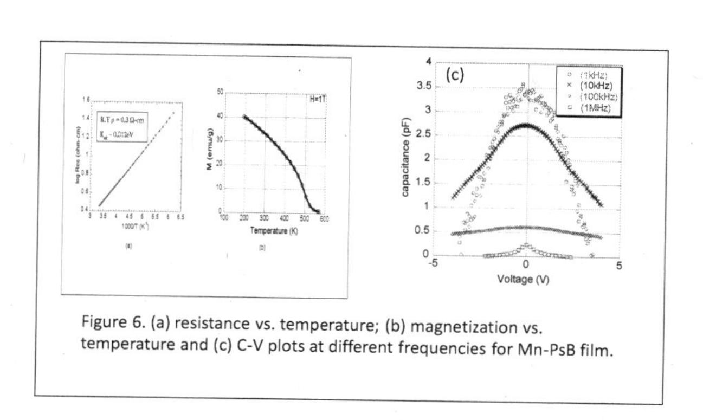

Figure 6. (a) resistance vs. temperature; (b) magnetization vs. temperature and (c) C-V plots at different frequencies for Mn-PsB film.

Ilmenite when doped with Mn^{+3} ions becomes ferrimagnetic with Curie points above room temperature[18]. Ilmenite with 45 atomic % Mn exhibits strong ferrimagnetic property with a well defined hysteresis loop. It forms solid solution with Mn_2O_3 for a wide range of concentration ranging from 0 to 0.45. For Mn concentrations of $x<0.2$ it is p-type and n-type for all concentrations for $x>0.2$. The band gap of ilmenite film grown by the PLD method is reported to be 3.55 eV

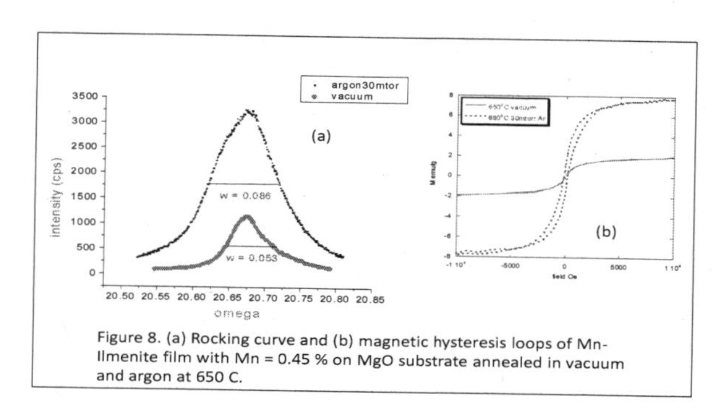

Figure 8. (a) Rocking curve and (b) magnetic hysteresis loops of Mn-Ilmenite film with Mn = 0.45 % on MgO substrate annealed in vacuum and argon at 650 C.

whereas it is 2.58 eV for bulk ilmenite. Like in the case of IH films here too the band gap appears to be dependent upon the environments of film growth as well as upon other processing parameters. From the rocking curve shown in Figure 8a we notice that the quality of film is highly dependent upon the processing atmosphere; for example when Ag gas is introduced in the PLD chamber and its pressure maintained at 30 mTorr then the FWHM (ω) = 0.086, whereas it is only 0.053 when a vacuum is maintained in the growth chamber. The value of (FWHM) is indicative of the crystalline quality of the film; the smaller the value the better is the quality of the film. However, the film grown under vacuum atmosphere (better FWHM) is weakly magnetic compared to the film processed with Ar in the chamber as seen in Figure 8b. Argon apparently enhances the magnetic ordering resulting in a stable hysteresis loop with a large value of the magnetic moment. The Curie point of this film was found to be 550 K and the magnetic moment at saturation corresponding to the magnetic field, H, equal to 10^4 Oe. An exhaustive treatment of semiconducting and magnetic properties of Mn-Il is given in[18].

POTENTIAL APPLICATIONS

The two terminal current-voltage (I-V) characteristics of IH, Mn-Il and PsB and Mn-PsB are found to be nonlinear which are highly responsive to external magnetic fields and biasing voltages. These properties can be exploited for developing tuned devices such as varistors, diodes, transistors etc. Since these oxides have excellent immunity to high level of exposure from neutrons, proton and heavy ions [13-17]. It is possible that these devices would be useful for applications in space as well as useful for other environments where these radiations might be present. It is known that widely used silicon based microelectronic devices are ill suited for space applications. Their performance level degrades drastically even terrestrially because of environmental radiations when the feature size reaches a certain limit such as submicron and nanometer scale. Wide band gap oxide based devices have advantage over silicon devices because of their high level of tolerance to terrestrial radiations.

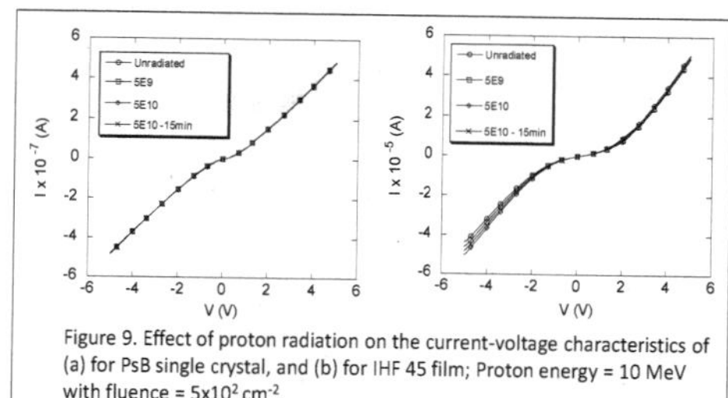

Figure 9. Effect of proton radiation on the current-voltage characteristics of (a) for PsB single crystal, and (b) for IHF 45 film; Proton energy = 10 MeV with fluence = $5x10^2$ cm^{-2}.

Figure 9 is just an example of radiation tolerance of modified iron titanate based devices. It shows the effect of proton radiation on the I-V characteristics of PsB (Figure 9a) and Mn-PsB (Figure 9b). Even for the proton energy of 10 MeV with fluence of $5x10^2$ cm^{-2} the I-V characteristics remains unaltered for their single crystals. Such immunity has been reported for the I-V characteristics of IH in multiple papers[13-17]. We have not studied the effect of radiation on the I-V behavior of Mn-Il but expect it to be highly tolerant to radiation. It is to be noted that though the I-V behavior of IH remains unaltered when exposed to radiations, the magnetic properties improve because of chemical ordering caused by protons[17].

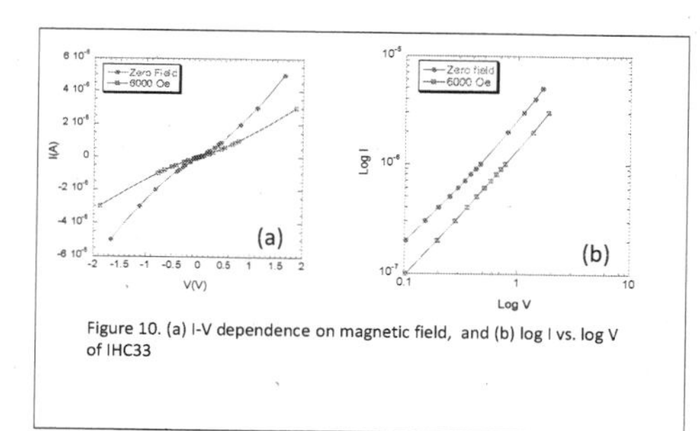

Figure 10. (a) I-V dependence on magnetic field, and (b) log I vs. log V of IHC33

The effect of magnetic field on the I-V characteristics is seen in Figure 10 for IHC45. As seen the nonlinear I-V curve shows a remarkable shift for a magnetic field of just 600 Oe, Figure 10a. In Figure 10b log I vs. log V is plotted for the same sample and we

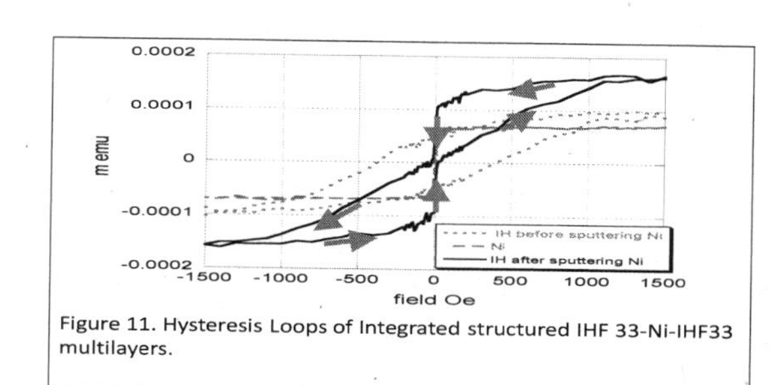

Figure 11. Hysteresis Loops of Integrated structured IHF 33-Ni-IHF33 multilayers.

see here that the dependence is linear as expected and the slope remains unchanged for the curves with and without a magnetic field. That is, the nonlinear coefficient, α (defined as ln

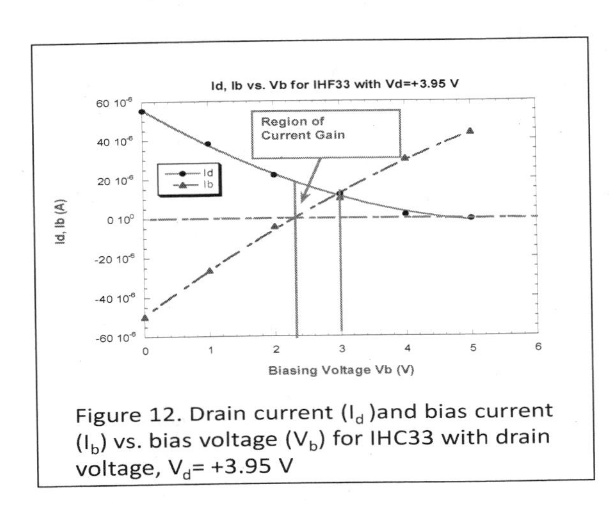

Figure 12. Drain current (I_d)and bias current (I_b) vs. bias voltage (V_b) for IHC33 with drain voltage, V_d= +3.95 V

I/lnV), which is the figure of merit of a varistor device remains unaltered. This means that varistors based on IHC 45 will operate satisfactorily in the presence of a magnetic field. This effect can be important for designing magnetic field sensors using IH and for varistor devices to operate satisfactorily in the presence of a magnetic field. We expect similar results for devices built using Mn-Il and Mn-PsB.

An integrated structure consisting of magnetic nickel film and an epitaxial IHF33 films on a single crystalline sapphire substrate was grown using the PLD method. Three magnetic layers were grown on top of each other; first the IH layer, then the Ni layer and finally the IH layer again. The selective switching of Ni and IHF films were achieved by carefully controlling the change in external magnetic field. Figure.11 shows the hysteresis loops of the three individual magnetic layers of the integrated structure: dotted curve represents the hysteresis loop of IHF33 film; the dashed curve is the hysteresis loop of Ni film grown over the IHF33 layer, and the black curve represents the loop of the integrated structure consisting of both IHF33 and Ni. One full cycle of switching is completed upon changing the field from +1500 Oe (+ maximum magnetization point for H = +1500 Oe) to the maximum magnetization in the negative direction at H = -1500 Oe and back again to +1500 Oe. The process can be cycled infinite number of times and controlled switching can be programmed and predicted precisely. Potential applications of such a structure are: a magnetic sensor for which sensor output would be the magnetic moment, and a magnetic sensor of the spin valve type where the output would be the change in resistance of magnetic states.

The effect of a biasing voltage was also studied on the two terminal I-V characteristics of IHC and IHF[16]. In that paper we showed that it was possible to generate positive and negative currents simply by changing the biasing voltage. The maximum number of such currents were found to be clustered between 2.5 and 4 V. This corresponds to the range for reported values of the band gap for IH[19]. It was also observed that the application of a biasing voltage changes the switching voltage of the varistor device with maximum slope corresponding to the 2.5 to 4 V range[16].

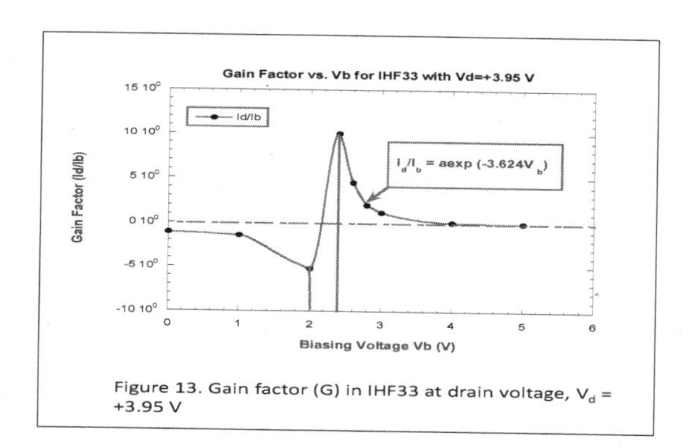

Figure 13. Gain factor (G) in IHF33 at drain voltage, V_d = +3.95 V

The biasing voltage (V_b) also causes power amplification in a very selective region of the biasing voltage. In Figure 12 it is shown how the drain current (I_d) and bias current (I_b) respond to the application of a biasing voltage, V_b. I_d decreases with increasing V_b and reaches its minimum value for $V_b = 5$ V. It never achieves a negative value. However, an opposite behavior is found for the bias current, I_b. It keeps on increasing with the biasing voltage and its sign switches from negative to positive at about 2.3 V (close to the band gap) and keeps on increasing. Between $2.3<V_b<3$ V power amplification takes place when $I_d > I_b$. This region is identified in Figure 12. From Figure 13 we see that the gain factor G (=I_d/I_b) reaches a maximum value at 2.3 V and decreases to its minimum value of zero following the exponential behavior given by:

$$G = a \exp(-3.624V_b) \qquad (1)$$

Between $V_b = 2.0$ V and $V_b = 2.3$ V G increases by a factor of 3 from -5 to +10. Such a gain of 300% in power is significant and can be used for many applications to boost the performance of a device including a handheld device.

CONCLUSIONS

All three magnetic-semiconductors in the family of modified iron titanates discussed here exhibit interesting magnetic, dielectric and nonlinear current-voltage characteristics that can be exploited for number of novel applications. The response of nonlinear I-V characteristics to an external magnetic field or a biasing voltage forms the basis for many applications. Integrated structures consisting of different magnetic layers can be used for selective magnetic switching opening the door for fabricating some unique sensors and spin valves. Also it is discussed that their I-V behavior remains practically unaltered even when these samples are exposed to high doses of radiations such as neutrons, protons and heavy ions. The high radiation immunity of these titanates make them particularly attractive for the fabrication of magnetically and electrically tuned devices and electronic components that can operate satisfactorily in radiation environments.

ACKNOWLEDGMENTS

The authors thank the National Science Foundation (Grant numbers: ECCS 1025395 and DMR 0213985), the Office of Naval Research (Grant number N00014-03-1-0358) and the Department of Energy (Grant DE-FG02-03er46039) for the support of this work. We also thank the research staff of the Center for Materials for Information Technology (MINT Center) and Central Analytical Facilities (CAF) at the University of Alabama, Tuscaloosa, AL for their collaboration and help during the course of this research. One of us (RKP) thanks his former students and associates in the Electronic Materials and Device Technology Laboratory (EMDTech) at the University of Alabama for conducting various parts of the research reported here. Our thanks also goes to DOE labs (Los Alamos National Lab and Brookhaven National Lab) as well as to the Cyclotron Center at Texas A&M University for allowing us to conduct radiation studies on our samples.

REFERENCES

[1] Suzanne A. McEnroe, B. C-S, Richard J. Harrison, Peter Robinson, Karl Fabian and Catherine McCammon: 2, (2007), 631
[2] Y. Ishikawa: J. Phys. Soc. Jpn. 12(10), (1957), 1083
[3] Y. Ishikawa: J. Phys. Soc. Jpn.13(1), (1958), 37
[4] Salah M. Bedair, John M. Zavada and Nadia El-Masry, IEEE Spectrum, 47, (2010).441
[5] R. P. Viswanathan et al., Solid State Communications, 92, (1994), 831
[6] L. Navarrete, J. Dou, D. M. Allen, R. Schad, P.Padmini, P. Kale and R. K. Pandey: J. Am. Ceramic Soc., 89(5), (2006), 1601
[7]Feng Zhou: "Processing, characterization, and device development of FeTiO3.Fe2O3 ceramic ad film for high temperature electronics", Dissertation, The University of Alabama, (2000).
[8]J. Dou, L. Navarrete, P. Kale, P. Padmini, R. K. Pandey, H. Guo, A. Gupta and R. Schad: J. Appl. Phys., 101, (2007), 053908
[9]J. Dou, L. Navarrete, R. Schad, P. Padmini, R. K. Pandey, H. Guo and A. Gupta: J. Appl. Phys., 103, (2008), 07D117.
[10]F. Zhou, S. Kotru and R. K. Pandey: Thin Solid Films, 33, (2002), 408
[11] F. Zhou, S. Kotru and R. K. Pandey: Materials Letters, 57, (2003), 2104
[12] P. Kale, P. Padmini, L. Navarrete, J. Dou, R. Schad and R. K. Pandey: J. Electronic Mats., 9, (2007), 1224
[13]P. Padmini, M. Pulikkathara, R. Wilkins, and R. K. Pandey: Appl. Phys. Lett.,82, (2003), 586
[14]P. Padmini, S. Ardalan-S, F. Tompkins, P. Kale, R. Wilkins and R. K. Pandey: J. Electronic Mats., 34(8), (2005), 1095
[15]R. K. Pandey, P. Padmini, L.F. Deravi, N. N. Patil, P. Kale, J. Zhong, J. Dou, L. Navarrete, R. Schad and M. Shamzuzzoha: IEEE Proceedings of the 8th. International Conference on Solid State and Integrated Circuit Technology (ICSICT 2006), ISBN: 1-4244-0160-5, Shanghai, China, 2, (2006), 992
[16] R. K. Pandey, P. Padmini, R. Schad, J. Dou, H. Stern, R. Wilkins, R. Dwivedi, W. J. Geerts, and C. O'Brien: J. Electroceramics, 22,(2009), 334.
[17]D. M. Allen, L. Navattete, J. Dou, R. Schad, P. Padmini, R. Kale, R. K. Pandey, S. Shojah-Ardalan and R. Wilkins: Appl. Phys. Lett. 85, (2004), 5092
[18] Pranoti Kale, "Processing and Evaluation of Ilmenite-Hematite Thin Films and Ceramic for Microelectronics and Spintronics", MS Thesis, The University of Alabama, (2005).
[19] Jian Dou, " Synthesis and Characterization of Ilmenite-Hematite Thin Films by Pulsed Laser Deposition", Dissertation, The University of Alabama, (2007).
[20] Hajime Hojo, Koji Fujita, Katsuhisa Tanaka and Kazuyuki Hiao, App. Phys. Lett., 89, (2006), 142503.

LONG-TERM CONVERGENCE OF BULK- AND NANO-CRYSTAL PROPERTIES

Sergei L. Pyshkin
Institute of Applied Physics, Academy of Sciences
Kishinev, Moldova

John M. Ballato
Center for Optical Materials Science & Engineering Technologies, Clemson University
Anderson, SC, USA

ABSTRACT

This paper continues a generalization of our observations on the improvement of properties from semiconductor GaP:N crystals prepared over 45 years ago and their convergence to the behavior of nanoparticles. We show that there are driving forces which, over time, result in the ordered redistribution of impurities and host atoms in a crystal. We observe a new type of the crystal lattice, where host atoms occupy their equilibrium positions, while impurities divide the lattice in the short chains of equal length in which the host atoms develop harmonic vibrations. This nearly half-centennial evolution of the GaP:N luminescence and its other optical and mechanical properties are interpreted as the result of both volumetrically ordered N impurities and the formation of an ordered bound exciton system. The highly ordered nature of this new host and excitonic lattices increases the radiative recombination efficiency and makes possible the creation of advanced non-linear optical media for optoelectronic applications. We demonstrate the considerable improvement of quality of GaP nanocrystals as the result of elaboration of an optimal for them nanotechnology.

Semiconductor nanoparticles were introduced into materials science and engineering mainly that to avoid limitations inherent to freshly grown semiconductors with a lot of different defects. The long-term ordered and therefore close to ideal crystals repeat behavior of the best nanoparticles with pronounced quantum confinement effect. These perfect crystals are useful for application in top-quality optoelectronic devices as well as they are a new object for development of fundamentals of solid state physics.

INTRODUCTION

Single crystals of semiconductors grown under laboratory conditions naturally contain a varied assortment of defects such as displaced host and impurity atoms, vacancies, dislocations, and impurity clusters. These defects result from the relatively rapid growth conditions and inevitably lead to the deterioration of the mechanical, electric, and optical properties of the material, and therefore to degradation in the performance of the associated devices.

Over time, driving forces such as diffusion along concentration gradients, strain relaxation associated with clustering, and minimization of the free energy associated with properly directed chemical bonds between host atoms result in ordered redistribution of impurities and host atoms in a crystal. Any attempt to accelerate these processes through annealing of GaP at increased temperatures cannot be successful because high-temperature processing results in thermal decomposition (due to P desorption) instead of improved crystal quality. Therefore successful thermal processing of GaP can only take place at temperatures below its sublimation temperature, requiring a longer annealing time. Evaluated in framework of the Ising model the characteristic time of the substitution reaction during N diffusion along P sites in GaP:N crystals at room temperature constitutes 15-20 years [1]. Hence, the observations of highly excited luminescence of the crystals made in the sixties and the nineties were then compared with the results obtained in 2009-2010 in closed experimental conditions.

The pure and doped GaP crystals discussed herein were prepared over 45 years ago. Throughout the decades they have been used to investigate electro- and photoluminescence (PL), photoconductivity, bound excitons, nonlinear optics, and other phenomena. Accordingly, it is of interest also to monitor the change in crystal quality over the course of several decades while the crystal is held under ambient conditions.

The long-term ordering of doped GaP and other semiconductors has been observed as an interesting accompanying process, which can only be studied in the situation when a researcher has a unique set of samples and the persistence to observe them over decade time scales.

More specifically, since 2005, we have analyzed the optical and mechanical properties of single crystalline Si, III–V (GaP [1], InP) semiconductors, and their ternary analog $CdIn_2S_4$, all of which were grown in the 1960s. Comparison of the properties of the same crystals has been performed in the 1960s, 1970s, 1980s, 1990s [1-9], and during 2000s [10–29] along with those of newly made GaP nanocrystals [14, 26-28] and freshly prepared bulk single crystals [20-24].

Jointly with Refs.[1-29] this paper is a generalization of the results on long-term observation of luminescence, absorption, Raman light scattering, and microhardness in the noted above bulk semiconductors in comparison with some properties of the best to the moment GaP nanocrystals . We show that the combination of these characterization techniques elucidates the evolution of these crystals over the course of many years, the ordered state brought about by prolonged room-temperature thermal annealing, and the interesting optical properties that accompany such ordering. We demonstrate that long-term natural stimuli improving perfection of our crystals prevail over other processes, which could lead to novel heterogeneous systems and new semiconductor devices with high temporal stability.

Note, that semiconductor nanoparticles were synthesized mainly that to avoid limitations inherent to freshly grown bulk semiconductors with a wide range of different defects. For instance, different defects of high concentration in freshly prepared GaP single crystals completely suppress any luminescence at room temperature due to negligible quantity of free path for non-equilibrium electron-hole pairs between the defects and their non-radiative recombination, while the quantum theory predicts their free movement in the field of an ideal crystal lattice. The long-term ordered and therefore close to ideal crystals demonstrate bright luminescence and stimulated emission repeating behavior of the best nanoparticles with pronounced quantum confinement effect. These perfect crystals due to their unique mechanical and optical properties are useful for application in top-quality optoelectronic devices as well as they are a new object for development of fundamentals of solid state physics and, nanotechnology and crystal growth.

RESULTS AND DISCUSSION

GaP Nanocrystals: Improvement of Quality and Properties

Since 2005 our efforts to prepare and improve the quality of GaP nanoparticles for light emissive devices have continued [14, 27-29]. While bulk and thin GaP films have been successfully commercialized for many years, its application in device nanocomposite structures for accumulation, conversion and transport of light energy has only received attention recently.

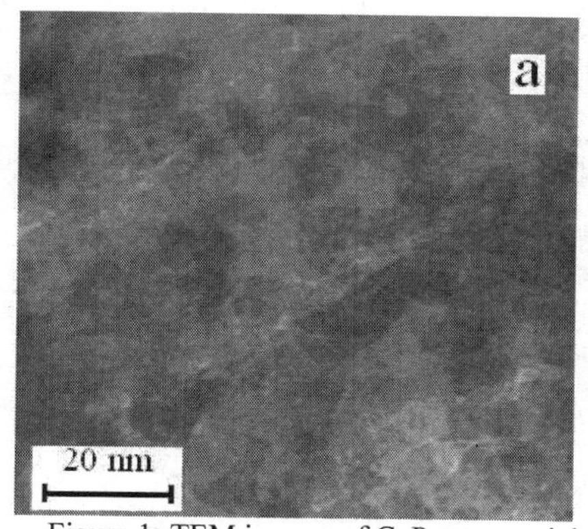

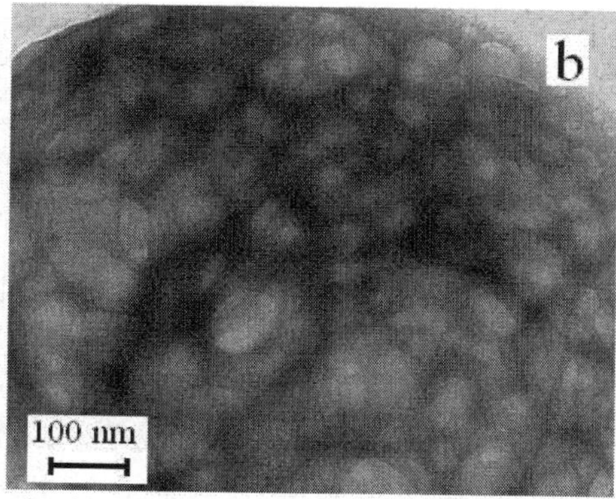

Figure 1: TEM images of GaP nanoparticles obtained by the aqueous synthesis. **a**. Thoroughly ultrasonicated and dried nanopowder. **b**. Initial clusters (the dimensions of the order of 100 nm).

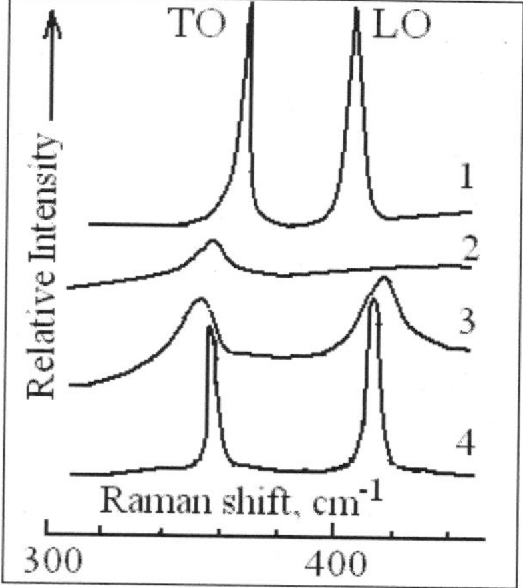

Figure 2: Raman light scattering from GaP nanoparticles of different treatment (spectra 2-4) and in comparison with perfect GaP bulk crystals (spectrum 1).
2. Not thoroughly treated powder of nanoparticles prepared using red phosphorus at 200°C.
3. Thoroughly treated GaP nanoparticles prepared using red phosphorus at 200°C.
4. Nanoparticles prepared on the base of yellow P by low temperature syntheses.

The best GaP nanoparticles prepared to-date have been based on white P under mild aqueous low temperature conditions [27]. The spectra of photoluminescence (PL) and Raman light scattering (RLS), X-ray diffraction (XRD) and electron microscopy (TEM) of the nanoparticles prepared under different conditions have been compared with each other as well as with those from bulk single crystals. After the relevant investigation of different regimes and components for hydrothermal reactions this type of the synthesis has been chosen as an optimal one. In preparation of the nanocomposite we used the fractions of uniform GaP nanoparticles having after a thorough ultrasonic treatment and a number of other operations a bright luminescence at room temperature in a broad band with the maximum, dependently on the concrete synthesis conditions, between 2.4 – 3.2 eV, while the value of the forbidden gap in GaP at room temperature is only 2.24 eV.

Note that according to our investigations of these conditions and data on nanoparticles characterization, only a combination of low temperature synthesis, using white P and thorough ultrasound treatment of the reaction products leads to the broad band and essential UV shift of luminescence that makes the obtained nanoparticles suitable for high quality light emissive nanocomposites.

Figure 1 shows the TEM images of GaP nanoparticles obtained by the aqueous synthesis. One can see GaP nanoparticles, having characteristic dimensions less than 10 nm. The washed, thoroughly ultrasonicated and dried nanopowder contains mainly single nanoparticles **(Fig. 1a)**, obtained from the initial clusters with the dimensions of the order of 100 nm **(Fig. 1b)**.

Figure 2 shows spectra of Raman light scattering from GaP nanoparticles prepared on the base of white or red P by mild aqueous synthesis at increased or low temperatures and ultrasonically treated.

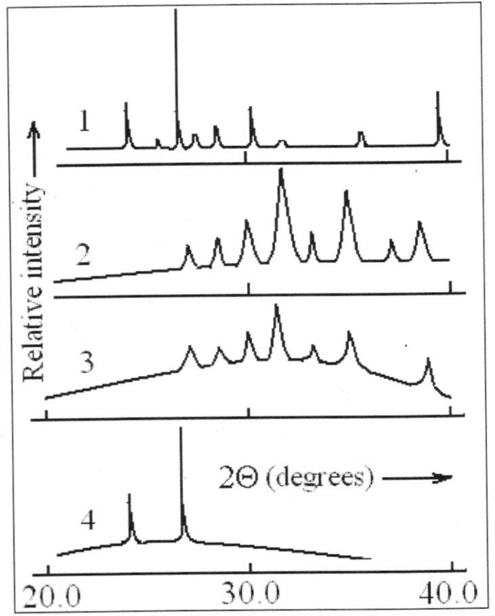

Figure 3 : X-ray diffraction from GaP nanoparticles.

1. White phosphorus, best performance of low temperature syntheses, well-treated powder. 2. White P, not the best performance and powder treatment. 3. Red phosphorus, the best result.

The characteristic Raman lines from the GaP nanoparticles prepared at low temperature using white P were narrow and intense **(Fig. 2, spectrum 1** and **4** respectively) whereas, they were weak and broad from the any nanoparticles prepared at high temperatures **(Fig. 2, spectra 2** and **3)**. The especially weak and broad spectrum exhibits not thoroughly washed powder (please see **spectrum 2**; spectra 1-3 are taken from [13]).

In **Figure 3** one can see X-ray diffraction from GaP nanoparticles prepared at different conditions using red or white phosphorus (**spectra 1-3**) in comparison with the diffraction from perfect GaP single crystal (**spectrum 4**). The nanoparticles obtained by low temperature aqueous synthesis using white phosphorus develop clear and narrow characteristic lines like those obtained from perfect GaP bulk single crystals taken from our unique collection of long-term (more than 40 years) ordered GaP single crystals (**Fig. 3, spectra 1** and **4**). Contrary to that, nanoparticles prepared on the base of red phosphorus or not in the best conditions show broad and weak characteristic lines (**Fig. 3, spectra 2** and **3**).

Initial results on luminescent properties of GaP nanoparticles [14] confirm the preparation of 10 nm GaP nanoparticles with clear quantum confinement effects but the luminescent spectrum was not bright enough and its maximum was only slightly shifted to UV side against the 2.24 eV forbidden gap at room temperature **(Fig. 4, spectrum 1)**. The nanoparticles obtained from the reaction with white P at low (125°C) temperature exhibit bright broad band spectra considerably shifted to UV side **(Fig. 4, spectrum 2, 3)**. Note that the original powder contains only a part of GaP particles with nearly 10 nm dimension, which develop quantum confinement effect and the relevant spectrum of luminescence, so the spectrum of luminescence consists of this band with maximum at 3 eV and of the band characterizing big particles with the maximum close to the edge of the forbidden gap in GaP **(Fig. 4, spectrum 2)**, but the thorough ultrasonic treatment gives an opportunity to get the pure fraction of nanoparticles with the **spectrum 3** having the maximum at 3 eV.

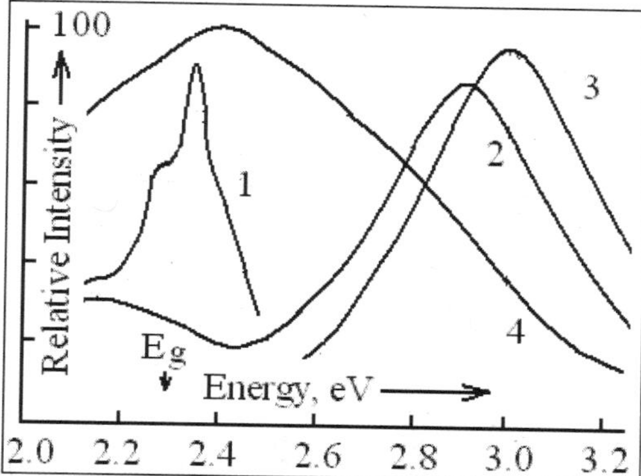

Figure 4 : Luminescence of GaP nanoparticles prepared at different conditions (spectra 1-3) and in comparison with the luminescence of perfect GaP bulk single crystals (4).

Comparison of Properties of Nanocrystals, Aged and Freshly-Grown Crystals

This work also continues monitoring of properties of GaP single crystals grown 45 years ago [1-29]. As we note, due to a significant number of defects and a highly intensive non-radiative recombination of non-equilibrium current carriers, initially luminescence of fresh undoped crystals could be observed only at the temperatures 80K and below. Now luminescence is clearly detected in the region from 2.0 eV and until 3.0 eV at room temperature (see **Figure 4**,

spectrum 4). Taking into account that the indirect forbidden gap is only 2.25 eV, it is suggested that this considerable extension of the region of luminescence to the high energy side of the spectrum as well as a pronounced increase of its brightness are connected with a very small concentration of defects, considerable improvement of crystal lattice, high transparency of perfect crystals, low probability of phonon emission at rather high temperature and participation of direct band-to-band electron transitions.

Comparing selected properties of recently measured GaP crystals (both doped and undoped) to those measured years ago allows a better understanding of their luminescence behavior. Luminescence and Raman light scattering spectra indicate that, over a period of about 25 years, the zero-phonon line A and its phonon replica LO, TO are narrower in their line-widths (**Figure 5**) in comparison with freshly prepared single crystals. The widths and the spectral positions of the excitonic photoluminescence (PL) lines at 15K both in freshly prepared and long-term ordered GaP:N crystals shown in **Figure 5** are in a good agreement with the known dependence of the line position on N concentration at ordered or chaotically distributed impurities (please see the details in [30]). Due to this dependence we observe a rather broad excitonic line in freshly prepared doped crystals (**Fig.1**, **spectrum 4**), but very narrow line (~ 1 meV which is equal to kT at 15K) in the same long-term ordered crystal.

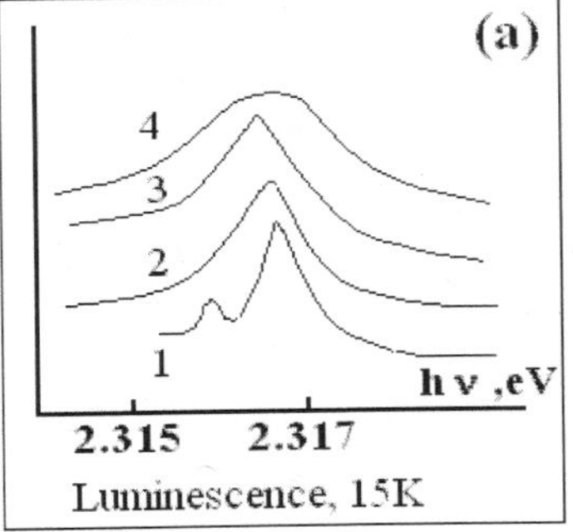

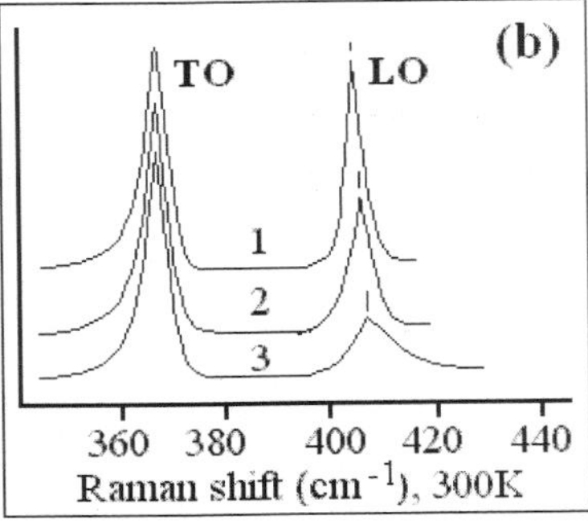

Figure 5. Evolution of the non-phonon line of bound exciton A in luminescence (a, excitation by Ar^+-laser at 488 nm, 15K) and its LO (longitudinal optical) and TO (transversal optical) phonon replica in Raman light scattering (b, Excitation by Ar^+-laser at 514 nm, 300K) as a function of N (nitrogen) concentration.
1-3: ordered crystals. 4: unordered. 1-4: $N_o = 10^{17}$; 10^{18}; 10^{19} and $10^{18}cm^{-3}$.

Fig. 5a, spectra 1-3 shows that the zero-phonon line A of single N impurity-bound excitons in the aged crystals shifts in spectral position with N impurities concentration according to the known dependence [20]:

$$\mathbf{E_{NN} = E_N - \beta r_{NN}^{-3}},$$

where the spacing $\mathbf{r_{NN}}$ is the distance between impurities, $\mathbf{E_N}$ is the A line position at $\mathbf{r_{NN}} \rightarrow \infty$, $\mathbf{E_{NN}}$ is the same at some non-zero nitrogen concentration and $\boldsymbol{\beta} = 13$ if $\mathbf{E_N}$, $\mathbf{E_{NN}}$ are measured in eV and $\mathbf{r_{NN}}$ in Å.

The narrow A line was fixed in our experiments and its shift (**Fig. 5a, spectra 1-3**) can be expected only in the case of equal distances between the impurities. The maintenance of the exact r_{NN} spacing facilitates the placement of the N impurities rather than the host P atoms in the nodes of the GaP crystal lattice. No shift in wavelength but, rather, a broad luminescence maximum is observed in case of the N impurities chaotically distributed in freshly grown crystals (**Fig. 5a, spectrum 4**). Therefore we can conclude that an ordered disposition of equally spaced N impurities exists in the aged crystals.

The difference in the present ordered state of the crystal lattice, with respect to the data obtained with the same crystals and conditions during 1989–1993 [9, 10] can be seen in **Figure 6a** and **b**, which shows the Raman spectra of pure GaP and GaP:N [17, 18]. One can see (**Fig. 6a, curves 3 and 4**) that, for the longitudinal optical (LO) phonon modes, the peak from the original samples is broad, weak, and shifts with impurity level. After 40 years (**Fig. 6a, curves 1 and 2**), the peak for heavily N-doped GaP is much more intense, has a more symmetric (Lorentzian) shape, narrower linewidth, and a spectral position that no longer depends on N concentration. These results are characteristic of harmonic vibrations in a more perfect lattice.

The LO phonon line is narrower in the aged doped crystal (**Fig. 6a, curve 2**) than in the undoped (pure) crystals (**Fig. 6a, curve 1**) and is also more intense than the TO phonon line. Similar results have been obtained for various impurities (N, Sm, and Bi) in spite of their maximum possible concentrations in GaP and different masses of impurity atoms and different types of substitution [17]. These results confirm the important role of the impurities that are periodically located in the host crystal lattice and result in the formation of the new perfect crystal lattice.

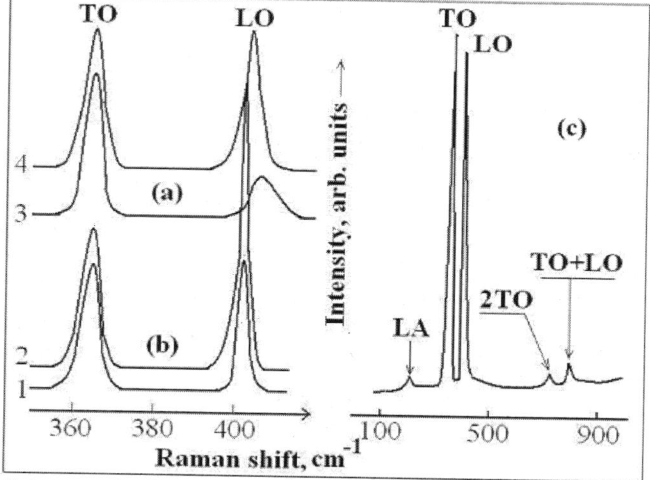

Figure 6. Raman light scattering in undoped GaP (a, b) and GaP:N (a, b, c) in 1989-1993 (a) and in 2006 (b, c) [17, 18].
1, 4 – pure GaP, 2,3 – GaP heavy doped by N. Spectra 3, 4 are taken from [10].

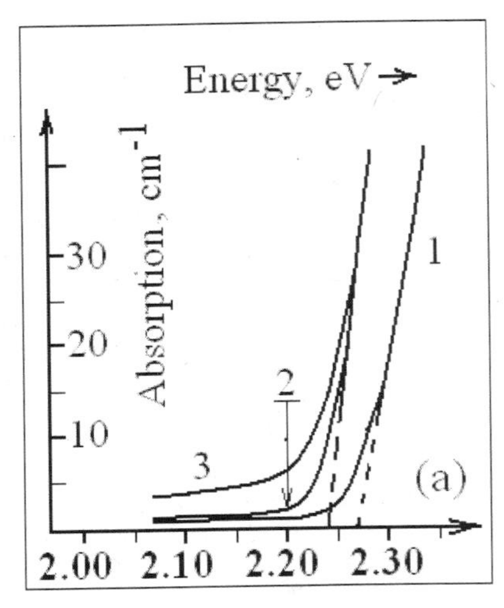

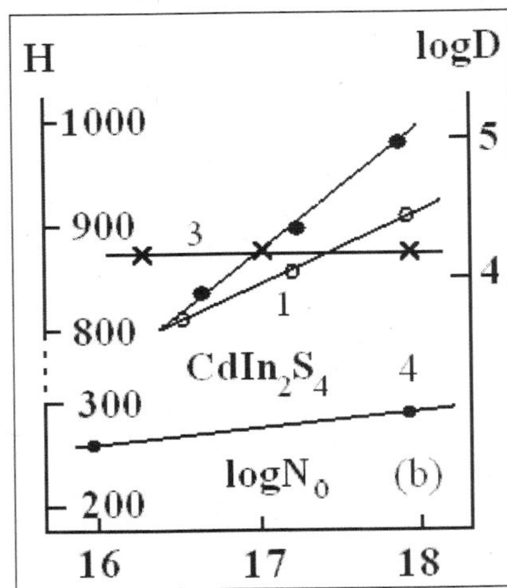

Figure 7. a. Absorption coefficient in long-term ordered (40 years) GaP:N (1) and in 25 year old (2) or freshly prepared undoped GaP (3). N concentration is 10^{18} cm^{-3}.
b. Microhardness H (Kg/mm^2) in fresh unordered GaP crystals as a function of N concentration (1); the same for 40 years ordered crystals (2); Density of dislocations (cm^{-2}) D as a function of N concentration (3); H in $CdIn_2S_4$ at different concentration of defects (4).

A new phenomenon observed in the Raman spectra, that has developed in the crystals over the course of 40 years, are the peaks denoted by us here as LA, 2TO and TO + LO (figure 6(c)). Note that the theory of Raman light scattering in GaP predicts the LO phonon to decay into two longitudinal acoustic (LA) phonons. LA phonons with a frequency LO/2 [31], and two-phonon processes of 2TO and TO + LO emission can also be observed in perfect crystals [4]. This observation of a multi-phonon process and a decay of LO phonon confirm the high quality of the host lattice, uniform impurity distribution, and as a consequence, low noise background in the Raman scattering.

Figures 5 - 7 demonstrate differences between the disordered, newly prepared single crystals, and crystals that have undergone ordering over either 25 or 40 years. As can be seen from **Fig. 5**, differences between the freshly prepared and 25-year ordered GaP crystals are observed in the line-widths and their spectral position with N concentration, which suggests an improvement in ordering of the host lattice and nearly equal spacing between impurities.

Figure 7 shows the absorption edge as a function of photon energy (**a**) and provides a comparison of microhardness and density of dislocations as a function of N concentration (**b**). The position of the absorption edge in freshly prepared GaP:N crystals does not depend on N impurity concentration and it coincides with the position for pure GaP crystals (**Fig. 7a, curves 2 and 3**). This N impurity creates the 21-meV energy level under the conduction band for bound excitons both in freshly prepared and 25-year ordered crystals. On the contrary, **curve 1, Fig. 7a** shows that the GaP:N crystals ordered for approximately 40 years demonstrate an increase in the forbidden gap and the clear shift of the absorption edge, which is expected to be proportional to the N concentration according to the Vegard law, similar to that in a dilute GaP-GaN solid solution.

Let us now discuss the influences of long-term ordering of the dopants and defects on the mechanical properties of GaP and its ternary analog $CdIn_2S_4$. The microhardness and dislocation density in these samples have been evaluated over many decades [32-36]. Taking into account the approximately 50 year timeframe makes an exact comparison difficult; therefore, only general trends are discussed here.

According to the classical point of view [37], good plasticity is determined by free movement of dislocations through the crystal under a mechanical load. Impurities act to pin the movement of dislocations. Therefore, the value of microhardness, H, in GaP should depend on the impurity concentration (**Fig. 7b, lines 1 and 2**). As can be seen, the relatively pure crystals have minimum microhardness. An increase in the impurity concentration in GaP crystals, with dopants such as N, Bi, N:Sm [1, 27], substituting for the host atoms (N, Bi) or occupying the interstitials in the crystal lattice (Sm), leads to an increase in microhardness for the long-term-ordered crystals (**Fig. 7b, lines 1 and 2**, respectively). The same behavior of microhardness for freshly prepared and long-term ordered single crystals was observed for mono-atomic Si and binary InP, pure and doped by different impurities.

A rather large difference in microhardness of the long-term-ordered highly doped crystal relative to the newly grown crystals can possibly be explained by the regular disposition of impurities. This regular disposition of impurities might create a more significant obstacle to dislocation movement than in the newly grown system, in which the impurities form clusters with large distances between them, permitting greater dislocation movement. Indeed, at the concentration of about N = $10^{18}cm^{-3}$ with the ordered disposition of the impurities in GaP crystal lattice, the 10 nm distance between the impurities is a serious obstacle for dislocation movement through this regular impurity net taking into account the dislocation density of the order of 10^4 cm^{-2} (please see **Fig. 7**) and the significant lateral extent of the dislocations. However, at the same concentrations of impurities and dislocation density, the distances between impurity clusters in disordered crystal can exceed 10 nm by a few orders of magnitude, resulting in a rather free dislocation movement and a corresponding decrease in microhardness.

Figure 7b, line 3, shows that the density of dislocations, D, in GaP does not depend on N concentration. Our observations show that neither the density of dislocations, D, in GaP nor the character of their distribution along the crystal changes with time (at least over nearly 50 years).

It is worth noting that the high-quality growth conditions for the initial GaP crystals and accurate control over temperature during growth significantly decreased the density of dislocations [36].

Let us discuss now the phenomena which can be observed only in luminescence of the perfect long-term ordered GaP and $CdIn_2S_4$ single crystals. The 45-year ordered GaP:N crystals demonstrate uniform luminescence from a broad excitonic band instead of the narrow zero-phonon line and its phonon replica in disordered and partly ordered (25-year-old) crystals due to the ordered crystals have no discrete impurity level in the forbidden gap. To the best of our knowledge, the transformation of a discrete level within the forbidden gap into an excitonic band (**Figs. 7a and 8b**) is observed for the first time. In this case, the impurity atoms regularly occupy the host lattice sites and affect the band structure of the crystals, which is now a dilute solid solution of GaP-GaN rather than GaP doped by occasionally located N atoms.

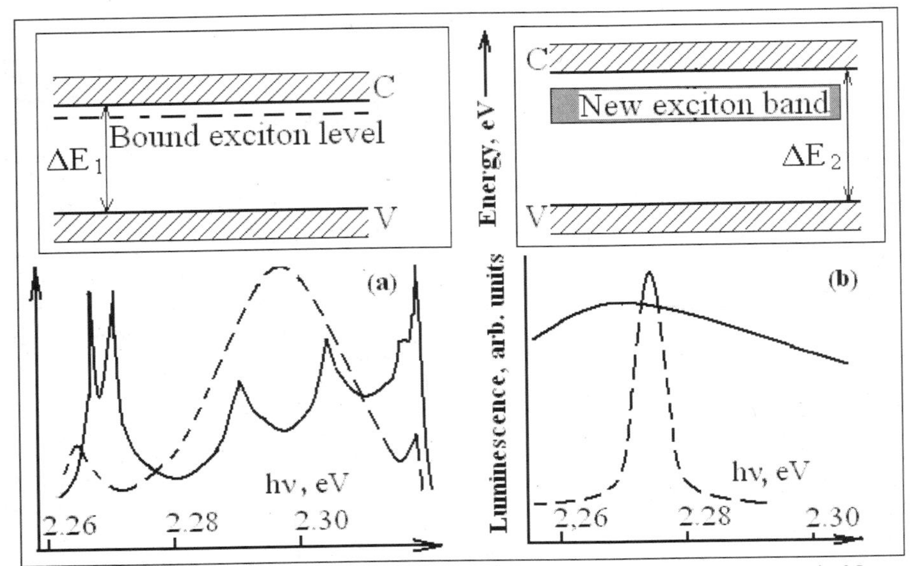

Fig. 8. Luminescent spectra and the view of the forbidden gaps (ΔE_1, ΔE_2) in the partly 25 year ordered (a) and perfect 40 year ordered (b) crystals GaP doped by N. The dotted lines correspond to highly optically excited crystals. C and V – the positions of the bottom of the conductance and the top of valence bands respectively.

As noted previously, the luminescence of fresh doped and undoped crystals could be observed only at temperatures below about 80 K. The luminescence band and lines were always seen at photon energies less than the value of the forbidden gap (2.3 eV). Now, after 45 years, luminescence of the long-term-ordered bulk crystals similar to the GaP nanocrystals [27, 28] is clearly detected in the region from 2.0 eV to 3.0 eV at room temperature [12-26]. We believe, in the long-term-ordered bulk crystals this considerable extension of the region of luminescence at 300 K to the high-energy side of the spectrum is due to: (a) a very small concentration of defects, (b) low contribution of nonradiative electron–hole recombination, (c) considerable improvement of crystal lattice, (d) high transparency of perfect crystals, and (e) low probability of phonon emission at indirect transition.

Note that increase of luminescence excitation in the case of partly ordered GaP:N (**Fig. 8a**) leads to a broad luminescence band as a result of bound exciton interaction [6], while in the case of perfectly ordered crystals (**Fig. 8b**) one can see an abrupt narrowing of the luminescence band, probably, due to stimulated emission in defect-free crystals. The density of power of the pumping laser at 4 ns and 50 mJ pulse duration and energy is usually enough for stimulated emission in a perfect GaP single crystal with natural faceting.

Earlier, in freshly prepared crystals we observed a clear stimulated emission from a GaP:N resonator at 80 K [38] as well as so called superluminescence from the GaP single crystals having natural faceting. Presently, our ordered crystals have a bright luminescence at room temperature that implies their perfection and very lower light losses. The narrowing of the luminescence band in Fig. 8b can probably be explained by stimulated emission at its initial stage of superluminescence. In our current presentations and papers [22, 23, 26] we demonstrate that the stimulated emission is developed even at room temperature by direct electron–hole recombination of an electron at the bottom of the conduction band with a hole at the top of the valence band and the LO phonon absorption.

Taking into account the above-mentioned results, a model for the crystal lattice and its behavior at a high level of optical excitation for 40-year-old ordered N-doped GaP (**Fig. 9**) can be suggested. At relevant concentrations of N, the anion sub-lattice can be represented as a row of anions where N substitutes for P atoms with the period equal to the Bohr diameter of the bound exciton in GaP (approximately 10 nm) (Fig. 6a). At some level of excitation, all the N sites will be filled by excitons, thereby creating an excitonic crystal (**Fig. 9b**) which is a new phenomenon in solid-state physics and a very interesting model for application in optoelectronics and nonlinear optics [7].

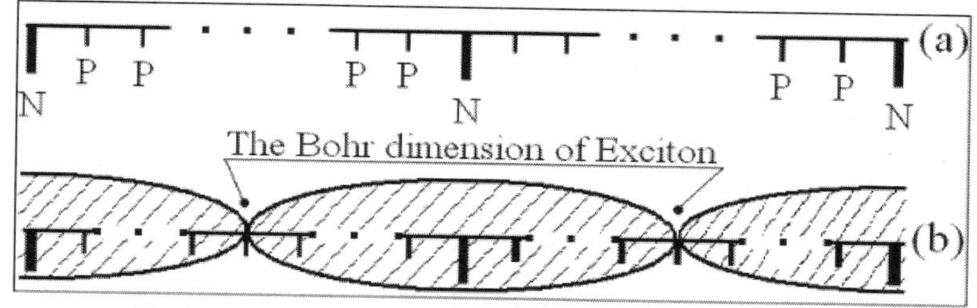

Fig. 9. The models of 40 year ordered GaP doped by N.
a. The new type of crystal lattice with periodic substitution of N atoms for the host P atoms.
b. The exciton crystal on the base of this lattice. The substitution period is equal to the Bohr diameter of exciton (~ 100Å) and optical excitation is enough for complete saturation of the N sublattice.

CONCLUSIONS

This study of GaP:N and $CdIn_2S_4$ brings a novel perspective to improving the quality of semiconductor crystals. The unique collection of pure and doped crystals of semiconductors grown in the 1960s provides an opportunity to observe the long term evolution of properties of these key electronic materials. During this almost half-centennial systematic investigation we have established the main trends of the evolution of optoelectronic and mechanical properties. It was shown that these stimuli to improve quality of the crystal lattice are the consequence of thermodynamic driving forces and prevail over tendencies that would favor disorder.

For the first time, to the best of our knowledge, we observe a new type of the crystal lattice where the host atoms occupy their proper (equilibrium) positions in the crystal field, while the impurities, once periodically inserted into the lattice, divide it in the short chains of equal length, where the host atoms develop harmonic vibrations. This periodic substitution of a host atom by an impurity allows the impurity to participate in the formation of the crystal's energy bands. It leads to the change in the value of the forbidden energy gap, to the appearance of a crystalline excitonic phase, and to the broad energy bands instead of the energy levels of bound excitons. The high perfection of this new lattice leads to the abrupt decrease of non-radiative mechanisms of electron-hole recombination, to both the relevant increase of efficiency and spectral range of luminescence and to the stimulated emission of light due to its amplification inside the well arranged, defect-free medium of the crystal. The further development of techniques for the growth of thin films and bulk crystals with ordered distribution of impurities and the proper localization of host atoms inside the lattice should be a high priority.

This long-term evolution of the important properties of our unique collection of semiconductor single crystals promises a novel approach to the development of a new generation of optoelectronic devices. The combined methods of laser assisted and molecular beam epitaxies [39-41] will be applied to fabrication of device structures with artificial periodicity; together with classic methods of crystal growth, can be employed to realize impurity ordering that would yield new types of nanostructures and enhanced optoelectronic device performance.

Note, that semiconductor nanoparticles were introduced into materials science and engineering mainly that to avoid limitations inherent to freshly grown semiconductors with a lot of different defects. Our long-term ordered and therefore close to ideal crystals repeat behavior of the best nanoparticles with pronounced quantum confinement effect. These perfect crystals are useful for application in top-quality optoelectronic devices as well as they are a new object for development of fundamentals of solid state physics.

ACKNOWLEDGEMENTS

The authors are very grateful to the US Dept. of State, Inst. of International Exchange, Washington, DC, The US Air Force Office for Scientific Research, the US Office of Naval Research Global, Civilian R&D Foundation, Arlington, VA, Science & Technology Center in Ukraine, Clemson University, SC, University of Central Florida, FL, Istituto di elettronica dello stato solido, CNR, Rome, Italy, Universita degli studi, Cagliari, Italy, Joffe Physico-Technical Institute, St.Petersburg State Polytechnical University, Russia, Institute of Applied Physics and Academy of Sciences of Moldova for support and attention to this protracted (1963-present time) research.

REFERENCES

1. Pyshkin S L, Radautsan S I et al., Processes of Long-Lasting Ordering in Crystals with a Partly Inverse Spinel Structure, *Sov. Phys. Dokl.*, **35(4)**, 301-304 (1990).
2. N.A. Goryunova, S.L. Pyshkin, A.S. Borshchevskii, S.I. Radautsan, G.A. Kaliujnaya, Yu.I. Maximov, and O.G. Peskov, Influence of Impurities and Crystallisation Conditions on Growth of Platelet GaP Crystals, *Growth of Crystals*, **8**, ed. N.N. Sheftal, New York, 68-72 (1969), Symposium on Crystal Growth at the 7th Int Crystallography Congress (Moscow, July 1966).
3. B.M. Ashkinadze, S.L. Pyshkin, A.I. Bobrysheva, E.V. Vitiu, V.A. Kovarsky, A.V. Lelyakov, S.A. Moskalenko, and S.I. Radautsan, Some Non-linear Optical Effects in GaP, *Proceedings of the IXth International Conference on the Physics of Semiconductors* (Moscow, July 23–29, 1968), **2,** 1189-1193 (1968).
4. S. Pyshkin, S. Radautsan, and V. Zenchenko, Raman Spectra of Cd-In-S with Different Cation-Sublattice Ordering, *Sov. Phys. Dokl.* **35 (12)**, 1064-67 (1990).
5. S.L. Pyshkin and A. Anedda, Preparation and Properties of GaP Doped by Rare-Earth Elements, *Proceedings of the 1993 Materials Research Society (MRS) Spring Meeting*, Symposium E, **301,**192-197 (1993).
6. S.L. Pyshkin and A. Anedda, Time-Dependent Behaviour of Antistructural Defects and Impurities in Cd-In-S and GaP, ICTMC-XI (Salford, UK, 1997), *Institute of Physics Conference Series, Ternary and Multinary Compounds*, **152**, Section E, 785-89 (1998).

7. S. Pyshkin and L. Zifudin, Excitons in Highly Optically Excited Gallium Phosphide, *J. Lumin.*, **9**, 302-308 (1974).
8. S. Pyshkin, Luminescence of GaP:N:Sm Crystals, *J. Sov. Phys. Semicond.*, **8**, 912-13 (1975).
9. S.L. Pyshkin, A. Anedda, F. Congiu, and A. Mura, Luminescence of the GaP:N Ordered System, *J. Pure Appl. Opt.*, **2**, 499-502 (1993).
10. S.L. Pyshkin (invited), Luminescence of Long-Time Ordered GaP:N, The 103rd ACerS Annual Meeting (Indianapolis, 2001), *ACerS Transaction series*, **126,** 3-10 (2002).
11. Sergei L. Pyshkin, Bound Excitons in Long-Time Ordered GaP:N, *Moldavian Journal of the Physical Sciences*, **1(3),** 14-19 (2002).
12. Sergei Pyshkin, John Ballato, Advanced Light Emissive Composite Materials for Integrated Optics, Symposium: The Physics and Materials Challenges for Integrated Optics - A Step in the Future for Photonic Devices, *Proc of the 2005 MS&T Conference*, Pittsburgh, TMS, 3-13 (2005)
13. J. Ballato and S.L. Pyshkin, Advanced Light Emissive Materials for Novel Optical Displays, Lasers, Waveguides, and Amplifiers, *Moldavian J. of Physical Sciences, 5(2),* 195-208 *(2006).*
14. S.L. Pyshkin, J. Ballato, G. Chumanov, J. DiMaio, and A.K. Saha, Preparation and Characterization of Nanocrystalline GaP, *Technical Proceedings of the 2006 NSTI Nanotech Conference,* **3**, 194-197 (2006).
15. S.L. Pyshkin, R.P. Zhitaru, and J. Ballato, Long-Term Evolution of Optical and Mechanical Properties in Gallium Phosphide , *Proceedings of the XVII St. Petersburg Readings on the Problems of Durability*, Devoted to the 90th Birthday of Prof. A.N. Orlov, **2,** 174-176 (2007).
16. S.L. Pyshkin, R.P. Zhitaru, and J. Ballato, Modification of Crystal Lattice by Impurity Ordering in GaP, Int. Symposium on Defects, *Proceedings of the MS&T 2007 Conference*, International Symposium on Defects, Transport and Related Phenomena (Detroit, MI, September 16–20, 2007), 303-310 (2007).
17. S.L. Pyshkin, J. Ballato, and G. Chumanov, Raman light scattering from long-term ordered GaP single crystals, *J. Opt. A. Pure Appl. Opt.*, **9**, 33-36 (2007).
18. S.L. Pyshkin, J. Ballato, M. Bass, and G. Turri (invited), Luminescence of Long-Term Ordered Pure and Doped Gallium Phosphide, TMS Annual Meeting, Symposium: Advances in Semiconductor, Electro Optic and Radio Frequency Materials (March 9–13, New Orleans, LA). *J. Electronic Materials,* **37(4)**, 388-395 (2008).
19. S. Pyshkin, and J. Ballato, Long-Term Ordered Crystals and Their Multi-Layered Film Analogues, *Proceedings of the 2008 MS&T Conference*, Pittsburgh Symposium on Fundamentals & Characterization, Session ''Recent Advances in Growth of Thin Film Materials'', 889-900 (2008).
20. S.L. Pyshkin, J. Ballato, M. Bass, G. Chumanov, and G. Turri, Time-dependent evolution of crystal lattice, defects and impurities in $CdIn_2S_4$ and GaP, *Phys. Stat. Sol.*, **C 6**, 1112-15 (2009).
21. S. Pyshkin, R. Zhitaru, J. Ballato, G. Chumanov, and M. Bass, Structural Characterization of Long Term Ordered Semiconductors , *Proceedings of the 2009 MS&T Conference*, International Symposium ''Fundamentals & Characterization,'' 698-709 (2009).
22. S. Pyshkin, J. Ballato, M. Bass, G. Chumanov, and G. Turri, Properties of the Long-term Ordered Semiconductors, *The 2009 TMS Annual Meeting and Exhibition, Suppl. Proc.*, (San Francisco, February 15–19, 2009), **3**, 477-484 (2009).
23. S. Pyshkin, J. Ballato, M. Bass, and G. Turri, Evolution of Luminescence from Doped Gallium Phosphide over 40 Years, *J. Electronic Materials,* **38(5)**, 640-646 (2009).
24. S. Pyshkin, J. Ballato, G. Chumanov, M. Bass, G. Turri, R. Zhitaru, and V. Tazlavan, Optical and Mechanical Properties of Long-Term Ordered Semiconductors, The 4th International

Conference on Materials Science and Condensed Matter Physics, Kishinev, Sept 23-26, 2008, *Moldavian Journal of the Physical Sciences*, **8(3-4)**, 287-295 (2009).
25. Sergei Pyshkin, John Ballato, Andrea Mura, Marco Marceddu, Luminescence of the GaP:N Long-Term Ordered Single Crystals, *Suppl. Proceedings of the 2010 TMS Annual Meetings* (Seattle, WA, USA, February, 2010, **3**, 47-54 (2010).
26. Sergei Pyshkin and John Ballato, Evolution of Optical and Mechanical Properties of Semiconductors over 40 Years, *J. Electronic Materials*, Springer, DOI: 10.1007/s11664-010-1170-z, **39(6)**, 635-641 (2010).
27. S. Pyshkin, J. Ballato, G. Chumanov, N. Tsyntsaru, E. Rusu, Preparation and Characterization of Nanocrystalline GaP for Advanced Light Emissive Device Structures, The 2010 Nanotech Conference (Anaheim, CA, June 21-24), *NSTI, NSTI-Nanotech 2010, www.nsti.org, ISBN 978-1-4398-3401-5*, **1**, 522-525 (2010).
28. S. Pyshkin, J. Ballato, I. Luzinov, B. Zdyrko (2010) Fabrication and Characterization of the GaP/Polymer Nanocomposites for Advanced Light Emissive Device Structures, The 2010 Nanotech Conference (Anaheim, CA, June 21-24), *NSTI, NSTI-Nanotech 2010, www.nsti.org, ISBN 978-1-4398-3401-5,* **1**, 772-775 (2010).
29. S. Pyshkin (Project Manager), Joint Moldova/US/Italy/France/Romania STCU (www.stcu.int) Project 4610 "Advanced Light Emissive Device Structures", 2009-2012
30. W.J. Allen, Phys. C: Solid State Phys. 1, 1136 (1968).
31. B. Bairamov B, Y. Kitaev, V.Negoduiko, and Z.Khashkhozhev, J. Sov. Phys. Solid State, **16**, 1323-25 (1975).
32. N.A. Goriunova, *J. All Union Chem. Soc.* 5, 522-527 (1960).
33. S.I. Radautsan, Yu.I. Maximov, V.V. Negreskul, and S.L. Pyshkin, Gallium Phosphide, Kishinev, Moldavian Academy of Sciences, *Shtiinza* (1969).
34. B.M. Pushkash, M.I. Valkovskaya, Yu.I. Maximov, and D.V. Martynko, Deformation of Crystals Under Influence of Localized Loading, Kishinev, Moldavian Academy of Sciences, *Shtiinza*, (1978).
35. M.I. Valkovskaya, B.M. Pushkash, and E.E. Maronchuk, Plasticity and Fragility of Semiconductors at the Tests for Microhardness, Kishinev, Moldavian Academy of Sciences, *Shtiinza*, 24-29 (1984).
36. S.L. Pyshkin, Preparation and Properties of Gallium Phosphide, Ph.D. thesis, *State University of Moldova* (1967).
37. C. Kittel, Introduction to Solid State Physics, *the 4th edition, New York: Wiley* (1978).
38. S.L. Pyshkin, Stimulated emission in gallium phosphide, Presented by Nobel Prize Laureate A.M. Prokhorov, *Sov. Phys. Dokl.*, **19**, 845-846 (1975)).
39. S.L. Pyshkin, Heterostructures $(CaSrBa)F_2$ on InP for Optoelectronics, Report to the US AFOSR/EOARD on the Contract No. SPQ-94-4098 (1995).
40. S.L. Pyshkin, V.P. Grekov, J.P. Lorenzo, S.V. Novikov, and K.S. Pyshkin, Reduced Temperature Growth and Characterization of $InP/SrF_2/InP(100)$ Heterostructure, Physics and Applications of Non-Crystalline Semiconductors in Optoelectronics, *NATO ASI Series, 3. High Technology*, **36**, 468-471 (1996).
41. S.L. Pyshkin, $CdF_2:Er/CaF_2/Si(111)$ Heterostructure for EL Displays, Report to the US AFOSR/EOARD on the Contract No. SPQ-97-4011 (1997).

INFLUENCE OF MAGNETIC FLUX DENSITY AND SINTERING PROCESS ON THE ORIENTED STRUCTURE OF C-AXIS-ORIENTED $Sr_2NaNb_5O_{15}$ PIEZOELECTRIC CERAMICS

Satoshi Tanaka[1], Takuma Takahashi[1], Tomoko Kawase[1], Ryoichi Furushima[1], Hiroyuki Shimizu[2], Yutaka Doshida[2] and Keizo Uematsu[1]

1 Dept. Materials Science and Technology, Nagaoka University of Technology
1603-1 Kamitomioka, Nagaoka Niigata, 9402188 JAPAN
2 General R&D Laboratories, Taiyo Yuden Co., Ltd
5607-2 Nakamuroda, Takasaki, Gumma, 3703347 JAPAN

ABSTRACT

Effect of magnetic field and subsequent sintering was examined on oriented structures of $Sr_2NaNb_5O_{15}$. Oriented direction was controlled by a rotating magnetic field of 0-10T. In specimen formed in a rotating magnetic field, c-axis of particles were oriented perpendicular to the plane of the rotation, since the diamagnetic susceptibility is larger along c-axis than other axis. The degree of orientation evaluated by XRD of green compact was about 0-0.65, and depended on the square of magnetic flux density. The degree of orientation increased upto 0.94 by sintering. Sample prepared in 5 T attaiend the moderate degree of orientation of 0.5 before sintering. After sintering, the microstructure observation indicated a well-oriented structure consisting of large grains over 10μm, in which longitudinal axes were oriented parallel to the rotating axis. Crystal-oriented ceramics shows electrical property 90pC/N at 10Tesla and is twice as high as 44pC/N at 0Tesla.

1. INTRODUCTION

Designing of texture has been studied actively to improve functions of ceramics drastically. New method consisting of forming in a high magnetic field followed by sintering is very effective to fulfill the objective for a variety of materials. Advantages of the technique include freedoms in the orientation direction and the choice of the raw powders[1-16]. High quality commercial powder can be applied without special treatment, allowing easy densification in the subsequent sintering. We have succeed to design highly textured structure for various materials such as alumina[6-8,12], titania[9], bismuth titanate family[10,11], and tungsten bronze systems.[16,17]

Ferroelectrics of tungsten-bronze type are very attractive as candidates for lead-free piezoelectric materials[14-23]. These materials require precise control of the textures, since the high piezoelectric performances occur only along the c-crystallographic axis. For manufacturing multi-layered piezoelectric devices, c-axis of particles must be oriented along the electrical field applied during application, or perpendicular to internal electrodes. Rotating strong magnetic field (RMF) orientation method is an excellent technique to align c-axis of tungsten bronze type materials. **Fig.1** shows the concept of RMF. The crystal axis of the highest diamagnetic susceptibility is oriented to the direction of rotation axis, that is, normal to the substrate or the

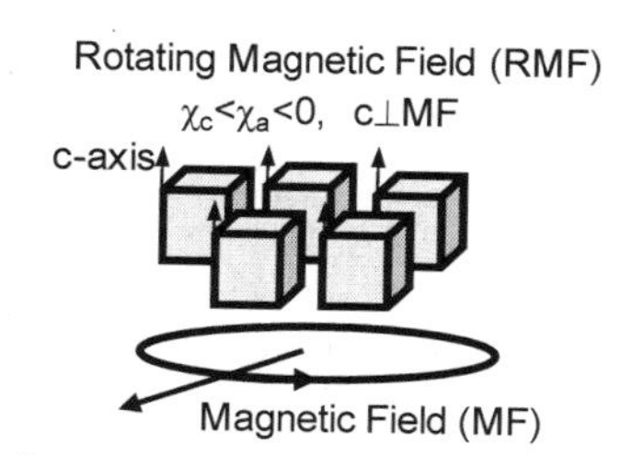

Figure 1 Oriented direction in a rotating magnetic field.

plane of rotating magnetic field. RMF has been successfully applied to produce c-axis oriented ZnO[13] and La-$Sr_2NaNb_5O_{15}$ [14]. The driving force of orientation is the minimization of magnetization energy in the magnetic field. The magnetic torque ***T*** for reducing the magnetization energy is expressed by the following equation,

$$T = \frac{1}{2\mu_0} \Delta\chi B^2 \left(\frac{4\pi r^3}{3} \right) \sin 2\theta \qquad (1)$$

where, $\Delta\chi$ is anisotropy of magnetic susceptibility, ***r*** particle radius and ***B*** magnetic flux density. The objectives of this study is to examine the influence of magnetic flux density and sintering on the orientation degree for the c-axis-oriented $Sr_2NaNb_5O_{15}$ (SNN) ceramics system.

2. EXPERIMENTAL

$Sr_2NaNb_5O_{15}$ (SNN) powder was synthesized by conventional solid-state reaction. Powders of sodium carbonate, strontium carbonate, and niobium oxide (Nakarai Co., Ltd.) were mixed with ethanol in a ball-mill with zirconia balls. The mixed powders were dried, and heated in an alumina crucible at 1200 °C for 3h. Crystalline phases were examined by X-ray diffraction analysis (MO3XHF22, Burker). The powder was also characterized by scanning electron microscopy (SEM) (5300LV, JEOL Co.).

The synthesized powder was mixed with distilled water and a poly acrylic acid dispersant by ball-milling to give slurry with solid loading of 30 vol%. A plastic mold filled with slurry ($5{\cdot}10^{-6}m^3$) was placed in a magnetic field (0-10T) of a superconducting magnet (TM-10VH10, Toshiba). The mold was set in the magnetic field (RMF) and kept to dry at the room temperature with or without the rotation at 30 rpm, as shown in **Fig.2**. The green compact, which had a disk shape (3mm thick and 25mm diameter), was heated at 5 °C/min to 1250°C, held for 6 h to prevent exaggerated grain growth, and then sintered at 1525°C for 2 h.

The density of the sample was measured from its weight and dimensions. The microstructure of the specimens was examined by SEM. Grain orientation was evaluated by XRD. The degree of orientation was calculated semiquantitatively with the Lotgering method[23,24]. The diffraction peaks used for calculations were in the range 20–60°. The degree of orientation was evaluated in terms of the Lotgering factor *F*, which is calculated by using the following equation[23].

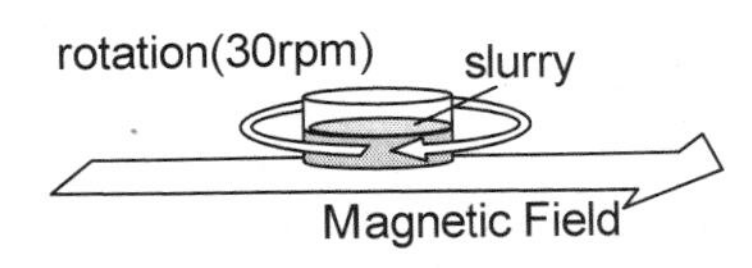

Figure 2 Illustration of setting of rotating magnetic field (RMF)

$$F = \frac{P - P_0}{1 - P_0} \qquad (2)$$

where $P_0 = \Sigma I_0(00l)/ \Sigma I_0(hkl)$ and $P = \Sigma I(00l)/ \Sigma I(hkl)$. I and I_0 are the intensities of each diffraction peaks in X-ray diffraction patterns as presented in ICDD data and those measured experimentally, respectively.

The piezoelectric constant was measured by d33 meter (ZJ-6B, Institute of Acoustic, Chinese Academy of Sciences). The specimens were poled parallel to the oriented direction under the electric field of 2.0-4.0 kV/mm for 15 min in silicone oil at 160°C, and silver electrodes were made on the major faces of these specimens.

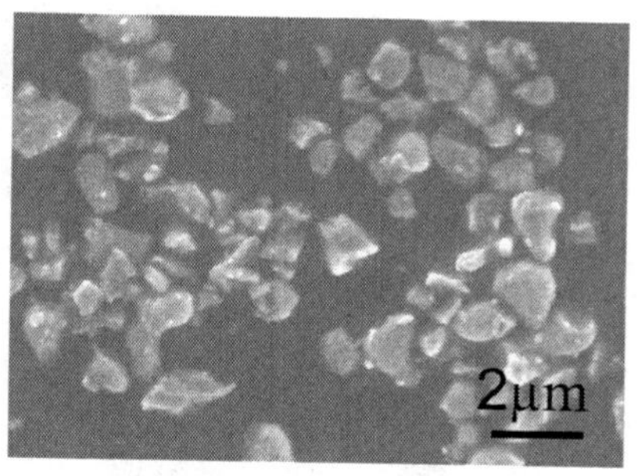

Figure 3 Synthesized $Sr_2NaNb_5O_{15}$ (SNN) particles.

3. RESULTS

Fig.3 shows a SEM micrograph of the $Sr_2NaNb_5O_{15}$ (SNN) particles. A majority of particles had isotropic or plate shapes. The mean particle size was about 1.6 μm. No agglomerates were found in the micrographs. Crystal phase was SNN only in the X-ray diffraction pattern.

Fig.4 shows XRD patterns of green body prepared with various magnetic fields 0-10 T. **Fig.5** shows the XRD patterns of sintered bodies, which were heat-treated at 1200 °C for 6h and at 1350°C for 2h. XRD patterns were taken on the top surface of the specimens. Peaks relevant to the c-planes of the crystal, such as *001* and *002*, were very strong in specimens prepared in

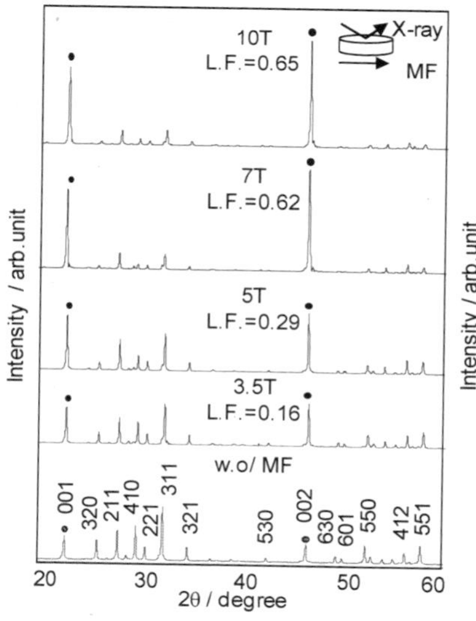

Figure 4 XRD patterns of green bodies formed in RMF with various magnetic flux densities.

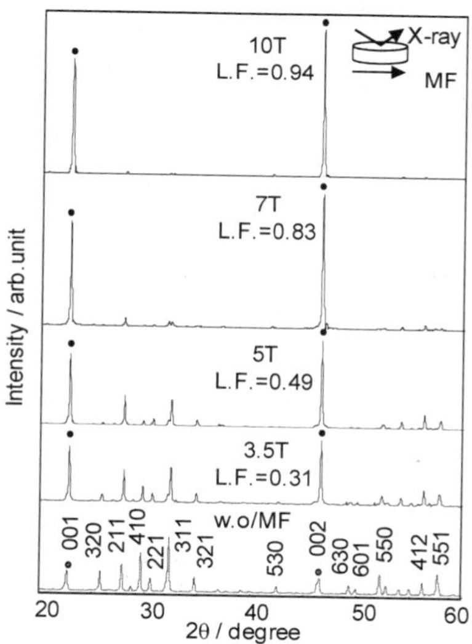

Figure 5 XRD patterns of sintered bodies. Magnetic flux densities (0-10T) are forming conditions before sintering.

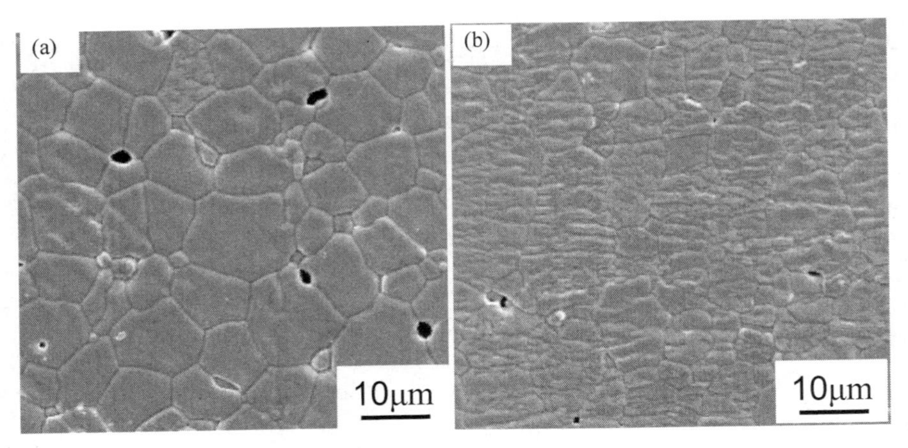

Figure 6 Microstructure of SNN ceramics heat treated at 1350°C for 2hrs
(a) top surface of sample, (b) cross section of sample

the magnetic field. Diffraction peaks such as *320*, *410* and *311* were very strong in the samples with 0 Tesla, but were absent in the specimen prepared with RMF. This showed that c-axis-oriented SNN polycrystalline ceramics were produced in the presence of the RMF. Their intensities increased with increasing magnetic flux density. The height of *00l* peaks increased in the sintering process.

Fig.6 shows SEM micrographs of the sintered specimens observed from the directions parallel and perpendicular to the rotating magnetic field (RMF). A significant difference was noted in the structures observed from two directions. Flat face of grains of c-faces, were observed in top surface of **Fig.6(a)**, whereas elongated grains with horizontal stripe of a-faces were observed in **Fig.6(b)**. The relative densities after sintering were 97.4% in the c-axis oriented specimen and 98.6% in the randomly oriented specimens.

Fig.7 shows piezoelectric constants of various oriented specimens. The piezoelectric constant increased up to 90pC/N with Lotgering factor.

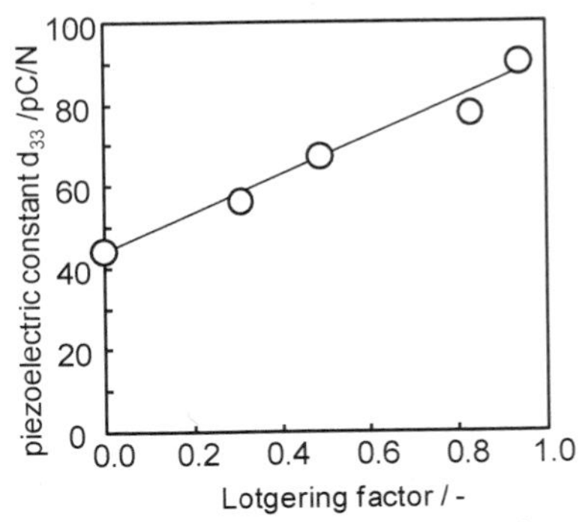

Figure 7 Piezoelectric constant d_{33} as a function of Lotgering factor.

4. DISCUSSIONS

The magnetic torque is proportional to the square of magnetic flux density as written in eq.(1). **Fig.8** shows the influence of the magnetic flux density on the orientation degree (Lotgering factor) of green and sintered bodies. The linear relationship was obtained in the green compact in the range of 0-50 $(\text{Tesla})^2$, (0-7 Tesla) in green body. The magnetic torque governs the orientation behavior for this range. The degree of orientation is saturated at high magnetic flux density 10 Tesla. Mechanism for this behavior is unclear at this stage. Presence of small agglomeration consisting of fine powders is a possible explanation to this result.

Fig.8 also shows that sintering processing contributes to development of oriented structure. The orientation degree, represented by the Lotgering factor, reached high value 0.94. Note that even the sample prepared in weak magnetic field 5 T attaiend the high degree of orientation of 0.5 after sintering. As shown in SEM micrographs (**Fig.6 (b)**), the large elongated grains with over 10 µm are observed in parallel to the rotation axis of the magnetic field. The horizontal stripes in each grain in oriented microstructure are also observed. The microstructure suggests that the oriented particles grow preferentially along the c-axis by involving small random particles during sintering. The processes of grain growth along to c-axis affect the morphologies in each grain as horizontal lines.

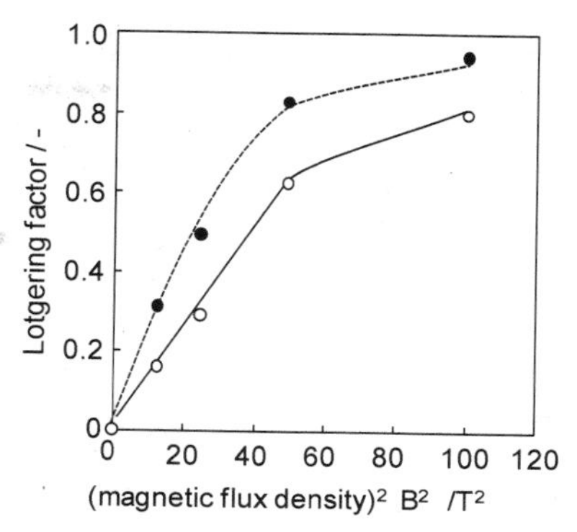

Figure 8 Lotgering factor as a function of square of magnetic flux density B^2, open circles: green body, closed circles: sintered body.

CONCLUSIONS

Effect of magnetic field and sintering process was examined on oriented structures of $Sr_2NaNb_5O_{15}$. Orientation direction was controlled by the rotating magnetic field in the range of 0-10 Tesla. The degree of orientation for *001* evaluated by XRD of green compact was about 0-0.65, which depended on the magnetic flux density. Oriented structures were developed with densification and grain growth by sintering at 1350°C. Microstructure observation indicated a well-oriented structure, in which longitudinal axes were oriented parallel to the rotating axis. Crystal-oriented ceramics shows improved electrical property.

REFERENCE

[1] P.De,Rango, M.Lees, P.Lejay, A.Sulpice, R.Tournier, M.Ingold, P.Germi and M.Pernet, Textur ing of magnetic materials at high temperature by solidification in a magnetic field, *Nature*, 349, 7 70-71 (1991).

[2] J.G.Noudema, J.Beilleb, D.Bourgaulta, D.Chateignerc and R.Tourniera, Bulk textured BiPbSrC aCuO (2223) ceramics by solidification in a magnetic field, *Physica C*, 264, 325-30 (1996)

[3] Y.Nakagawa, H.Yamasaki, H.Obara and Y.Kimura, Superconductiong properties of grain-oriented samples of $YBa_2Cu_3O_y$, Jpn,J.Appl. Phys. 28, 4, L547-50 (1989)

[4] M.H.Zimmerman, K.T.Faber, E.R.Fuller Jr., K.L.Kruger and K.J.Bowman, Texture assessment of magnetically processed iron titanate, *J.Am.Ceram.Soc.*, **79**, 1389-93 (1996)

[5] M.H.Zimmerman, K.T.Faber and E.R.Fuller Jr, Forming textured microstructures via the gelcast ing technique, *J.Am.Ceram.Soc.*, **80,** 2725-2729 (1997)

[6] K.Uematsu, T.Ishikawa, D.Shoji, T.Kimura and K.Kitazawa, Japan-Patent 3556886

[7] T.S.Suzuki, Y.Sakka and K.Kitazawa, Orientation amplification of alumina by colloidal filtration in a strong magnetic field and sintering, *Adv. Engineering Mater.*, **3**, 44-46 (2001)

[8]A.Makiya, D.Shoji, S.Tanaka, N.Uchida, T.Kimura and K.Uematsu, Grain oriented microstructure made in high magnetic field, *Key Engineering Mater.*, **206-213**, 445-48(2002).
[9]A.Makiya, K.Kusumi, S.Tanaka and K.Uematsu, Particle oriented titana ceramics prepared in a magnetic field, *J.Euro.Ceram.Soc.*, **27**, 797-99 (2007)
[10]A.Makiya, D.Kusano, S.Tanaka, N.Uchida, K.Uematsu, T.Kimura, K.Kitazawa and Y.Doshida, Particle oriented bismuth titanate ceramics made in high magnetic field, *J. Ceram. Soc. Japan*, **1 11**, 702-4 (2003)
[11]Y.Doshida, K.Tsuzuku, H.Kishi, A.Makiya, S.Tanaka, K.Uematsu and T.Kimura, Crystal-orien ted $Bi_4Ti_3O_{12}$ ceramics fabricated by high-magnetic-field method, *J. J. Appl. Phys.*, **43**, 6645-48 (2004)
[12]Y.Sakka and T.S.Suzuki, Textured development of feeble magnetic ceramics by colloidal processing under high magnetic field, *J.Ceram.Soc.Jpn.*, **113,** 26-36 (2005)
[13]S.Tanaka, A.Makiya, Z.Kato, N.Uchida, T.Kimura and K.Uematsu, Fabrication of c-axis orientated polycrystalline ZnO using a rotating magnetic field and following sintering, *J.Material Res.*, **21,** 703-7 (2006)
[14]Y.Doshida, H.Kishi, A.Makiya, S.Tanaka, K.Uematsu and T.Kimura, Crystal-oriented La-substituted $Sr_2NaNb_5O_{15}$ ceramics fabricated using high-magnetic-field method, *J. J. Appl. Phys.*, **45**, 7460-64 (2006)
[15]S.Tanaka, A.Makiya, T.Okada, T.Kawase, Z.Kato and K.Uematsu, C-axis orientation of $KSr_2Nb_5O_{15}$ by using a rotating magnetic field, *J.Am.Ceram.Soc*, **90**, 3503-06 (2007)
[16]S.Tanaka, T.Takahashi, R.Furushima, A.Makiya and K.Uematsu, Fabrication of c-axis-oriented potassium strontium niobate ($KSr_2Nb_5O_{15}$) ceramics by a rotating magnetic field and electrical property, *J.Ceram.Soc.Jpn*, **118**, 722-725 (2010)
[17]Y.Saito, H.Takao, T.Tani, T.Nonoyama, K.Takatri, T.Homma, T.Nagaya and M.Nakamuram, Lead-free piezoceramics, *Nature*, **432**, 84-87 (2004)
[18]R. R. Neurgaonkar, W. W. Ho, W. K. Cory, W. F. Hall and L .E. Cross, Low and high frequenc y dielectric properties of ferroelectric tungsten bronze $Sr_2KNb_5O_{15}$ crystals, *Ferroelectrics*, **51**, 1 85-91 (1984)
[19]K.Matsuo, R.J.Xie, Y.Akimune and T.Sugiyama, Preparation of lead free $Sr_{2-x}Ca_xNaNb_5O_{15}$ ba sed piezoceramics with tungsten bronze structure, *J. Ceram. Soc. Japan*, **110**, 491-4 (2002)
[20]R.R.Neurgaonkar, J.R.Oliver, W.K.Cory, L.E.Cross and D.Viehland, Piezoelectricity in tungste n bronze crystals, *Ferroelectrics*, **160**, 265-6 (1994)
[21]C.Duran, S.Trolier-McKinstry and G..L.Messing, Fabrication and Electrical properties of textur ed $Sr_{0.47}Ba_{0.47}Nb_2O_6$ ceramics by templated grain growth, *J. Am. Ceram. Soc.*, **83**, 2203-13(2000)
[22]C.Duran, S.Trolier-McKinstry and G.L.Messing, Dielectric and piezoelectric properties of textured $Sr_{0.47}Ba_{0.47}Nb_2O_6$ ceramics by templated grain growth, *J. Mater. Res.*, **17**, 2399-2409(2002)
[23]F.K.Lotgering, Topotactial reactions with ferromagnetic oxides having hexagonal crystal struct ure, *J. Inorg. Nucl. Chem.*, **9,** 113-23 (1959)
[24]J.L.Jones, E.B.Slamovich and K.J.Bowman, Critical evaluation of the Lotgering degree of orientation texture indicator", *J. Mater. Res.*, **19**, 3414-22 (2004)

SINTERING OF DEFECT-FREE $BaTi_{0.975}Sn_{0.025}O_3/BaTi_{0.85}Sn_{0.15}O_3$ FUNCTIONALLY GRADED MATERIALS

S. Marković and D. Usković
Institute of Technical Sciences of the Serbian Academy of Sciences and Arts
Belgrade, Serbia

ABSTRACT

Shrinkage anisotropy could be one of the reasons of deformation during densification of functionally graded materials (FGMs). In this study we calculated coefficient of shrinkage anisotropy during sintering of BTS2.5/BTS15 FGMs by different heating rates of 2, 5, 10 and 20 °/min. We found that concentration gradient in BTS2.5/BTS15 FGMs implies a gradient of anisotropy coefficient. Moreover, we calculated effective activation energy for sintering of BTS2.5/BTS15 FGMs and compared with those for BTS2.5 and BTS15 graded layers. The values of 359.5 and 340.5 kJ/mol were obtained for BTS2.5 and BTS15 graded layers, respectively, while for the sintering of entire BTS2.5/BTS15 FGMs the value of 460 kJ/mol was obtained. The difference between the effective activation energy could be attributed to potential insulator interlayer formed between layers during uniaxial pressing. Impedance spectroscopy was used to determine the electrical characteristics of BTS2.5/BTS15 FGMs sintered by different heating rates, as well as to determine influence of concentration gradient on intrinsic and microstructural features. Activation energies of BTS2.5/BTS15 FGMs, separately for grain interior and grain boundary, were calculated. It was established that the activation energy deduced from grain-interior conductivity kept the intrinsic properties (0.71−0.73 eV) and is not influenced by tin/titanium concentration gradient, neither by heating rate. However, activation energy for grain boundary conductivity (1.05−1.24 eV) is determined by microstructural development (density and average grain size) which is a consequence of both concentration gradient developed during sintering processes and heating rate. Finally, electrical properties of BTS2.5/BTS15 FGMs sintered by different heating rates were correlated with their microstructure.

INTRODUCTION

Functionally graded materials (FGMs) belong to an interesting class of materials in which it is possible to obtain a gradient of properties that cannot be attained in any spatially-homogenous materials. In particular, continuous changes in the properties such as chemical composition, grain size or porosity result in the gradient of mechanical, thermal and/or electrical features of graded materials. Over the years, FGMs have found applications in various functional materials, like piezoelectric ceramics,[1] thermoelectric semiconductors,[2] and biomaterials.[3]

It is well known from previous studies that barium titanate stannate ($BaTi_{1-x}Sn_xO_3$, BTS) FGMs have a relatively high dielectric permittivity in a wide temperature range.[4,5] Furthermore, their electrical characteristics (dielectric permittivity, position of $\varepsilon_{r,max}$ at the temperature scale, width of Curie temperature intervals, grain boundary resistivity, etc.) can be tailored by modifying tin/titanium concentration gradient.[4,5] Thus, BTS FGMs are promising candidates for the manufacture of ceramic capacitors, bending actuators,[6] microwave phase shifters,[7] as well as sensors.[8]

An important processing goal in the design and processing of FGMs which will be used as devices in electronic industry is to obtain high-quality microstructure with desired grain size and density, free from any form of deformations. The easiest and the cheapest method of FGM fabrication is powder processing, i.e. by uniaxially pressing and sintering. However, this technique could yield significant problems of components' macrostructure. Precisely, during the sintering process, graded layers with different chemical compositions and/or average particles size, specific surface area, etc., show different shrinkage rates, final extents of shrinkage and densities. This phenomenon could lead to excessive shape distortion, warping, delamination, development of cracks and microstructural damage in the sintered FGMs, which have detrimental influence on their electrical properties. Therefore, it is necessary to find optimal sintering conditions (heating rate and sintering temperature) to achieve high-quality FGMs free from any form of deformation, with microstructure required for good electrical properties.

In our previous investigation, during calculation of effective activation energy for sintering of BTS2.5 and BTS15 graded layers in BTS2.5/BTS15 FGMs we noticed that the shrinkage of the graded layers in FGMs is different.[9] In particular, shrinkage depends on the tin content and the heating rate, where BTS2.5 graded layer in FGMs always reaches higher values of shrinkage than BTS15 graded layer. Furthermore, the percent of shrinkage of the graded layers decreases with the increase in heating rate during sintering of FGMs. These results confirm that the kinetics of densification of graded layers within the FGMs depends on heating rate. Hence, it is very important to apply optimal sintering heating rate to prepare FGMs with ideal microstructure.

Since shrinkage anisotropy could be one of the reasons of compact deformation, here, we calculated coefficient of shrinkage anisotropy for BTS2.5/BTS15 FGM during sintering by different heating rates. Moreover, we calculated effective activation energy for sintering of BTS2.5/BTS15 FGMs and compared with those for BTS2.5 and BTS15 graded layers. Finally, we correlated electrical properties of BTS2.5/BTS15 FGMs, sintered by different heating rates, with their microstructure.

EXPERIMENTAL PROCEDURE

The initial BTS powders ($BaTi_{1-x}Sn_xO_3$, x=0.025 and 0.15, denoted as BTS2.5 and BTS15, respectively) were prepared by a conventional solid state reaction between $BaCO_3$, TiO_2, and SnO_2 at 1100 °C during 2 h. A complete preparation procedure is described elsewhere.[4] The phase structure of the BTS powders was determined using a powder X-ray diffractometer (Philips PW 1050). Theoretical density of BTS2.5 and BTS15 powders, tin content, and average crystallite size were calculated according to the Rietveld analysis on the XRD data. The average particle size (d_{50}) was measured by laser particle size analyzer, Mastersizer 2000 (Malvern Instruments Ltd., UK). The average green density of the BTS monomorph layers, pressed in the same conditions as further FGMs, was measured according to Archimedes' principle. The average green density values of 63.5 and 65% of theoretical density were estimated for BTS2.5 and BTS15, respectively.

The BTS2.5/BTS15 FGMs were fabricated by the powder stacking method. The BTS powders with different compositions were stacked sequentially in die; they were uni-axial pressed into cylindrical compacts (∅ 4 mm and $h \approx 2$ mm) under a pressure of 300 MPa. In order to investigate the influence of heating rate on the shrinkage anisotropy, microstructure and electrical properties of BTS2.5/BTS15 FGMs, sintering experiments were performed in heating microscope (E. Leitz, Wetzlar, Germany). The experiments were performed in air up to 1420 °C, using a heating rate of 2, 5, 10 and 20 °C/min. The diameters and height of the recorded samples were measured on appropriate images (Figure 1a) of considerable enlargement. Figure 1b shows schematically the graded structure of the investigated FGMs. The changes of dimensions, height and diameters (d_{bottom} and d_{top}, denoted as diameters of graded layers BTS2.5 and BTS15, respectively, see Figure 1b) were photographed during sintering process at appropriate time intervals.

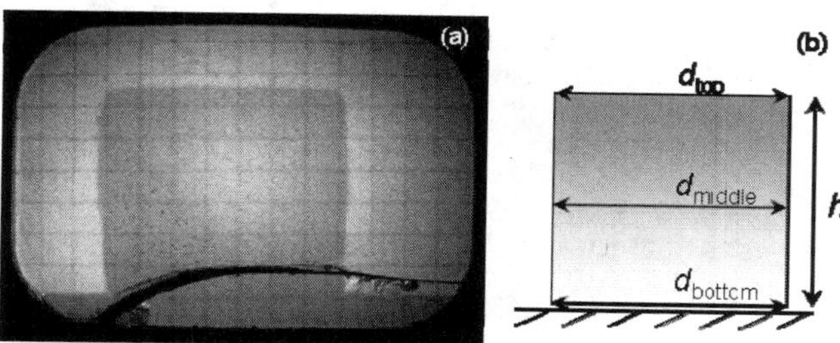

Figure 1. (a) Photograph of sintered cylindrical sample as observed in heating microscope, and (b) scheme of graded sample with marked dimensions measured during sintering. Colored scheme: graded concentration of BTS2.5/BTS15 FGM, from BTS2.5 (bottom, white) to BTS15 (top, dark gray).

Gamry EIS300 Impedance Analyzer was used for electrical characterization of FGMs. Measurements were done in frequency interval 1 Hz-100 kHz, in air atmosphere, during cooling from 320 to 25 °C; the applied voltage was 100 mV. As electrodes, high conductivity silver paste was applied onto both sides of the samples, parallel to the layers. The impedance data were fitted by software *Z*-View2 (version 2.6 demo).

The morphology of the BTS powders were analyzed by field emission scanning electron microscopy (FE-SEM, Supra 35 VP, Carl Zeiss). Before the analysis, the powders were dispersed in ethanol, filtered, and carbon coated. The microstructure of the FGMs was studied by scanning electron microscopy (SEM model JSM 5300, operating at 30 kV). Before SEM measurements, the FGM samples were cut perpendicularly with respect to the layers; the cross section surfaces were gold coated.

RESULTS AND DISCUSSION

The characteristics of the starting powders influenced the quality of the final ceramics as well as their electrical properties. In particular, for the preparation of dense ceramics with optimal microstructure and good electrical properties, sub-micro or nano-sized powders with uniform particle size are required.[10,11] The main characteristics of used BTS powders are listed in Table I, while their morphology is shown in Figure 2.

Table I. Characteristics of BTS powders.

	BTS2.5	BTS15
Stoichiometry	$BaTi_{0.975}Sn_{0.025}O_3$	$BaTi_{0.85}Sn_{0.15}O_3$
Theoretical density (g/cm^3)	6.08	6.23
Crystallite size (nm)	69.4	65.0
Average particle size (nm)	330	340
Average green density (% T.D.)	63.5	65.0

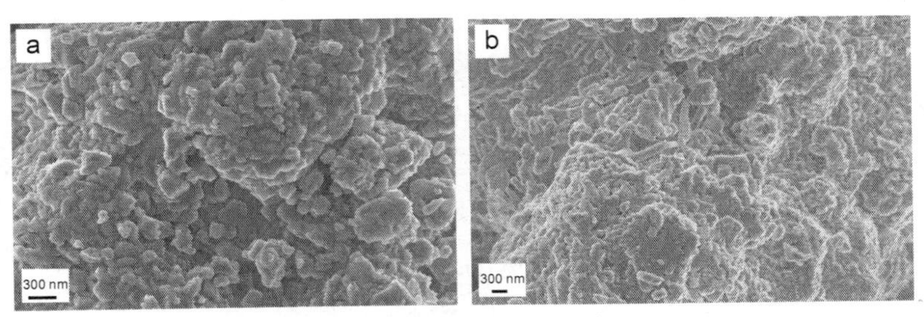

Figure 2. FESEM images of: (a) BTS2.5, and (b) BTS15 powders.

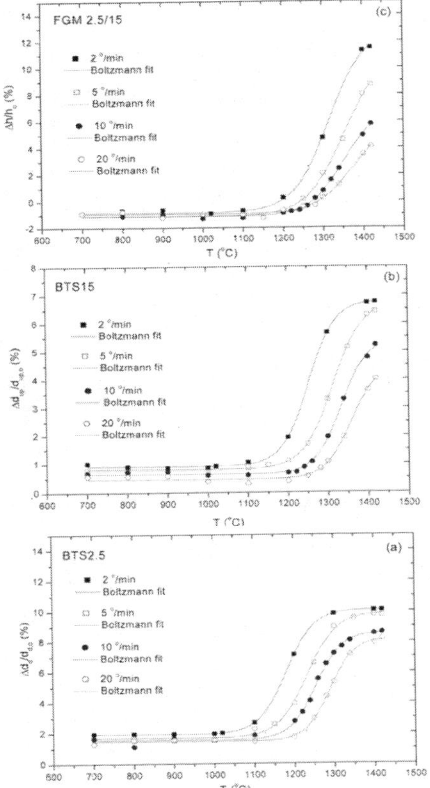

Figure 3 Shrinkage curves for: (a) bottom diameter, (b) top diameter, and (c) height.

The sintering shrinkage of BTS2.5/BTS15 FGMs was recorded in the heating microscope; axial (h) and radial (d) dimensions were measured. From the experimental data for h and d, and using equation (1), the percentage of shrinkage was calculated for the height, as well as for the bottom and the top diameters:

$$shrinkage\ (\%) = \frac{\Delta l}{l_o} \times 100 \qquad (1)$$

where Δl ($=l_o - l_i$) denotes the difference between the initial value of height or diameter l_o at time t_o and the values l_i at time t_i. The calculated values of shrinkage were used for the determination of the samples' sintering behavior, including shrinkage anisotropy.

The densification of BTS layers in FGMs is represented by shrinkage curves of the height, and both diameters, of the bottom and top layer, of the samples *versus* temperature, Figure 3 (a-c). It is noticed that the shrinkage curves are relatively similar for all the measured dimensions. The shrinkage is greater for bottom layer than for top layer, which is a consequence of tin content. Increase of tin content decreases the shrinkage. Furthermore, for top layer, the shrinkage interval is shifted toward higher temperatures for ~ 100 °C. For both diameters, the shrinkage is the most intensive during heating by 2 °/min (~ 10% for bottom layer, and ~ 7% for top layer), and the least intensive during sintering by 20 °/min (~ 8% for bottom layer, and ~ 4% for top layer). The axial shrinkage begins at 1200 °C, whereby densification is not completed up to 1420 °C; the highest attained axial shrinkage is ~ 11%, Figure 3c. From these results it is obvious that $\Delta h \neq \Delta d$, which implies anisotropic densification.

During the years a series of phenomenological hypotheses have been proposed as an explanation for the existence of shrinkage anisotropy, including the effect of gravity, residual stresses, varying density distribution in compacts, and preferred orientation of the crystal lattice.[12] Also, the origin of the anisotropic shrinkage was attributed to the orientation of elongated particles in the compacts in addition to the non-uniform particle packing density in uniaxially pressed compacts.[13,14] A larger sintering shrinkage in the h direction than in the d direction occurs when elongated particles align with their longest axis perpendicular to the direction of uniaxial pressing within the plane parallel to the direction of uniaxial pressing and randomly oriented within the plane normal to the uniaxial pressing.[14]

A convenient way of quantifying shrinkage anisotropy during sintering is the calculation of anisotropy factor k.[15] In cylindrical compacts, k determines the extent of shrinkage anisotropy relating axial and diametrical shrinkage. The anisotropy factor was calculated using equation (2):

$$anisotropy\,factor = k = \frac{\Delta d}{\Delta h} = \frac{d_o - d_i}{h_o - h_i} \tag{2}$$

where d_o and d_i denote the diameter, whereas h_o and h_i denote the height, at initial time t_o and at time t_i, respectively.

The values of k calculated for the shrinkage data at 1420 °C are listed in Table II. It can be seen that the anisotropy factor varies a lot depending on both stoichiometry of graded layer and heating rate. At first, the shrinkage anisotropy in barium titanate materials can be explained by the use of powders consisting of anisotropic crystallites elongated in [100] direction, as well as by their non-uniform packing during uni-axially pressing. As it is emphasized above, average particle size, their morphology, as well as the nature of agglomerates were important to the densification process and anisotropy factor.

Table II. Anisotropy coefficient for bottom (k_1), top (k_2), and average (k_3) diameter.

Heating rate (°/min)	k_1 $(=\frac{d_{b,o} - d_{b,i}}{h_o - h_i})$	k_2 $(=\frac{d_{t,o} - d_{t,i}}{h_o - h_i})$	k_3 $(=\frac{d_{av,o} - d_{av,i}}{h_o - h_i})$
2	0.88	0.59	0.73
5	1.12	0.74	0.93
10	1.50	0.91	1.20
20	2.05	0.98	1.49

Here, we find that for BTS2.5 powder the slower heating rates (2 and 5 °/min) yield almost isotropic shrinkage ($k \approx 1$). Contrarily, the heating rate that resulted in isotropic shrinkage of BTS15 powder is 20 °/min. Since the morphology of the used BTS powders, as well as their average particle size, are almost equal; different anisotropy factor of BTS2.5 and BTS15 could be explained by different tin content i.e. different theoretical density. Increasing of tin content in BTS powders required increasing of sintering temperature to obtain full density ceramics. Second explanation of different anisotropy factor of the bottom and the top layers could be the direction of gravity during sintering; in particular, weightier BTS15 graded layer was on the top and exerted extra pressure on BTS2.5 graded layer during sintering process. Hence, according to presented values of k it can be concluded that concentration gradient promotes gradient of anisotropy coefficient.

In our previous study[9] we used the concept of master sintering curve (MSC)[16] and diameter shrinkage data to estimate the effective activation energy for sintering of BTS2.5 and BTS15 graded layers in BTS2.5/BTS15 FGMs. The values of 359.5 and 340.5 kJ/mol were obtained.[9] Here, since a large anisotropy shrinkage was noticed, we used the axial shrinkage data to estimate activation energy for sintering of entire BTS2.5/BTS15 FGMs; the value of 460 kJ/mol was obtained. This difference between the effective activation energy, of ~ 100 kJ/mol, could be attributed to potential insulator interlayer formed between graded layers during uniaxial pressing.

Furthermore, we used impedance spectroscopy to determine the electrical characteristics of BTS2.5/BTS15 FGMs sintered by different heating rates, as well as to distinguish the grain-interior and grain boundary resistivity of the FGMs. The experimental data for total ionic conductivity were fitted as a function of temperature T following the Arrhenius law:

$$\sigma = (\sigma_o/T)\exp(-E_a/k_BT) \tag{3}$$

where E_a is the activation energy for ionic migration, k_B is the Boltzmann constant, and σ_o, the pre-exponential factor, is a constant related to the density of charge carriers. Total ionic conductivity can be separated on grain-interior ($\sigma_{gi} = 1/\rho_{gi}$) and grain boundary ($\sigma_{gb} = 1/\rho_{gb}$) contribution. Arrhenius plots of grain-interior and grain boundary conductivity, for BTS2.5/BTS15 FGMs, are shown in Figure 4. All obey the Arrhenius law and therefore activation energies (separately E_{gi} and E_{gb}) can be estimated from the slopes of these diagrams. Values of activation energies determined by least square fitting of specific conductivity data are listed in Table III.

It is known that E_{gi} represents intrinsic properties of materials; here, E_{gi} varies in the range from 0.71 to 0.73 eV and is not influenced by tin/titanium concentration gradient or by applied heating rate. E_{gb} varies in the range of 1.05 to 1.24 eV. The highest E_{gb} of 1.24 eV was for FGM heated with 10 °/min, while the smallest E_{gb} of 1.05 eV was FGM heated with 20 °/min. It is known that E_{gb} depends on macro- and micro-structural development; air gaps on the contact between grains act as potential barriers for oxygen ion transport, while macro-structural defects allowed leakage currents through the components. Here, it can be deduced that all of main parameters (shrinkage anisotropy, density, average grain size) that influence E_{gb} depend on heating rate. Consequently, E_{gb} of FGMs is determined by heating rate in a very complex way.

The associated relaxation times for grain interior, τ_{gi}, and grain boundary, τ_{gb}, were also studied. Relaxation time is defined as $\tau = (2\pi f_{max})^{-1}$, where f_{max} is the frequency at the top of the Z'' *versus* Z' arcs.[17] The relaxation time is a geometry independent parameter; also it does not depend on microstructure. Instead, τ depends only on the intrinsic conductivity of material.[18] We found that for the studied FGMs the calculated values of bulk relaxation process, τ_{gi}, for temperatures interval from 320 to 230 °C, fall in the range of 10^{-6} to 10^{-4} s. Besides, the calculated values for grain boundary relaxation process τ_{gb} are from 10^{-3} to 10^{0} s. In Figure 4 (b) Arrhenius plots of $\ln\tau$ *versus* 1000/T are shown separately for grain interior (bottom) and grain boundary (top). It is found that grain boundary relaxation times are about three orders of magnitude larger than those of the grain interior. Naturally, in the grain boundary structure the time spent in the relaxation process is longer. So, it can be emphasized that there is no significant difference between bulk or grain interior relaxation processes in FGMs.

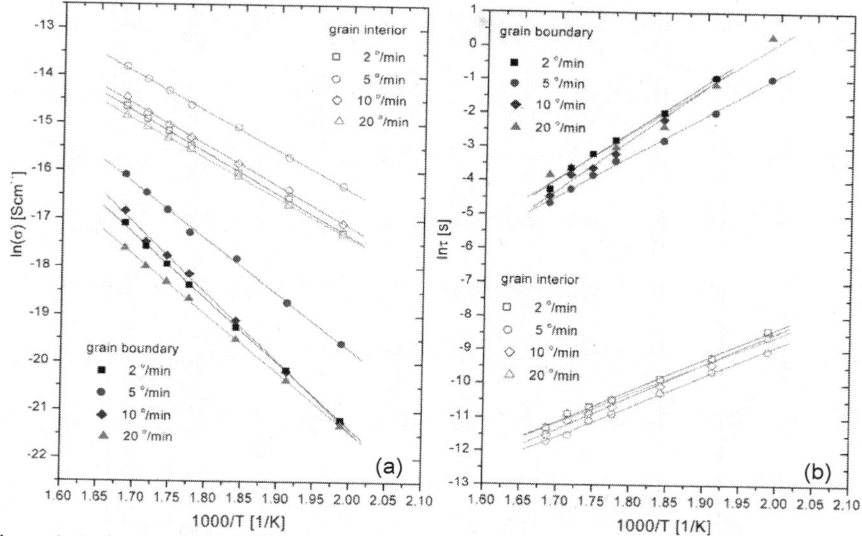

Figure 4. Arrhenius plots of: (a) ln(σ) versus 1000/T, and (b) lnτ versus 1000/T. Contributions of grain interior and grain boundary are separated.

Table III. Electrical and microstructural characteristics of FGMs.

Heating rate (°/min)	E_{gi} [eV]	E_{gb} [eV]	$D_{50\ b.l.}$ [μm]	$D_{50\ t.l.}$ [μm]	ρ [g/cm³]
2	0.73±0.02	1.16±0.02	18.5	16.0	5.35
5	0.71±0.02	1.14±0.01	18.0	13.0	5.10
10	0.72±0.02	1.24±0.04	18.0	11.0	4.90
20	0.71±0.01	1.05±0.02	12.0	10.0	4.50

In the final stage of this study, we studied the influence of heating rate on FGMs' microstructure, which is correlated with electrical properties. Figure 5 shows microstructure of BTS2.5 and BTS15 graded layers in BTS2.5/BTS15 FGMs prepared by different heating rate during sintering processes. It is obvious that heating rate affected FGMs' microstructure; it impacts final density as well as average grain size. Increasing of the heating rate decreases both

FGMs density and average grain size. Estimated values of average grain sizes (for bottom and top layers) are shown in Table III.

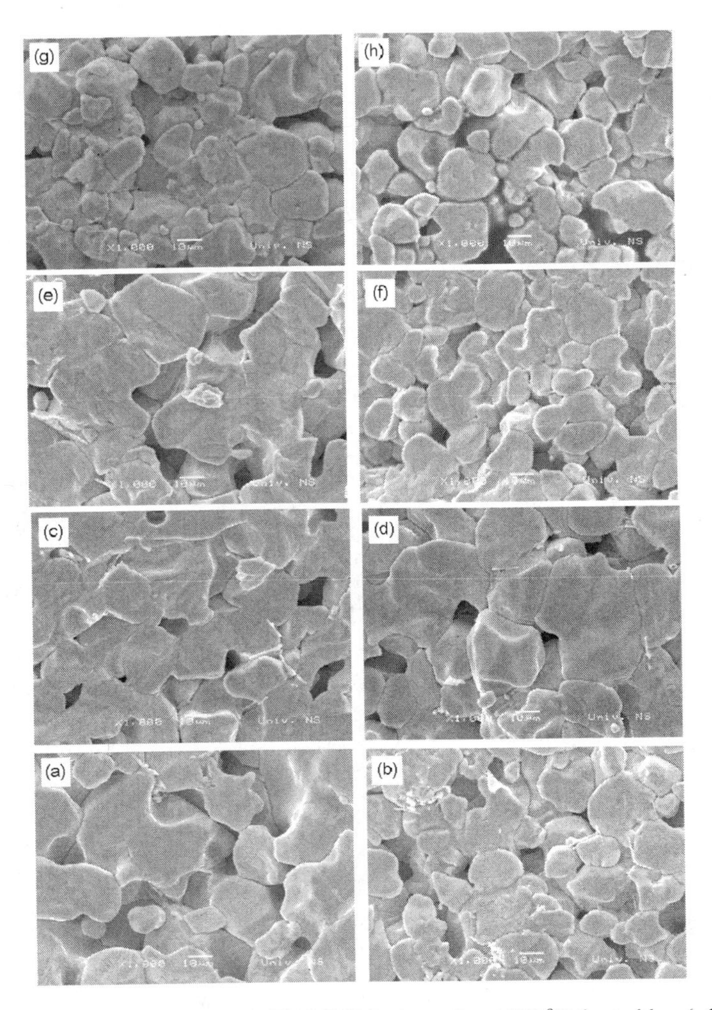

Figure 5. Microstructure of BTS2.5/BTS15 FGMs sintered at 1420 °C, heated by: (a,b) 2 °/min; (c,d) 5 °/min; (e,f) 10 °/min, and (g,h) 20 °/min. Left side – BTS2.5 layers, right side – BTS15 layers.

CONCLUSION

During sintering of BTS2.5/BTS15 FGMs anisotropic densification occurs, the extent of which depends on heating rate. A large concentration gradient in BTS2.5/BTS15 FGMs implies a gradient of anisotropy coefficient; hence, it is very important to choose a heating rate to give moderate anisotropy.

The concept of master sintering curve (MSC) was used to estimate the effective activation energy for sintering of BTS2.5 and BTS15 graded layers as well as whole BTS2.5/BTS15 FGMs. The values of 359.5 and 340.5 kJ/mol were obtained for BTS2.5 and BTS15 graded layers, respectively; while for sintering of entire BTS2.5/BTS15 FGMs the value of 460 kJ/mol was obtained. This difference of ~ 100 kJ/mol in the effective activation energy could be attributed to a potential insulator interlayer formed between graded layers during uniaxial pressing.

Furthermore, impedance spectroscopy was used to determine the electrical characteristics of BTS2.5/BTS15 FGMs sintered by different heating rates, as well as to determine influence of concentration gradient on intrinsic and microstructural features. Activation energies of BTS2.5/BTS15 FGMs, separately for grain interior and grain boundary, were calculated. It was established that the activation energy deduced from grain-interior conductivity (0.71–0.73 eV) kept the intrinsic properties of BTS materials, hence, it is not influenced by concentration gradient neither by heating rate. Contrarily, activation energy for grain boundary conductivity (1.05–1.24 eV) is influenced by macrostructural (shrinkage anisotropy) and microstructural development (density and average grain size), which is furthermore correlated with concentration gradient and heating rate on very complex way. These results confirmed that by tailoring of heating rate during sintering of BTS2.5/BTS15 FGMs their electrical features can be tailored, too.

ACKNOWLEDGEMENTS

The Ministry of Science and Technological Development of the Republic of Serbia provided financial support under Grant No. 142006. The FESEM results were obtained at Jožef Stefan Institute in Ljubljana, Slovenia, owing to the bilateral cooperation program between the Republic of Serbia and the Republic of Slovenia. The authors would like to thank to Prof. Dr. Danilo Suvorov and Dr. Srečo Škapin for their support.

REFERENCES

1. C.C.M. Wu, M. Kahn, and W. Moy, Piezoelectric ceramics with functional gradients: A new application in material design, *J. Am. Ceram. Soc.*, **79** (3), 809–812 (1996).
2. M. Koizumi, Recent progress in FGMs research in Japan, *Int. J. of SHS*, **6** (3), 295–306 (1997).
3. C. Chenglin, Z. Jingchuan, Y. Zhongda, W. and Shidong, Hydroxyapatite–Ti functionally graded biomaterial fabricated by powder metallurgy, *Mater. Sci. Eng. A*, **271**, 95–100 (1999).
4. S. Marković, M. Mitrić, N. Cvjetićanin, and D. Uskoković, Preparation and properties of $BaTi_{1-x}Sn_xO_3$ multilayered ceramics, *J. Eur. Ceram. Soc.*, **27**, 505–509 (2007).
5. S. Marković, Č. Jovalekić, Lj. Veselinović, S. Mentus, and D. Uskoković, Electrical properties of barium titanate stannate functionally graded materials, *J. Eur. Ceram. Soc.*, **30**, 1427–1435 (2010).
6. R. Steinhausen, A. Kouvatov, H. Beige, H.T. Langhammer, and H.-P. Abicht, Poling and bending behavior of piezoelectric multilayers based on $Ba(Ti,Sn)O_3$ ceramics, *J. Eur. Ceram. Soc.*, **24**, 1677–1680 (2004).

7. W. Xiaoyong, F. Yujun, and Y. Xi, Dielectric relaxation behavior in barium stannate titanate ferroelectric ceramics with diffused phase transition, *Appl. Phys. Lett.*, **83**, 2031–2033 (2003).
8. E. Müller, Č. Drašar, J. Schilz, and W.A. Kaysser, 'Functionally graded materials for sensor and energy applications'. *Mat. Sci. Eng. A*, **362**, 17–39 (2003).
9. S. Marković, and D. Uskoković, The master sintering curves for $BaTi_{0.985}Sn_{0.025}O_3/BaTi_{0.85}Sn_{0.15}O_3$ functionally graded materials, *J. Eur. Ceram. Soc.*, **29**, 2309–16 (2009).
10. P. Pinceloup, C. Courtois, A. Leriche, and B. Thierry, Hydrothermal synthesis of nanometer-sized barium titanate powders: Control of barium/titanium ratio, sintering, and dielectric properties, *J. Am. Ceram. Soc.*, **82**, 3049–56 (1999).
11. Z. Zhao, V. Buscaglia, M. Viviani, M. T. Buscaglia, L. Mitoseriu, A. Testino, M. Nygren, M. Jonhsson, and P. Nanni, Grain-size effects on the ferroelectric behavior of dense nanocrystalline $BaTiO_3$ ceramics, *Phys. Rev. B*, **70**, 024107-1–8 (2004).
12. E.A. Olevsky, and R.M. German, Effect of gravity on dimensional change during sintering - I. Shrinkage anisotropy, *Acta Mater.*, **48**, 1153–1166 (2000).
13. A. Shui, N. Uchida, and K. Uematsu, Origin of shrinkage anisotropy during sintering for uniaxially pressed alumina compacts, *Powder Tech.*, **127**, 9–18 (2002).
14. A. Shui, Z. Kato, N. Uchida, and K. Uematsu, Sintering deformation caused by particle orientation in uniaxially and isostatically pressed alumina compacts, *J. Eur. Ceram. Soc.*, **22**, 311–316 (2002).
15. A.R. Boccaccini, V. Adell, C.R. Cheeseman, and R. Conradt, Use of heating microscopy to assess sintering anisotropy in layered glass metal powder compacts, *Adv. Appl. Ceram.*, **105** (5), 2532–3540 (2006).
16. H. Su, and D.L. Johnson, Master sintering curve: A practical approach to sintering, *J. Am. Ceram. Soc.*, **79**, 3211–17 (1996).
17. L. Bucio, E. Orozco, and E.A. Huanosta-Tera, Relaxation and conductivity behaviour in the compounds: $FeRGe_2O_7$ (R = Pr, Tb), *J. Phys. Chem. Solids*, **67**, 651–658 (2006).
18. P. Sooksaen, I.M. Reaney, and D.C. Sinclair, Crystallization and dielectric properties of borate-based ferroelectric $PbTiO_3$ glass-ceramics, *J. Electroceram.*, **19**, 221–228 (2007).

APPLICATIONS OF HIGH-THROUGHPUT SCREENING TOOLS FOR THERMOELECTRIC MATERIALS

[1] W. Wong-Ng, [1]H. Joress, [1]J. Martin, [1]Y. Yan, [2]J. Yang, [1]M. Otani, [3]E. L. Thomas, [1]M.L. Green, and [1]J. Hattrick-Simpers
[1]MML, NIST, Gaithersburg, MD 20899
[2]GM Research Center, Warren, MI 48090
[3]Air Force Research Laboratory, WPAFB, OH 45433

ABSTRACT

The increased research and development on thermoelectric materials in recent years has been driven primarily by the need for improved efficiency in the global utilization of energy resources. To facilitate the search for new thermoelectric materials, we have developed a high-throughput thermoelectric screening system for combinatorial thin films. This screening system is comprised of two tools, one for measuring the Seebeck coefficient and resistance via an automated multiprobe apparatus, and another for measuring thermal effusivity by a frequency domain thermoreflectance technique. Using these systems, we are able to make measurements for calculating the thermoelectric power factor ($S^2\sigma$, where S = Seebeck coefficient, σ = electrical conductivity) and thermal conductivity of over 1000 sample-points within 6 hours for each instrument. This paper provides examples of our current studies on both oxide and intermetallic materials.

INTRODUCTION

Due to the high gas price and negative effect of green house gas emission in recent years, there are desperate needs to have alternate energy capabilities and environmentally friendly technologies. Thermoelectric materials have demonstrated the potential for widespread applications in areas of waste heat recovery and solid-state refrigeration. The efficiency and performance of thermoelectric power generation or cooling is related to the dimensionless figure of merit (ZT) of the thermoelectric materials, given by $ZT = S^2\sigma T/\kappa$, where T is the absolute temperature, S is the Seebeck coefficient, σ is the electrical conductivity ($\sigma = 1/\rho$, ρ is electrical resistivity), and k is the thermal conductivity [1]. ZT is directly related to the performance of a thermoelectric material and is the reference by which these materials are judged. Thermoelectric materials with desirable properties (i.e., high ZT >>1), are characterized by high electrical conductivity, high Seebeck coefficient, and low thermal conductivity.

Until about 10 years ago only a small number of materials have been found to have practical industrial thermoelectric applications because of generally low thermoelectric efficiencies. Increased attention to research and development of thermoelectric materials has been partly due to the dramatic increase of the ZT values of materials being discovered in bulk and thin film form [2, 3]. Additional novel materials include quantum well films [4] and quantum dot films [5] that have been reported to yield ZT as high as 2.5.

The goal of this paper is to summarize our recent efforts on the development of high-throughput thermoelectric metrology, including a tool suite for screening the Seebeck coefficient, resistance, and thermal conductivity of combinatorial films. Demonstration of the applications of these tools on the combinatorial composition-spread films on cobalt oxide systems and the study of the homogeneity of bulk inter-metallic compounds will be given.

A SUITE OF SCREENING TOOLS FOR COMBINATORIAL FILMS[1]

For large-scale applications of thermoelectric technology, continued efforts to identify novel materials and to optimize the properties of existing materials are crucial. The combinatorial state-of-the-art synthesis approach is an efficient technique to systematically investigate the thermoelectric properties as a function of composition in complex multi-component systems. This method typically involves fabricating film libraries with compositions varying between two or three different end-member materials on a substrate using physical vapor deposition techniques, followed by evaluation of the film libraries with high-throughput screening tools [6-10]. We have developed a high-throughput screening system for thermoelectric material exploration using combinatorial films prepared with a continuous spread of compositions. This system consists of a power factor screening tool and a thermal conductivity screening tool [11].

Power Factor Screening tool

The high-throughput power factor screening tool that we developed can be used to measure electrical resistance and the Seebeck coefficient [12, 13]. Measurements are fully automated by a computer (Fig 1). The salient features of the power factor screening tool consist of a probe to measure Seebeck coefficient and electric conductivity, an automated translation stage to scan the film in the *x-y-z* directions, and various voltage measuring instruments. The measurement probe consists of four gold-plated spring probes as sample contacts, a heater to generate temperature differences between two of the spring probes, two thermometers to measure the temperature of these probes, two insulators, and two copper plates. To achieve accurate Seebeck coefficient measurements, a spring probe is placed directly on each copper plate, and the other spring probe is attached on the insulator. The four probes can be arranged either in a square array or in a colinear fashion. A thermometer is attached on each copper plate and a heater is attached on one of the copper plates. Electrical conductivity is measured by the conventional 4-probe van der Pauw method [14] (Fig. 2). All Seebeck coefficient measurements were conducted at room temperature and at $\Delta T = 4.1$ K. It takes about 20 seconds to measure both electric conductivity and Seebeck coefficient for each sample point. This probe allows us to measure electric conductivity and Seebeck coefficient of over 1000 sample points within 6 hours.

In order to estimate the accuracy of our screening tool, we compared the Seebeck coefficients of several bulk and film samples using our screening probe and a traditional, one-sample-at-a-time system (PPMS, Quantum Design[a], with the Thermal Transport Option at room temperature). As seen in Fig.3, measurements using these two instruments are in excellent agreement on a number of representative samples, including films and bulk materials. These results suggest that the scanning probe is reliable and sufficiently accurate as a screening tool for Seebeck coefficients of combinatorial samples.

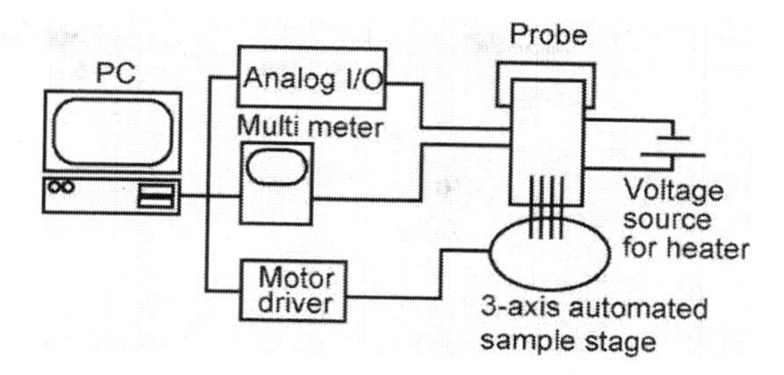

Fig. 1. Schematic diagram of the power factor screening system

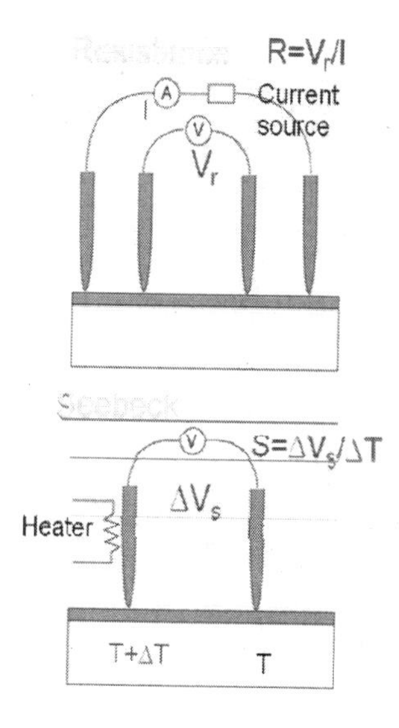

Fig. 2. Schematic of the screening probe to measure electrical conductivity and Seebeck coefficient.

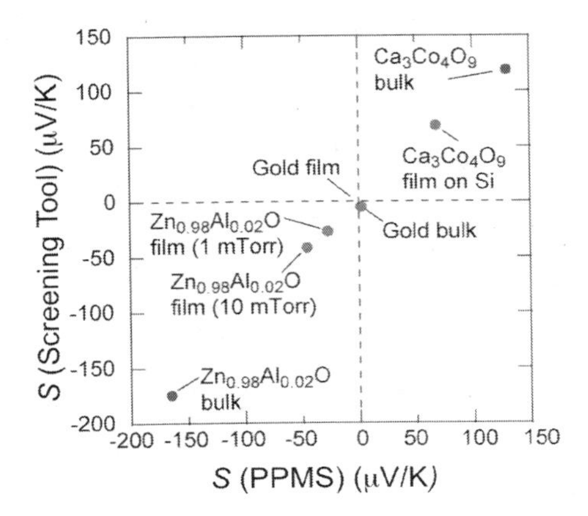

Fig 3. Comparison between Seebeck coefficients measured on several samples, including bulk and film, with our screening tool and a thermal transport option of physical property measurement system (PPMS). The bulk samples are gold foil, polycrystalline $Zn_{0.98}Al_{0.002}O$, and polycrystalline $Ca_3Co_4O_9$. The film sample fabricated by PLD are gold film, $Ca_3Co_4O_9$ film, and two $Zn_{0.98}Al_{0.002}O$ films grown at 1.3 and 13 Pa.

We demonstrated the high-throughput screening of Seebeck coefficient and electrical conductivity using the ternary (Ca, Sr, La)$_3Co_4O_9$ composition-spread films using $Ca_3Co_4O_9$, $Ca_2LaCo_4O_9$, and $Ca_2SrCo_4O_9$ as targets [12, 13]. The composition-spread films were fabricated with a pulsed laser deposition system by the continuous-composition-spread technique. Figure 4 gives the results of the power factor screening of this system (depicted as a conventional ternary diagram). It is clear, from our results, that the power factor reach a maximum between the Sr-rich region and the La-rich region. Substitution of the trivalent La^{3+} for the divalent Ca^{2+} is believed to decrease the hole concentration, leading to an increased Seebeck coefficient in the La-rich region. On the other hand, substitution of Sr^{2+} for Ca^{2+} leads to an increase of electrical conductivity and an insignificant change of Seebeck coefficient. This is caused by the substitution of a larger divalent Sr^{2+} cation for a smaller divalent Ca^{2+} cation which does not change the carrier concentration but it does change the carrier mobility by lattice deformation.

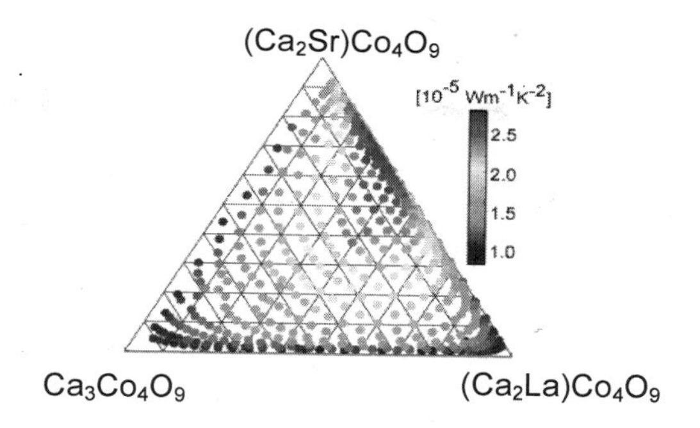

Fig. 4. Power factor of the composition-spread $(Ca_{1-x-y}Sr_xLa_y)_3Co_4O_9$ film ($0 < x < 1/3$ and $0 < y < 1/3$).

We further demonstrated the applications of the screening tool with a quaternary system by extending the ternary system by adding the Mg-cobaltite component. The chemical formula for the 4 targets are $Ca_3Co_4O_9$, $(Ca_2La)Co_4O_9$, $(Ca_2Sr)Co_4O_9$, and $(Ca_2Mg)Co_4O_9$. The quaternary diagram of the composition-spread $(Ca_{1-x-y-z}Sr_xLa_yMg_z)_3Co_4O_9$ film ($0 < x < 1/3$, $0 < y < 1/3$, and $0 < x < 1/3$) was constructed using eleven ternary films. We were able to make these 11 films (about 6000 data points) in about one week. The Seebeck coefficient diagram of this composition-spread film is shown in Fig. 5a. The Seebeck coefficient variation with valence is shown in Fig. 5b. Similar to the previous example, as the valence of the cations increases, the Seebeck coefficient increases because doping with the La^{3+} cation decreases the hole concentration, leading to also a decrease of electrical conductivity.

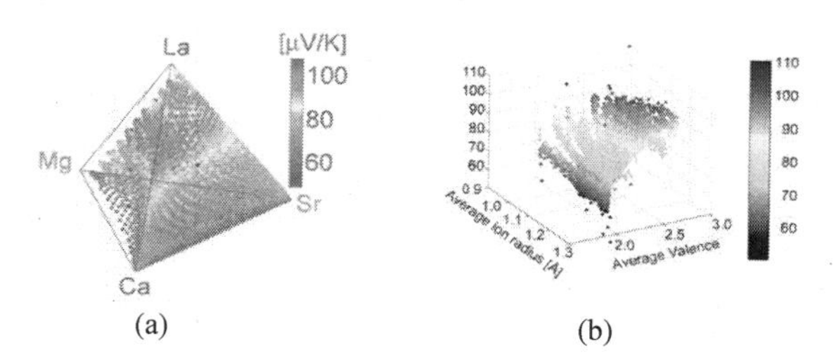

Fig. 5a. Seebeck coefficient of the composition-spread $(Ca_{1-x-y-z}Sr_xLa_yMg_z)_3Co_4O_9$ film ($0 < x < 1/3$, $0 < y < 1$, and $1/3$, $0 < z < 1$); Ca=$Ca_2CaCo_3O_9$, Sr= $Ca_2SrCo_3O_9$, La=$Ca_2LaaCo_3O_9$, Mg=$Ca_2MgCo_3O_9$,. Fig. 5b, Seebeck coefficient variation with valence.

Additionally, the screening tool has been used to evaluate the homogeneity of three industrially important bulk samples. The first example is screening the homogeneity of a single crystal of S-substituted Bi_2Te_3 (Fig. 6a). The structure is rhombohedral and can be best described as having quintuple layers of atoms (Te-Bi-Te-Bi-Te) stacked along the *c*-axis [15]. The structure can be considered as having quasi 2D electron layers with van der Waals planes in between each group of the quintuple layers. We have studied the crystal structure at three different locations of this 4.5 cm long crystal, and have screened the Seebeck coefficient along the long axis of the crystal.

The crystal was found to have a concentration gradient along the long-axis. The structure result shows the S and Te concentration are approximately 50:50, but slightly Te-rich (Fig. 6b). In all three compositions that we studied, the Te site in the middle of the quintuple layer in Bi_2Te_3 is occupied by only S, whereas the gap position of Te is occupied by mixed S/Te. The Te has the highest concentration in position #1 ($Bi_2(Te_{1.58(1)}S_{1.42(1)})$), followed by position #2 ($Bi_2(Te_{1.53(1)}S_{1.47(1)})$), and the least in position #3 ($Bi_2(Te_{1.52(2)}S_{1.48(2)})$). In other words, there is slightly more Te in the gap Te/S sites in position #1 than those in positions #2 and #3 .This correlates well with the Seebeck coefficient values (Fig. 7), namely, the higher the Te concentration, the higher the Seebeck coefficient.

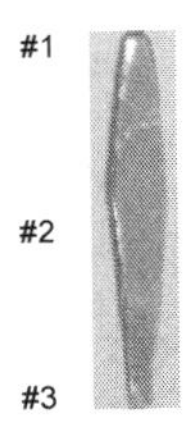

Fig. 6a. Dimension of the $Bi_2(Te,S)_3$ crystal (about 4.5 cm long)

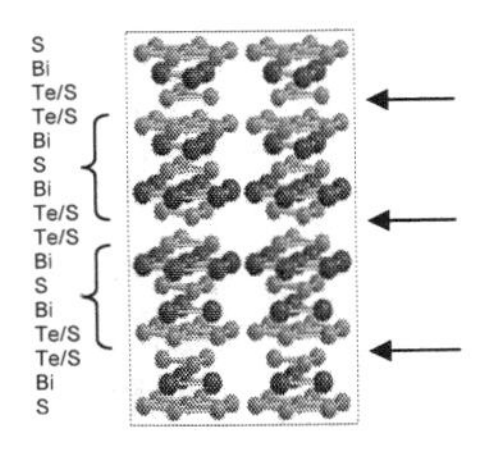

Fig. 6b. Structure of $Bi_2(Te,S)_3$ (← indicates the position of the van der Waals planes)

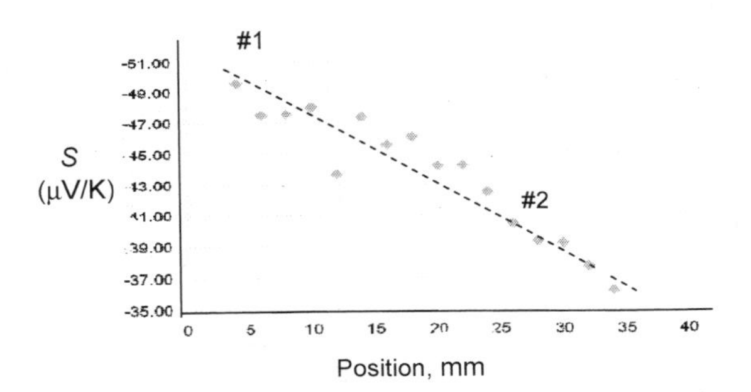

Fig. 7. Seebeck coefficient as a function of distance along the c-axis of the crystal

The second example (in the form of rectangular bars of approximately 2.5 mm x 2.5 mm x 10 mm) is a Type-I clathrate, $Ba_8(Ni,Pt,Pd,Zn)Ga_{13}Ge_{46}$ [16] (Fig. 8), and the third example is a double-filled skutteruide $(Ba,Yb)Sb_4Co_{12}$ [17, 18] (Fig. 9). $Ba_8(Ni,Pt,Pd,Zn)Ga_{13}Ge_{46}$ has an "open" structure that can host guest atoms inside the crystallographic voids. The "rattling" Ba atoms inside the large Ge/Ga "cages" scatter thermal phonons and thus reduce thermal conductivity. The measurement positions are parallel to the thermal gradient. Each data point was collected after a 30 second stabilization period. A total of 13 scanned data points on $Ba_8(Ni,Pt,Pd,Zn)Ga_{13}Ge_{46}$ gives an average

Seebeck coefficient value of -67 μV/K with a 4.5 % (3.0 μV/K) standard deviation. The $(Ba,Yb)Sb_4Co_{12}$ (a skutterudite) sample is also in the form of rectangular bars of approximately 2.5 mm x 2.5 mm x 10 mm. $(Ba,Yb)Sb_4Co_{12}$ has a cubic structure that consists of six 4-membered Sb-rings that are almost parallel to the cell edges. The two large voids in the unit cell can be filled with "rattling" atoms (Ba and Yb ions). Based on the screening of 16 data points, the average Seebeck coefficient for $(Ba, Yb)Sb_4Co_{12}$ is –128 μV/K with a 2 % (2.4 μV/K) standard deviation. Therefore we conclude that both $Ba_8(Ni,Pt,Pd,Zn)Ga_{13}Ge_{46}$ and $(Ba, Yb)Sb_4Co_{12}$ are essentially homogenous samples, as no systematic change in the Seebeck coefficient was observed as a function of position.

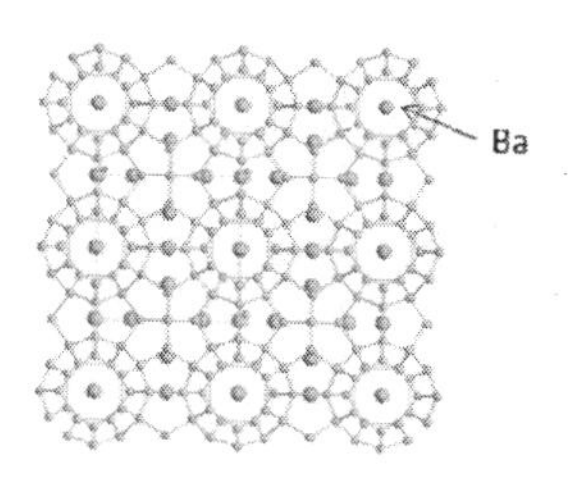

Fig. 8. Planar view of the Type-I clathrate crystal structure showing the two different polyhedra (20-atom cage and 24-atom cage) that form the unit cell.

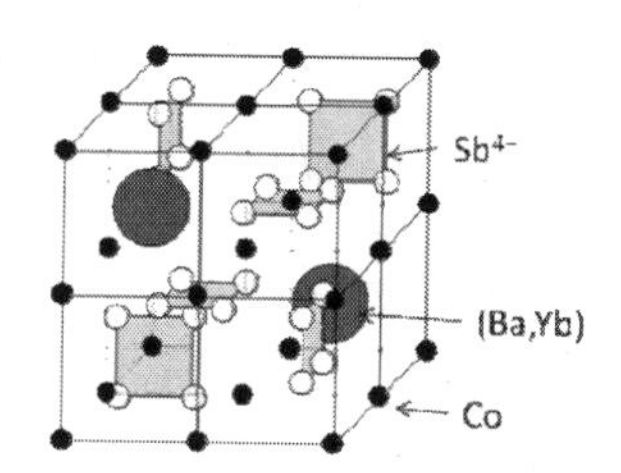

Fig. 9. Structure of the skutterudite $(Ba, Yb)Sb_4Co_{12}$ showing the Sb_4 rings and the voids in the unit cell where the mixed (Ba, Yb) ions reside.

Thermal conductivity screening tool (Frequency-domain thermoreflectance technique)

A scanning thermal effusivity measurement system using the frequency domain thermoreflectance technique has been developed [11]. The thermoreflectance technique is based on the relationship between the change in the optical reflection coefficient of a material and the change in its temperature through periodic heating. By using a metal for the film layer, this method can measure thermal effusivity of ceramics, metals, glass, and plastics. The thermal effusivity measurement system can rapidly and locally (10 micrometer spot size) measure the thermal effusivity of combinatorial (composition-spread) films.

Figure 10 gives a schematic of the scanning thermal effusivity frequency system that includes two laser diodes, a compensating network, a voltage source heater, and an *x-y-z* axis automated sample stage driven by a motor driver. The sample, which is a thermoelectric film or a bulk sample with a smooth surface, is first coated with a thin molybdenum layer (typically about 100 nm thick), then is locally heated by an intensity-modulated heating laser; the thermal response of the film is detected by the reflected beam of a second (probe) laser. The reflected probe signal is detected by a balance detector. The amplitude and phase difference between signal and reference signal from the pulse generator are obtained by a lock-in amplifier.

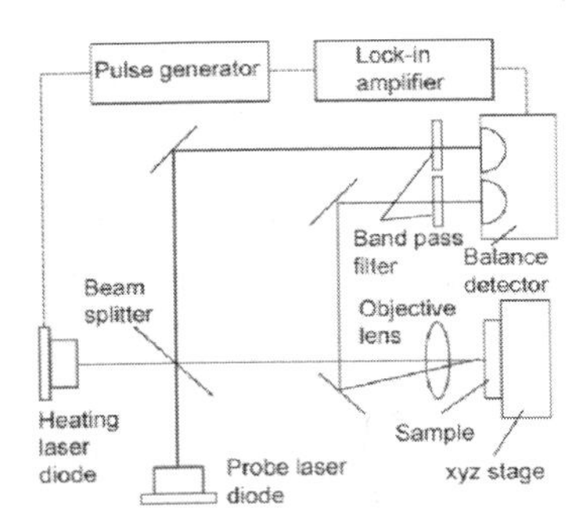

Fig. 10. Schematic diagram of the thermal effusivity screening system (frequency domain thermoreflectance technique)

The thermal effusivity b can be derived from the phase lag, δ, between the thermoreflectance and the heating laser signals. Thermal effusivity is related to thermal conductivity as

$$b = (\kappa c \rho)^{1/2}, \qquad (1)$$

where c is the specific heat, ρ is the density, and κ is thermal conductivity. By choosing the modulation frequency of 1 MHz, we obtained a calibration curve of the phase lag of five bulk samples of which the thermal effusivity values are known: SiO_2, $SrTiO_3$, $LaAlO_3$, Al_2O_3, and Si. The linear dependence of the phase lag on thermal effusivity of these five compounds can be expressed in the following equation,

$$b = -402.34\ \delta + 26352, \qquad (2)$$

where b is the thermal effusivity. If the δ value for an unknown material can be measured experimentally, one can estimate b using equation (2).

We successfully estimated the thermal conductivity value for a 800nm/(100nm Mo) thick conventional $Ba_2YCu_3O_{6+x}$ film on a $SrTiO_3$ substrate. Using equation (1), the thermal effusivity was estimated to be 1370 $Js^{1/2}M^{-2}K^{-1}$ ($\delta = 62.1$ °). The thermal conductivity of the $Ba_2YCu_3O_{6+x}$ film was then determined (c= 430 $J.kg^{-1}K^{-1}$[19], and ρ= 6.38 $g.cm^{-3}$ [20]) to be 12.0 $Jm^{-1}K^{-1}s^{-1}$, which agrees reasonably well (within 10 %) with the reported value of 12.9 $Jm^{-1}K^{-1}s^{-1}$ [19].

To assess the application of the tool for 2-dimensional screening, we determined the δ values of a Si sample in the *yz*-direction (with a 100 nm-thick Mo metallic film on the surface). Using equation (2) with the c, ρ values of 0.713 $Jg^{-1}K^{-1}$ [21] and 2.3290 $g\text{-}cm^{-3}$ [21], respectively, and the measured δ values, we obtained the thermal effusivity and thermal conductivity values of the Si sample in an area of approximately 1 mm x 1 mm in increments of 100 μm. The resulting average κ is (165 ± 15) $Wm^{-1}K^{-1}$, which is within 8 % of the literature reported value of 156 $Wm^{-1}K^{-1}$ [21]. Using this tool, we plan to study the thermal properties of combinatorial film systems in the near future.

FUTURE DEVELOPMENTS

Our continuing development of the power factor screening tool for the thermoelectric combinatorial films focuses on the improvement of its performance. There are three limitations to the current screening tool configuration [12, 13]. First, it is difficult to measure materials with low electrical conductivity (i.e., $\leq 10^{-1}$ $(\Omega cm)^{-1}$ for a 200 nm thin film) and low value of Seebeck coefficient ($\approx$5 μV/K) due only to limitations of the homemade current source and voltage meter. Although these limitations do not hamper our ability to screen thermoelectric materials, we could extend our measurement capability by upgrading those two components. Second, the spatial resolution of our screening tool is about 2 mm, corresponding to the distance between two spring probes. If the electrical conductivity of the sample drastically changed over a very short length scale, measurements at a sample point might be affected by its neighbors. We plan to improve this resolution. Third, measurements can be carried out only at room temperature at present. We are currently developing an improved instrument that is capable of high temperature up to 600 °C) measurement.

[1] Certain commercial equipment, instruments, or materials are identified in this paper in order to specify the experimental procedure adequately. Such identification is not intended to imply recommendation or endorsement by the National Institute of Standards and Technology, nor is it intended to imply that the materials or equipment identified are necessarily the best available for the purpose.

REFERENCES

[1] T.M. Tritt, Thermoelectrics Run Hot and Cold, Science, **272**, 1276-1277 (1996).

[2] K.F. Hsu, S. Loo, F. Guo, W. Chen, J.S. Dyck, C. Uher, T. Hogan, E.K. Polychroniadis, and M.G. Kanatzidis, Cubic $AgPb_mSbTe_{2+m}$: Bulk Thermoelectric Materials with High Figure of Merit, Science, **303**, 818-821 (2004).

[3] R. Venkatasubramanian, E. Siivola, T. Colpitts, and B. O'Quinn, Growth of one-dimensional Si/SiGe heterostructures by thermal CVD, Nature, **413**, 597-602 (2001).

[4] S. Ghamaty, and N.B. Eisner, "Development of Quantum Well Thermoelectric films, Proceedings of the 18th International Conference on Thermoelectrics, Baltimore, MD, pp. 485-488 (1999).

[5] M.S. Dresselhaus, G. Chen, M.Y. Tang, R.G. Yang, H. Lee, D.Z. Wang, Z.F. Ren, J.P. Fleurial, and P. Gogna, Enhanced Thermopower in PbSe Nanocrystal Quantum Dot Superlattices, Adv. Mater., **19**, 1043-1053 (2007).

[6] R. Funahashi, S. Urata, and M. Kitawaki, Exploration of n-type oxides by high throughput screening, Appl. Surf. Sci., **223**, 44-48 (2004).

[7] T. He, J.Z. Chen,T.G. Calvarese, M.A. Subramanian, Thermoelectric properties of $La_{1-x}A_xCoO_3$ (A = Pb, Na), Solid State Sciences, **8**, 467-469 (2006).
[8] R. B. Van Dover, L. F. Schneemeyer, R. M. Fleming, Discovery of a useful thin-film dielectric using a composition-spread approach, Nature, **392**, 162 (1998).
[9] T. Fukumura, M. Ohtani, M. Kawasaki, Y. Okimoto, T. Kageyama, T. Koida, T. Hasegawa, Y. Tokura, H. Koinuma, Rapid construction of phase diagram for doped Mott insulators with a composition-spread approach, Appl. Phys. Lett., **77**, 3426-3428 (2000).
[10] H. M. Christen, S. D. Silliman, and K. S. Harchavardhan, Continuous compositional-spread technique based on pulsed-laser deposition and applied to the growth of epitaxial films, Rev. Sci. Instrum., **72**, 2673-2678 (2001).
[11] M. Otani, E. Thomas, W. Wong-Ng, P. K. Schenck, K.-S. Chang, N. D. Lowhorn, M. L. Green, and H. Ohguchi, A High-throughput Screening System for Thermoelectric Material Exploration Based on a Combinatorial Film Approach, Jap. J. Appl. Phys, **48**, 05EB02 (2009).
[12] M. Otani, N.D. Lowhorn, P.K. Schenck, W. Wong-Ng, and M. Green, A high-throughput thermoelectric Screening Tool for Rapid Construction of Thermoelectric Phase Diagrams, Appl. Phys. Lett., **91**, 132102-132104 (2007).
[13] M. Otani, K. Itaka, W. Wong-Ng, P.K. Schenck, and H. Koinuma, Development of high-throughput thermoelectric tool for combinatorial thin film libraries, Appl. Surface Science, **254**, 765-767 (2007).
[14] L.J. van der Pauw, Philips Tech. Rev., A Method of Measuring the Resistivity and Hall Coefficient on Lamellae of Arbitrary Shape, **20**, 220-224, (1958).
[15] P.M. Lee and L. Pincherle, "The Electronic Band Structure of Bismuth Telluride," Pro. Phys. Soc., **81**, 461 (1963); P.W. Lange, Naturwissenschaften **27** 133 (1939).
[16] C. Uher, Skutterudites: Prospective Novel Thermoelectrics in Semiconductors and Semimetals, Semiconductors & Semimetals, **69**, 139-253 (2001).
[17] X. Shi, H. Kong, C.-P. Li, C. Uher, J. Yang, J.R. Salvador, H. Wang, L. Chen, and W. Zhang, Low thermal conductivity and high thermoelectric figure of merit in n-type $Ba_xYb_yCo_4Sb_{12}$ double-filled skutteruide, Appl. Phys. Lett. **92**, 182101 (2008).
[18] X. Shi, J.R. Salvador, J. Yang, and H. Wang, Thermoelectric Properties of *n*-Type Multiple-Filled Skutterudites, J. Electronic Mater. **38**(7), 930-933 (2009).
[19] T. Yagi, N. Taketoshi, and H. Kato, Distribution analysis of thermal effusivity for sub-micrometer YBCO thin films using thermal microscope, Physica C, **412-414**, 1337-1342 (2004).
[20] W. Wong-Ng, R.S. Roth, L.J. Swartzendruber, L.H. Bennett, C.K. Chiang, F. Beech, and C.R. Hubbard, X-ray powder characterization of $Ba_2YCu_3O_{7-x}$, Adv. Ceram. Mater., Vol. **2**(3B), Ceramic Superconductors, ed. W. J. Smothers, Am. Ceram. Soc., Westerville, OH, p. 565 (1987).
[21] Robert Hull, Editor, Properties of Crystalline Silicon, INSPEC, The Institution of Electrical Engineers, London, p. 165 (1999).

Developments in High Temperature Superconductors

ALTERING SELF-ASSEMBLY OF SECOND PHASE ADDITIONS IN $YBa_2Cu_3O_{7-x}$ FOR PINNING ENHANCEMENT

Francisco J. Baca, Paul N. Barnes*, Timothy J. Haugan, Jack L. Burke
Air Force Research Laboratory, Propulsion Directorate
Wright-Patterson AFB, Ohio, USA
*Army Research Laboratory, Sensors and Electron Devices Directorate
Adelphi, Maryland, USA

C.V. Varanasi
Army Research Office
Research Triangle Park, North Carolina, USA

Rose L. Emergo, Judy Z. Wu
Department of Physics and Astronomy, University of Kansas
Lawrence, Kansas, USA

ABSTRACT

Different mechanisms can be used to provide additional pinning enhancement to $YBa_2Cu_3O_{7-x}$ (YBCO) films by altering the growth of nanoparticulate and nanocolumnar additions. This can be done, for example, by chemically altering the additions such as through Ca-doping of the secondary phase additions or by simply by the use of engineered substrates. Our experimental results presented here confirm that the microstructure and directional alignment of both BZO and BSO nanorods in YBCO thin films is directly influenced by the vicinal angle of the substrate. Above a 15° vicinal angle, nanorods begin to orthogonally realign parallel to the YBCO a-b planes, enhancing the J_c(H||a-b) for magnetic fields to several tesla. Finally, the issue of tailored pinning is discussed for nanocolumnar additions that allows an emphasis on either the low-field or high-field performance via varying amounts of the BZO or BSO pinning material.

INTRODUCTION

Regardless of the particular means of adding secondary phases as pinning centers in high temperature superconductors (HTS), the pinning enhancement is largely dependent on the geometry of the particular pinning centers, such as in $YBa_2Cu_3O_{7-x}$ (YBCO) thin films. There have been both nanoparticulate dispersions of nonsuperconducting additions[1-3] and nanorod pinning centers using $BaZrO_3$ (BZO), $BaSnO_3$ (BSO) or the more recent Ba_2YNbO_6 (BYNO) or YBa_2NbO (YBNO)[4-9]. Both improved the critical current density (J_c) in YBCO providing either an isotropic pinning enhancement or preferentially enhancement in the c-axis direction, respectively. Either way, each can provide an overall angular improvement.

Although these pinning additions do increase the J_c of the YBCO superconductor based on the particular geometry of the inclusions, this may not be a complete perspective of how the additions affect J_c. For example, it is known that chemical interactions between the doping additions and the YBCO can degrade the T_c of the sample with an attended reduction in the J_c performance. This can limit the amount of some additions such as BZO where excessive amounts heavily degrade the J_c in these samples.[10] Others can be more forgiving such as BSO[8]. A chemical interaction can be beneficial such as the minute doping of YBCO with deleterious rare earths.[11] In the particular case of Y_2BaCuO_5 (Y211) additions, Ca-doping of this secondary phase provides an added increase to the $J_c(H)$ of YBCO.[12] Y211 doped with Ca and deposited into YBCO resulted in short-ranged self-assembly of the nanoparticles into short nanorods.

The self-assembled nanorods formed by pulsed laser deposition of materials such as BZO and BSO form highly correlated vortex pinning sites in YBCO. While these defects significantly enhance $J_c(H)$, the improvement is highly anisotropic with respect to the direction of the applied magnetic field. Therefore, understanding how to modify the pinning enhancement, e.g. columnar alignment or vortex pinning, is critical to controlling the $J_c(H, \theta)$. BZO pinning tends to provide enhancement generally, but in particular for the c-axis field orientation. However, the method that it is introduced can dramatically change that behavior. The focus of this paper is to show how vicinal substrates, previously demonstrated to splay the nanocolumnar growth of BZO in YBCO,[13] to more broadly induce splaying of the nanocolumnar structure in both BZO and BSO. This indicates that the splaying is a more general result of the use of vicinal substrates.

EXPERIMENTAL

Flux pinning is studied in YBCO thin films deposited by pulsed laser deposition (PLD). Details of the film deposition conditions and process parameters are given elsewhere.[13-14] In short, PLD targets were made by solid-state reaction of high purity precursor powders finely ground together in an appropriate stoichiometric ratio. These powders were reacted and formed into targets of dimensions 1 inch diameter and 1.5 cm thick, using the standard solid-state sintering methods. Polished $LaAlO_3$ (LAO) and $SrTiO_3$ (STO) (100) substrates were used for growth of the films. Deposition parameters included a KrF laser λ=248 nm wavelength, 3 J/cm^2 laser fluence, 25 nm pulse length. A post-deposition anneal was conducted for the films at 500 °C and 1 atmosphere of oxygen. Film thicknesses were in the range of 0.2 to 0.3 µm. YBCO films with 2-4 vol % BZO and 9.5 vol % BSO were deposited via single-target PLD with a repetition rate of 4-8Hz. The substrates were held at 780-800 °C (heater block temperature) during the deposition, with an O_2 partial pressure of 300 mTorr. For the vicinal substrates, deposited films were grown on STO substrates with miscut angles of 0°, 5°, 10°, 15° and 20°.

The superconducting transition temperature (T_c) was measured using an AC susceptibility technique with the amplitude of the magnetic sensing field strength, H, varied from 0.025-2.2 Oe, at a sensing frequency of approximately 4 kHz. Magnetic J_c measurements were made with a vibrating sample magnetometer (VSM) in magnetic field strengths of 0 to 9 T, and a ramp rate of 0.01 ($T \cdot s^{-1}$). The J_c of the square-shaped samples was estimated using a simplified Bean model $J_c = 15\Delta M/R$, where M is the magnetization/volume from M-H loops, and R is the radius of volume interaction. The critical current was measured in liquid nitrogen at 77.2 K. For transport current measurements, the samples were patterned by photolithography, and for the critical currents of YBCO films on miscut STO substrates were measured in the direction parallel to the vicinal steps via a four-point probe.

Scanning electron micrographs were taken with a FEI Sirion High Resolution Microscope in an ultra-high resolution mode using a through-lens-detector (TLD), with magnifications up to 100 kX. To study the microstructural properties by transmission electron microscopy (TEM), cross sections were prepared using focused ion beam systems. Micrographs of TEM were taken with a Phillips CM-200 and a FEI Titan TEM operating at 200 and 300 kV, respectively. Care was taken to cut the cross-sectional foils perpendicular to the vicinal steps so that the effects of the modulated surface could be directly observed. Films were also characterized using X-ray diffraction.

RESULTS AND DISCUSSION

This section shows 1) how BZO or BSO nanocolumn splaying or even reorientation from c-axis alignment to ab-plane alignment can be accomplished, and 2) how the in-field behavior of the columnar pinning can be affected based on how the BZO/BSO is introduced.

BZO nanocolumn reorientation

In typical non-vicinal growth, BZO and BSO additions form arrays of columnar defects (nanorods) that align generally along the YBCO c-axis.[4-8] The formation of these nanorods and their alignment within the YBCO matrix is considered to depend on several microstructural and growth conditions. Among the conditions favorable for columnar growth is the preferential nucleation of precipitates given the local strain distribution, and the accompanied alignment of misfit dislocations.[6,10,15-16] However, it is possible to take advantage of the modulation provided by the stepped surface of a vicinal substrate as a means of tuning the microstructure during growth. By initiating growth on a vicinal substrate, the modulated surface may be expected to influence the strain distribution within the YBCO matrix. Indeed, experiments have shown that the growth of pure YBCO on vicinal $SrTiO_3$ (STO) produces antiphase boundaries resulting from the misalignment of the unit cell with the height of a vicinal step.[17-18] These induced microstructural changes would then be expected to also interact with the formation and alignment of added nanorods. We have previously shown some of the microstructural and accompanied flux pinning effects for vicinal angles ranging from 0° to 20° with inclusions of BZO,[13] and now show some preliminary results for YBCO + BSO thin films here.

The microstructural effects on BZO nanorod formation become increasingly pronounced as the vicinal angle is increased.[13,19] At low vicinal angles, ~0 – 10°, BZO nanorods are formed and grow generally along the YBCO c-axis, as typically observed in non-vicinal films. However, closer examination reveals that the 5° substrate roughly doubles the angular splay of the BZO nanorods to 9.7 ± 3.1° from the c-axis when compared to the non-vicinal film.[13] The interaction of the increased strain induced by a 10° vicinal substrate and the added BZO produces a YBCO film marked by a high concentration of planar defects, while the nanorods maintain nearly the same splay observed at 5°.[13,20] While the lower vicinal angles show an effect of the added microstructural strain, a more distinct change occurs at higher angles of ~15 – 20°. At these angles the BZO nanorods undergo an orthogonal realignment such that they grow parallel to the YBCO a-b planes instead of the c-axis.

Fig. 1 shows transmission electron microscopy (TEM) cross-sections of 15° and 20° vicinal films with 2 vol % BZO. As seen in this figure, the nanorods form with their long dimension aligned with the a-b planes, and inclined by ~15 – 20° from the substrate. A small number of c-axis aligned nanorods are also visible in the 15° sample (marked by arrows), showing the realignment is not complete. The c-axis oriented nanorods were not observed at 20°, indicating the transition to a-b alignment is most preferable above 15°. Also of note are the 10.8 ± 1.5 nm diameters of the a-b aligned BZO nanorods at 20°, which increased from 5.2 ± 0.5 nm in the non-vicinal c-axis nanorods.[13]

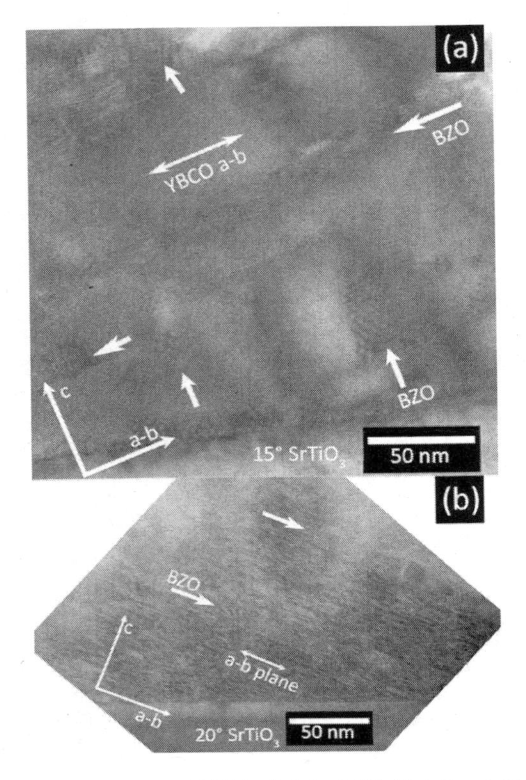

Figure 1. Cross-sectional TEM of vicinal YBCO + 2 vol. % BZO. (a) 15° vicinal STO substrate, where both c-axis and a-b nanorods are observed (marked by vertical and horizontal arrows, respectively). (b) On a 20° vicinal STO substrate the BZO nanorods align with the YBCO a-b planes. Scale bars are 50 nm.

It can be speculated that an observed increase in the dislocation spacing at the YBCO/BZO interface for a film on a 10° miscut substrate may indicate an increasing interfacial energy that makes c-axis alignment less favorable with increasing vicinal angle. The spacing between misfit dislocations is predicted to depend on the lattice constants, a_1 and a_2, of the interfacing materials: $D = a_{avg}^2 / |a_1 - a_2|$.[21] Therefore, the reduced lattice misfit between YBCO and BSO (6.7%, vs. 8.2% for YBCO/BZO) may also influence the nanorod alignment when modulated by the vicinal substrate. Indeed a similar orthogonal realignment takes place for YBCO + 4.5 vol % BSO films at 20°. In that case, the alignment of the BSO nanorods parallel to the a-b planes shows that the strain modulated films provide a mechanism for directional control over a range of misfit values. Additionally, since the internal strain distribution appears to play a role in the preferential alignment direction, we may also expect that the dopant concentration would also be influential.

The transport properties were accordingly influenced by the modified microstructure, as illustrated by the $J_c(H)$ dependence plotted in Fig. 2. $J_c(H)$ for H || c is shown in Fig. 2(a), and clearly reflects the modified defect structure of the vicinal films. It shows that up to ~4.7 T, there is an increase of $J_c(H||c)$ in the 5° sample. This improvement in the low-to-moderate applied magnetic field range may be attributed to a reduction of vortex hopping from the increased splay in the BZO nanorods at this vicinal angle.[20] Similar increases have been observed in ion-irradiated YBCO for doses less than the matching field.[22] However, as the vicinal angle is increased the increase in $J_c(H||c)$ reduced to lower field intensities, indicating the reduction in BZO nanorods aligned in this direction. Specifically, at 10° an increased J_c is measured up to ~1.3 T, while at 20° there is a slight increase of $J_c(H||c)$ up to ~0.6 T. These $J_c(H||c)$ behaviors are consistent with the observed defect structure, as the alignment of the nanorods is directed away from the c-axis of the YBCO. This is also verified by the $J_c(H||a\text{-}b)$ curves shown in Fig. 2(b), where an overall increase is seen for the vicinal samples through the measured range of 5 T. The largest increase in $J_c(H||a\text{-}b)$ at ~0.5 T in the 20° sample is probably due to the fully a-b aligned BZO nanorods. However, given their larger diameter and smaller number density at this vicinal angle, the greatest increase in J_c occurs at a lower value of applied magnetic field.

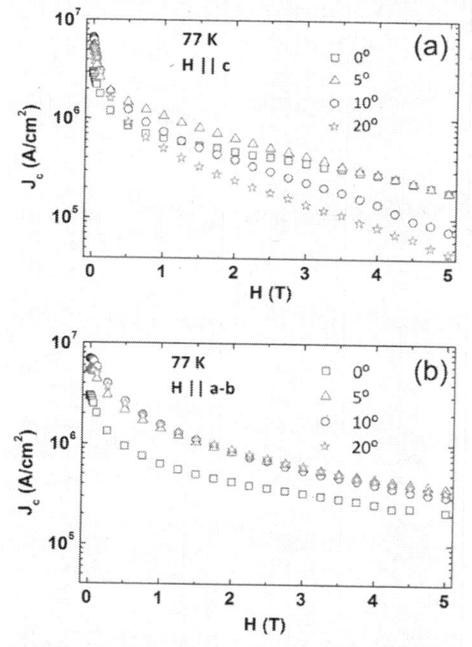

Figure 2. Transport critical current density vs. applied magnetic field for YBCO + 2vol. % BZO grown on 0° to 20° vicinal STO substrates. $J_c(H)$ is plotted for: (a) H || c and (b) H || a-b at 77 K.

Modification of In-field Performance

Of interest here is the difference in the in-field effective performance for the reported columnar pinning. Reports by other groups have shown that BZO and BSO pinning in YBCO provide fairly uniform enhancement to the plain YBCO J_c (H) curve; see, for example, the work of Mele et al. and Goyal et al.[5,7] This is in stark contrast to the behavior previously reported by Varanasi et al.[8,14] and that as grown by Baca et al.,[20] see Fig. 3. In the case of these reports from our group, the low field enhancement is not as large as that reported by the other representative groups; however, the high-field performance is significantly greater than those same reports. This is indicative that differences in how the pinning material was incorporated into the lattice bears a significant on the relative pinning provided by the material, in this case BZO and BSO.

The particular difference may be related to the amount of material inclusion and not just the deposition parameters or method of incorporation. For example, Varanasi et al has shown that variations in amount of pinning material added can directly impact the high-field behavior.[14] The work recently presented by Wee et al also indicates the same[23]--at higher concentrations of BSO, the low-field performance of the YBCO+BSO becomes less, but the high-field performance is further enhanced. Of course too little BSO or BZO material or too much does not provide the desired effects. This certainly does not mean that one is better than the other; it all depends on whether one desires a greater enhancement in the low-field regime of the high-field regime. The primary point made here is that both options are available simply by modifying the amount or manner in which it is incorporated. Fig. 4 shows the data at 65 K.

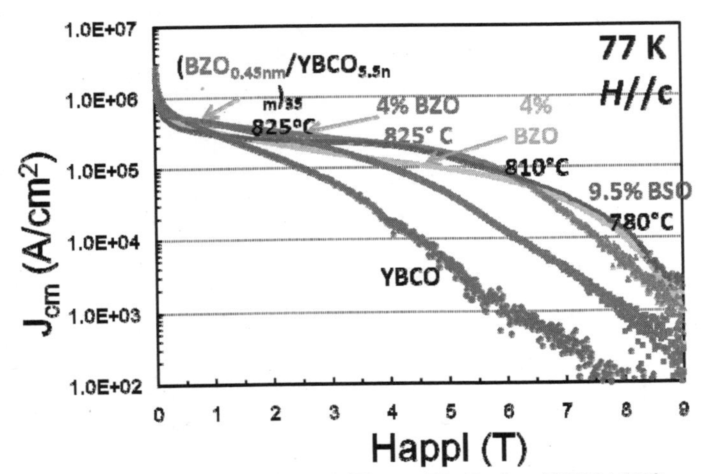

Figure 3. Magnetic Jc (H) at 77 K for YBCO+BZO and YBCO+BSO pinning compared to YBCO. The pinning demonstrated here shows exceptional high-field enhancement at the expense of some low-field J_c performance. The doping levels shown in the figure are in vol. %.

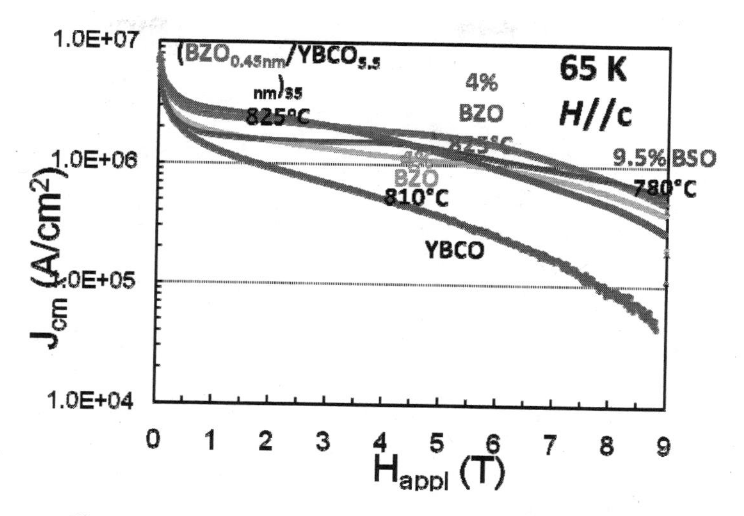

Figure 4. Magnetic Jc (H) at 65 K for YBCO+BZO and YBCO+BSO pinning compared to YBCO.

CONCLUSION

In summary, this paper addressed the means to provide specialized enhancement of pinning by splaying of the nanocolumns formed with both BZO and BSO additions to YBCO via vicinal substrates. We have also shown that the microstructure and direction of alignment of BZO and BSO nanorods in YBCO thin films is directly influenced by the vicinal angle of the substrate on which they were grown. Above a 15° vicinal angle, the BZO nanorods begin to orthogonally realign parallel to the YBCO a-b planes. Increasing to 20°, the nanorods become fully aligned in the a-b direction, from which the J_c(H||a-b) is enhanced for magnetic fields through 5 T. The same is observed for the BSO nanorods, which also undergo a realignment parallel to the a-b planes at a vicinal angle of 20°. This indicates the reduced strain field associated with the lower misfit between BSO and YBCO still favors a-b aligned nanorod formation. Lastly, pinning can be tailored in the nancolumnar growth to emphasize the low-or high field performance depending on the desired application. This effect can likely be accomplished through larger additions of the BZO or BSO material. It is worthy of noting that a higher concentration of BZO or BSO may also be expected to influence the strain distribution due to the increased overlap between individual nanorod strain fields.

REFERENCES

[1]T. J. Haugan, P. N. Barnes, R. Wheeler, F. Meisenkothen, and M. Sumption, Addition of nanoparticle dispersions to enhance flux pinning of the type II high temperature superconductor $YBa_2Cu_3O_{7-x}$, *Nature*, **430**, 867-870 (2004).

[2]J. Hanisch, C. Cai, R. Huhne, L. Schultz, and B. Holzapfel, Formation of nanosized $BaIrO_3$ precipitates and their contribution to flux pinning in Ir-doped $YBa_2Cu_3O_{7-\delta}$ quasi-multilayers, *Appl. Phys. Lett.*, **86**, 122508 (2005).
[3]N. Long, N. Strickland, B. Chapman, N. Ross, J. Xia, X. Li, W. Zhang, T. Kodenkandath, Y. Huang, and M. Rupich, Enhanced in-field critical currents of YBCO second-generation (2G) wire by Dy additions, *Supercond. Sci. Technol.*, **18**, S405 (2005).
[4]J. L. MacManus-Driscoll, S. R. Foltyn, Q. X. Jia, H. Wang, A. Serquis, L. Civale, B. Maiorov, M. E. Hawley, M. P. Maley, D. E. Peterson, Strongly enhanced current densities in superconducting coated conductors of $YBa_2Cu_3O_{7-x}$+$BaZrO_3$, *Nature Mater.*, **3**, 439-443 (2004).
[5]A. Goyal, S. Kang, K. J. Leonard, P. M. Martin, A. A. Gapud, M. Varela, M. Paranthaman, A. O. Ijaduola, E. D. Specht, J. R. Thompson, D. K. Christen, S. J. Pennycook, F. A. List, Irradiation-free, columnar defects comprised of self-assembled nanodots and nanorods resulting in strongly enhanced flux-pinning in $YBa_2Cu_3O_{7-\delta}$ films, *Supercond. Sci. Technol.*, **18**, 1533-1538 (2005).
[6]C. V. Varanasi, P. N. Barnes, J. Burke, L. Brunke, I. Maartense, T. J. Haugan, E. A. Stinzianni, K. A. Dunn, and P. Haldar, Flux pinning enhancement in $YBa_2Cu_3O_{7-x}$ films with $BaSnO_3$ nanoparticles, *Supercond. Sci. Technol.*, **19**, L37-L41 (2006).
[7]P. Mele, K. Matsumato, T. Horide, A. Ichinose, M. Mukaida, Y. Yoshida, and S. Horii, Enhanced high-field performance in PLD films fabricated by ablation of YSZ-added $YBa_2Cu_3O_{7-x}$ target, *Supercond. Sci. Technol.*, **20**, 244 (2007).
[8]C.V. Varanasi, J. Burke, L. Brunke, H. Wang, J.H. Lee, and P.N. Barnes, Critical current density and microstructure variations in $YBa_2Cu_3O_{7-x}$ + $BaSnO_3$ films with different concentrations of $BaSnO_3$, *J. Mater. Res.*, **23**, 3363-3369 (2008).
[9]S.H. Wee, A. Goyal, Y.L. Zuev, C. Cantoni, V. Selvamanickam, and E.D. Sprecht, Formation of Self-Assembled, Double-Perovskite, Ba_2YNbO_6 Nanocolumns and Their Contribution to Flux-Pinning and J_c in Nb-Doped $YBa_2Cu_3O_{7-\delta}$ Films, *Appl. Phys. Express*, **3**, 023101 (2010).
[10]P. Mele, K. Matsumoto, T. Horide, A. Ichinose, M. Mukaida, Y. Yoshida, S. Horii, R. Kita, Ultra-high flux pinning properties of $BaMO_3$-doped $YBa_2Cu_3O_{7-x}$ thin films (M = Zr, Sn), *Supercond. Sci. Technol.*, **21**, 032002 (2008).
[11]P.N. Barnes, J.W. Kell, B.C. Harrison, T.J. Haugan, C.V. Varanasi, M. Rane, F. Ramos, Appl. Phys. Lett. **89**, 012503 (2006).
[12]P.N. Barnes, T.J. Haugan, F.J. Baca, C.V. Varanasi, R. Wheeler, F. Meisenkothen, and S. Sathiraju, Minute doping with deleterious rare earths in $YBa_2Cu_3O_{7-\delta}$ films for flux pinning enhancements, *Physica C*, **469**, 2029-2032 (2009).
[13]F. J. Baca, P. N. Barnes, R. L. S. Emergo, T. J. Haugan, J. N. Reichart, J. Z. Wu, *Appl. Phys. Lett.*, **94**, 102512 (2009).
[14]C.V. Varanasi, J. Burke, L. Brunke, H. Wang, J.H. Lee, P.N. Barnes, *J. Mater. Res.*, **23**, 3363-3369 (2008).
[15]J. P. Rodriguez, P. N. Barnes, C. V. Varanasi, *Phys. Rev. B*, **78**, 052505 (2008).
[16]Y. F. Gao, J. Y. Meng, A. Goyal, G. M. Stocks, *JOM*, **60**, 54-58 (2008).
[17]T. Haage, J. Zegenhagen, J. Q. Li, H.-U. Habermeier, M. Cardona, Ch. Jooss, R. Warthmann, A. Forkl, H. Kronmuller, *Phys. Rev. B*, **56**, 8404-8418 (1997).
[18]J. L. Maurice, J. Briático, D. G. Crété, and J. P. Contour, O. Durand, *Phys. Rev. B*, **68**, 115429 (2003).
[19]F. J. Baca, R. L. Emergo, J. Z. Wu, T. J. Haugan, J. N. Reichart, P. N. Barnes, *IEEE Trans. Appl. Supercond.*, **19**, 3371-3374 (2009).

[20]R. L. Emergo, F. J. Baca, J. Wu, T. Haugan, P. Barnes, Generating splayed $BaZrO_3$ nanorods in $YBa_2Cu_3O_{7-\delta}$ films, unpublished results.
[21]J. M. Woodall, G. D. Pettit, T. N. Jackson, C. Lanza, K. L. Kavanagh, J. W. Mayer, *Phys. Rev. Lett.*, **51**, 1783-1786 (1983).
[22]L. Civale, L. Krusin-Elbaum, J. R. Thompson, R. Wheeler, A. D. Marwick, M. A. Kirk, Y. R. Sun, F. Holtzberg, C. Field, *Phys. Rev. B*, **50**, 4102-4106 (1994).
[23]S.H. Wee, A. Goyal, Y.L. Yuri, C. Cantoni, and S. Cook, "Tuning Flux-Pinning and Critical Current Density for $YBa_2Cu_3O_{7-\delta}$ Coated Conductors via $BaZrO_3$ Compositional Control," presented at the Materials Research Society Spring Meeting, San Francisco, CA (2009).

ELECTRICAL PROPERTIES OF $Hg_{0.8}Tl_{0.2}Ba_2Ca_{n-1}Cu_nO_{2n+2+\delta}$ FOR (n=1-5) HTSC SYSTEM

Ghazala Y. Hermiz and Ebtisam Kh AL-Beyaty*
University of Baghdad, College of Science
*University of Diyala, College of Education

ABSTRACT

Samples of high temperature superconductor $Hg_{0.8}Tl_{0.2}Ba_2Ca_{n-1}Cu_nO_{2n+2+\delta}$ for (n=1-5) system has been prepared using solid state thermochemical reaction method by two steps. The present work has studied the influences of preparation method and the effect of n on the transition temperature (T_c). The value of T_c enhanced with the increase of n up to 3, after that there was a reduction of T_c for n = 4 and 5. The highest transition temperature was 125 K for (Hg-Tl)-1223 samples sintered at 860°C in capsule for 13 h under 0.5 GPa. Annealing the samples in the flow of oxygen exhibited an enhancement of the crystallites improving superconducting behavior and thus increasing the T_c of the doped samples, but decreasing T_c above optimum doped.

INTRODUCTION

The discovery of Hg-based cuprates high temperature superconductors (HTSC), with general formula $HgBa_2Ca_{n-1}Cu_nO_{2n+2+\delta}$ (n=1----8) where n is the number of consecutive Cu-layers, show the highest transition temperature into the superconducting state (T_c), both at normal pressure (for instance $HgBa_2Ca_2Cu_3O_{8+\delta}$ has T_c =135K) and high pressure (for $HgBa_2Ca_2Cu_3O_{8+\delta}$ T_c=166K at P=23 GPa) [1] . That is why these compounds are promising candidates for a number of possible applications. However, the scope of Hg-based superconductors application is not wide, because of difficulties in reproducible synthesis of samples containing only one superconducting phase, the toxicity of several substances that may by formed during the synthesis, render chemical instability of the cuprates.

Liu et al. [2] reported that substitution of Hg in $HgBa_2Ca_2Cu_3O_{8+\delta}$ (1223) by 5-10 % Tl considerably enhances the volume fraction of the superconducting phase while T_c remains unchanged at about 134 K. They reported that the addition of small amounts of Tl enhances the solid state reaction rate and greatly increases the superconducting volume fraction. They also suggested that the (Hg,Tl)-1223 system of superconducting compounds will allow for the practical fabrication and development of superconducting wires, tapes, thin films, powders, and devices with transition temperatures higher than 130 K.

The partial substitution of Tl to this high T_c mercury-based oxide was studied by Dai et al. [3] they followed a procedure as a nominal $Ba_2Ca_2CuO_7$ precursor was prepared first, and then mixed with HgO and Tl_2O_2 to form $Hg_{0.8}TL_{0.2}Ba_2Ca_2Cu_3O_{8.33}$. The pressed rods are put inside quartz tube, evacuated sealed by a hot metal at 860°C for 400 minutes in annealed O_2. The transition temperature for this compound is about 138K.

The highest transition temperature within the complex homologous series $HgBa_2Ca_{n-1}Cu_nO_{2n+2+\delta}$ (n=1-7) was found by Peacock et al.[4] The third member, i.e. Hg-1223, is possessing the critical temperature (T_c) of 135K and the first member of this family i.e., Hg-1201, having T_c up 97K.

Giri et al. [5] reported investigations on the effect of substitution of Bi and Tl at the Hg sites in the oxygen deficient HgO_δ layer of $HgBa_2Ca_2Cu_3O_{8+\delta}$ cuprate superconductor. They prepared the samples by the two-step reaction process and observed that the as-grown $HgBi_{0.2-x}Tl_xBa_2Ca_2Cu_3O_{8+\delta}$ (with x = 0, 0.05, 0.1, 0.15, and 0.2) corresponds to the 1223 phase. It has been found that the T_c varies with the average cationic size of the dopant cations. The optimum T_c of ~131K has been found at x=0.05 with a highest stability.

Giri et. al. [6] studied the influence of simultaneous doping of Tl and Bi on microstructures and critical current density (J_c) of $HgBa_2Ca_2Cu_3O_{8+\delta}$. They found the phase induce interesting microstructure variants in the form of long period polytypoid (LPP) like structure embodying native defect substructures. They also observed that $(HgTl_{0.2-x}Bi_x)Ba_2Ca_2Cu_3O_{8+\delta}$ with (x = 0.05, 0.15, 0.15) phases have superconducting onset transition temperature (T_c) of about ~ 133 K and the highest J_c at 5K has been found to vary from 6.2×10^6 - 2.6×10^6 Acm^{-2} for the $(HgTl_{0.1}Bi_{0.10})Ba_2Ca_2Cu_3O_{8+\delta}$ phase.

Hase et. al. [7] reported the synthesis of $HgBa_2Ca_{n-1}Cu_nO_{2a+2+\delta}$ phases with (n = 3, 4) and the relationship between their structural and physical properties. They calculated Fermi velocities V_f and Fermi surface volume .From measurements of Fermi surfaces they found 0.93-1.11 holes in Hg-1223 and 0.90-1.14 holes in Hg-1234 and an anisotropy defined by the ratio V_f^x/V_f^y is similar to that in $TlBa_2Ca_2Cu_3O_9$.

Zaleski et al .[8] studied a model of n-layer high-temperature cuprates of homologous series like $HgBa_2Ca_{n-1}Cu_nO_{2+2n+\delta}$ to explain the dependence of the critical temperature $T_c(n)$ on the number n of Cu-O planes in the elementary cell.

In this paper, we studied the role of CuO_2 layer playing in the properties of superconductor material. This could be observed by the variation of n of the $HgBa_2Ca_{n-1}Cu_nO_{2a+2+\delta}$ system.

EXPERIMENT

The samples were prepared by thermochemical solid state reaction. Appropriate amounts of the powder materials $CaCO_3$, CuO and $BaCO_3$ were mixed together, the mixture were grounded and reground many times to produce an appropriate fine powder and then calcined in air at 800°C for 20-24 h in two stages. The calcined powder was reground again after the mixing of HgO and Tl_2O_3 with it and pressed into disc-shaped pellets under 0.5 and 0.9GPa.

Two sets of samples (P= 0.5, P=0.9) GPa were put into a sealed quartz tube, and then evacuated using a rotary pump to a pressure of 10^{-2} mbar, the third set was pressed under 0.9GPa.and put in an alumina crucible.

All the samples under 0.9 GPa were placed in a tube programmable furnace. The temperature of the furnace was set to rise at a rate 200°C/h up to 600°C and thereafter the rate was 100°C/h up to 860°C and held at this temperature for 20 h. Finally, the furnace was cooled to a room temperature by the same rate of heating. The samples in the sealed quartz tube pressed under 0.5 GPa were placed in a tube programmable furnace, then repeat the program above but it was held at 860°C for 10h, 13 h and 20h.

All the pellets were placed in an alumina boat and heated at a rate 120°C/h, when the temperature became 450°C an oxygen gas started flow until the furnace temperature become 650°C.It is held at this temperature about two hours. After that time the flow oxygen was stopped and the furnace was cooled to room temperature by the same rate of heating. Iodometric titration was used to find the oxygen content (δ) in the samples. Four probe dc methods at temperature range (77-300) K was used to measure the resistivity (ρ) and to determine the critical temperature

RESULTS AND DISCUSSION

Figure 1 represents the quantitative testing for the elements of the compound $Hg_{0.8}Tl_{0.2}Ba_2Ca_{n-1}Cu_nO_{2n+2+\delta}$ by using x-ray fluorescent (XRF) method, for the samples A, B, C and D prepared under 0.9 GPa for 20h at 860°C ,where A and C samples were sintered in capsule(quartz tube) while B and D samples were sintered in air. From this figure we found the differences in the quantity of elements such as the Hg and Tl of the compounds prepared by the two methods. The samples prepared in capsule were better than the samples sintered in air. Such results are in agreement with those reported by Xu et al. [9] They found that the prepared samples under high compressed pressure in capsule will prevent the escape of Hg from the mixture.

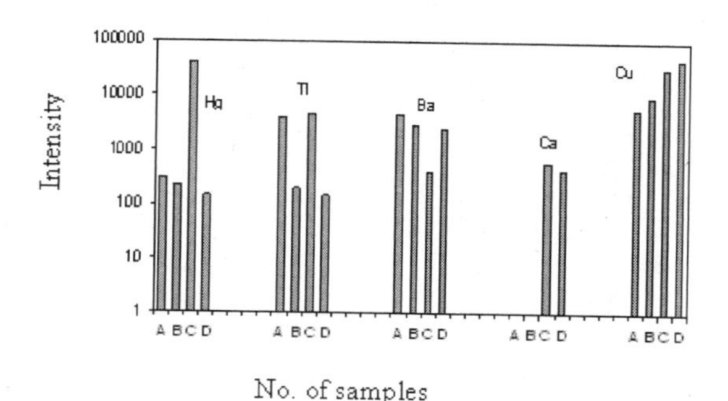

Figure 1. XRF for Hg-1201 and Hg-1245 and compound at different preparation techniques were A is n=1 in capsule, B is n=1 in air, C is n=5 in capsule, D is n=5 in air

In order to clarify the effect of sintering time on the resistivity and transition temperature (T_c), the prepared samples of $Hg_{0.8}Tl_{0.2}Ba_2Ca_2Cu_3O_{8+\delta}$ under 0.5GPa inside capsule were sintered at 860°C for different times 10,13and 20h,their T_c are shown in figure 2 It is clear from this figure that the critical temperature T_c increases from 95K to 125K with the increasing of the sintering time from 10h to 13h, while the samples sintered for 20h did not reach to zero-resistivity even at the boiling point of liquid nitrogen but it behaved like superconductor albeit it showed a transition temperature at about 101K after annealing with oxygen.

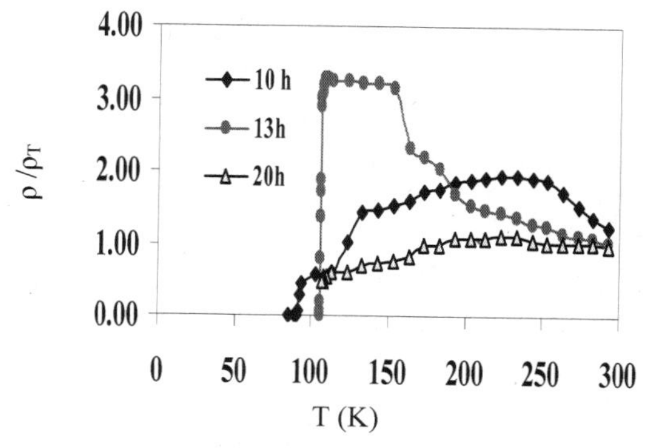

Figure 2. Normalized resistivity vs. temperatures for various compositions of $Hg_{0.8}Tl_{0.2}Ba_2Ca_2Cu_3O_{8+\delta}$ sintered in capsule at T_s=860°C for 10h,13h,20h under pressure 0.5GPa.before flow of oxygen.

The increase in the T_c value is mainly due to the strong grain links due to the increase of the contact areas between the grains during the sintering process time. When the sintering time increased up to 20h a decrease in the T_c value was observed. The decrease can be attributed to amorphous phase formation at the grain boundaries due to volatile mercury during the prolonged sintering treatment.

The synthesis of (Hg-Tl) samples fabricated under vacuum pressure in sealed quartz tube pressed under 0.5GPa and 0.9 GPa at 860°C for 13 h and 20h.respectively have been done .The prepared samples were dark and homogenous. Plots of normalized resistivity (ρ/ρ_T) versus temperature measurement for the as-synthesized as well as annealed samples with oxygen are shown in figure 3 and 4.

Before the flow of oxygen as shown in figures 3a&4a ,the samples have good superconductor behavior and the resistivity dropped nearly sharp but at different onset transition temperatures i.e. 100, 120, 135, 98 and 110K for **n**=1,2,3,4 and 5 respectively for samples prepared under 0.5GPa while the samples under 0.9 GPa have $T_{c\ (onset)}$ within a rang of (94 -103) K for **n** is equal (1-5) .However, not all of the compositions reach $T_{c\ zero}$ except $Hg_{0.8}Tl_{0.2}Ba_2Ca_1Cu_2O_{6+\delta}$ and $Hg_{0.8}Tl_{0.2}Ba_2Ca_2Cu_3O_{8+\delta}$ that prepared under 0.5GPa and $Hg_{0.8}Tl_{0.2}Ba_2Ca_1Cu_2O_{6+\delta}$ and $Hg_{0.8}Tl_{0.2}Ba_2Ca_3Cu_4O_{10+\delta}$ that prepared under 0.9GPa, their transition temperatures are 111 K, 125 K, 120 and 104 K respectively.

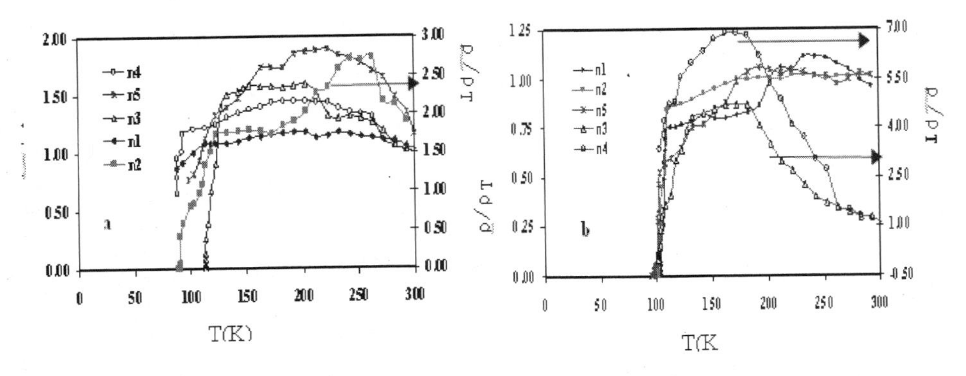

Figure 3. Normalized resistivity vs. temperatures for various compositions $Hg_{.8}Tl_{.2}$ $Ba_2Ca_{n-1}Cu_nO_{2n+2+\delta}$ sintered at T_s=860°C for13h under pressure 0.5 GPa in capsula. (a) without flow of oxygen,(b) with flow oxygen.

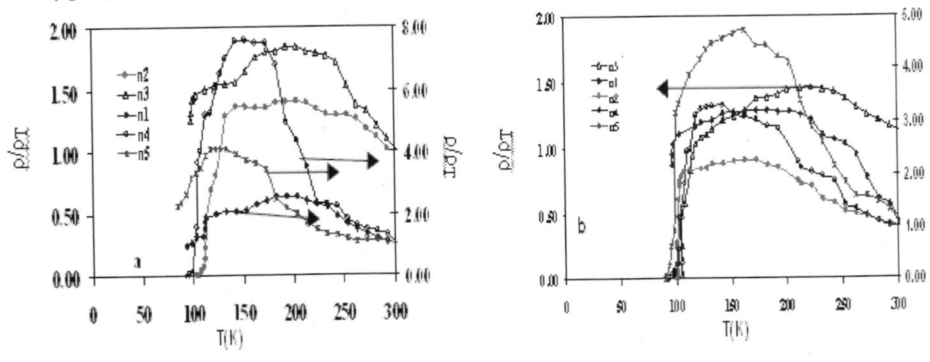

Figure 4.Normalized resistivity vs. temperatures for various compositions of $Hg_{.8}Tl_{.2}$ $Ba_2Ca_{n-1}Cu_nO_{2n+2+\delta}$ sintered at T_s=860°C for 20h under pressure 0.9 GPa. in capsula (a) without flow of oxygen, (b) after flow oxygen.

Most of the samples, after annealing with oxygen, see Figs.3b&4b, have a good response ,their superconducting transition temperatures are 106,110,115,108,and 102K under 0.5GPa and 98,107,113,106,and 101K under 0.9GPa, for n equal to 1,2,3,4 and 5 respectively. On the other hand we found a decrease of transition temperature for samples that previously behaved as superconductor (before flow of oxygen) .as shown in Table (I,II) and figure 3b&4b.Our results are in agreement with that obtained by Giri et al .[5] they found annealing with oxygen leads to a decrease in T_c from 131 K to 127 K for Hg $Bi_{.15}Tl_{0.5}Ba_2Ca_2Cu_3O_{8+\delta}$.

Table I. Exhibit values of δ, υ and T_c for the composition $Hg_{0.8}Tl_{0.2}Ba_2Ca_{n-1}Cu_nO_{2n+2+\delta}$ sintered in capsule at T_s= 860°C for 13 h under pressure 0.5GPa.without and with flow of oxygen.

P (GPa.)	n	Without Flow of Oxygen			With Flow of Oxygen.		
		T_c (K)	δ	υ	T_c (K)	δ	υ
0.5	1	onset 100	-0.093	2.185	106	0.233	2.467
	2	111	0.118	2.118	110	0.305	2.305
	3	125	0.111	2.074	115	0.489	2.326
	4	onset 98	0.314	2.157	onset 108	0.778	2.389
	5	onset 110	0.694	2.274	102	0.890	2.356

Another feature observed for samples after annealing was that the $T_c(n)$ dependence for Hg-based curate has a parabolic shape as shown in figure 5, T_c values increase from Hg-1201 to Hg-1212 and finally reach their maximum for the Hg-1223.Strating from the fourth member of the series ,the T_c values go down (e.g., for Hg1234 and Hg-1245), it could be related to increase of external CuO_2 layers distortion as a result of increasing O-atoms.

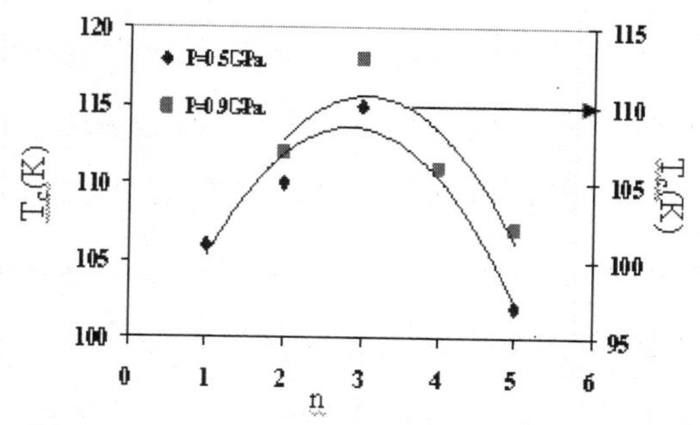

Figure 5. Transition temperature (T_c) as a function of n for $Hg_{0.8}Tl_{0.2}Ba_2Ca_{n-1}Cu_nO_{2n+2+\delta}$ sintered in capsule with flow of oxygen for 13h,20 under pressure 0.5,0.9 GPa respectively.

The observed decrease for higher homologues was explained [10] as a result of insufficient charge carrier concentration, in other words, as if these samples were underdoped. However, synthesis of Hg-1234 and Hg-1245 samples are in overdoped state refuted this hypothesis. However, we don't exclude the hypothesis that for n>3, the charge carriers concentration decreased, due to holes decreasing, since the HTSc is a structural distortion property.

Table II. Values of δ,υ and T_c for the Composition $Hg_{0.8}Tl_{0.2}Ba_2Ca_{n-1}Cu_nO_{2n+2+\delta}$ with Different n ,Sintered in Capsule and Air at T_s= 860°C for 20 h under pressure 0.9GPa. without and with flow of oxygen.

		Sintering in Capsule					
		Without Flow of Oxygen			With Flow of Oxygen		
P(Gpa.)	n	Tc(K)	δ	ν	Tc(K)	δ	ν
	1	Onset 92	-0.414	1.172	Onset 98	0.446	2.892
	2	120	0.573	2.572	107	0.941	2.676
	3	Onset 98	-0.585	1.61	115	0.926	2.263
	4	104	0.609	2.348	106	0.815	2.48
0.9	5	Onset 85	-0.378	1.849	102	0.936	2.345
		Sintering in AIR					
	1	Onset 93	-0.364	1.272	99	-0.12	1.78
	2		-0.749	1.251		0.397	2.397
	3		-0.532	1.646		0.829	2.553
	4		-0.164	1.918		0.878	2.439
0.9	5		-0.378	1.849		1.298	2.519

In order to clarify the effect of the preparation technique on superconducting behavior and transition temperature, another set of $Hg_{0.8}Tl_{0.2}Ba_2Ca_{n-1}Cu_nO_{2n+2+\delta}$ under 0.9 GPa, were sintered in air for 20h with and without annealing with oxygen. The normalized resistivity curves of these samples show that all the samples start like semiconductor in the range (300-150) K then the normalized resistivity decreases with the decrease of temperature within the range (80-150) K, but the curves didn't reach the zero resistivity except for (Hg-Tl)-1201 which has a good superconductor behavior with transition temperature equal to 96K after annealing with oxygen as shown in Fig.(6).This value is nearly equal to that obtained by Peacock et al [4] for the same phase (T_c = 97K).

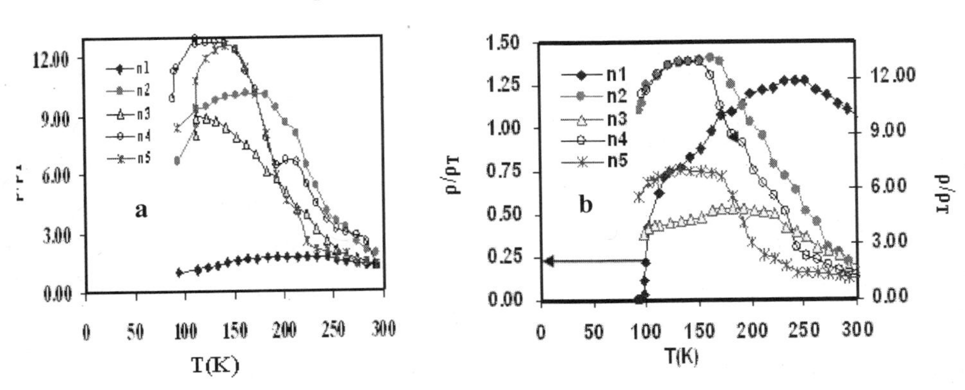

Figure 6 Normalized resistivity vs. temperatures for various compositions of $Hg_{.8}Tl_{.2}Ba_2Ca_{n-1}Cu_nO_{2n+2+\delta}$ sintered at T_s=860° C for 20h under pressure 0.9 GPa. (a) In air with no flow of oxygen and (b) with flow of oxygen

It is well known that the variation of oxygen amount in high-T_c superconductors (HTSc) is exclusively due to changes of oxygen content in the reservoir layer (CuO_2) and its effect on the crystalline phase of the superconductor, thus the doping is a delicate and has decisive rule on the superconducting properties of HTSc system as in $HgBa_2Ca_{n-1}Cu_nO_{2n+2+\delta}$.

The diffusion of oxygen through the superconductor material is suggested to be a guideline for annealing treatment intended to optimize the oxygen content and superconductor properties of Hg-based .Our data showed that the oxygen deficiency δ increased continuously with flow of oxygen, as shown in Table (I,II) and figure 7 for samples prepared in capsule under 0.5 GPa and 0.9 GPa.

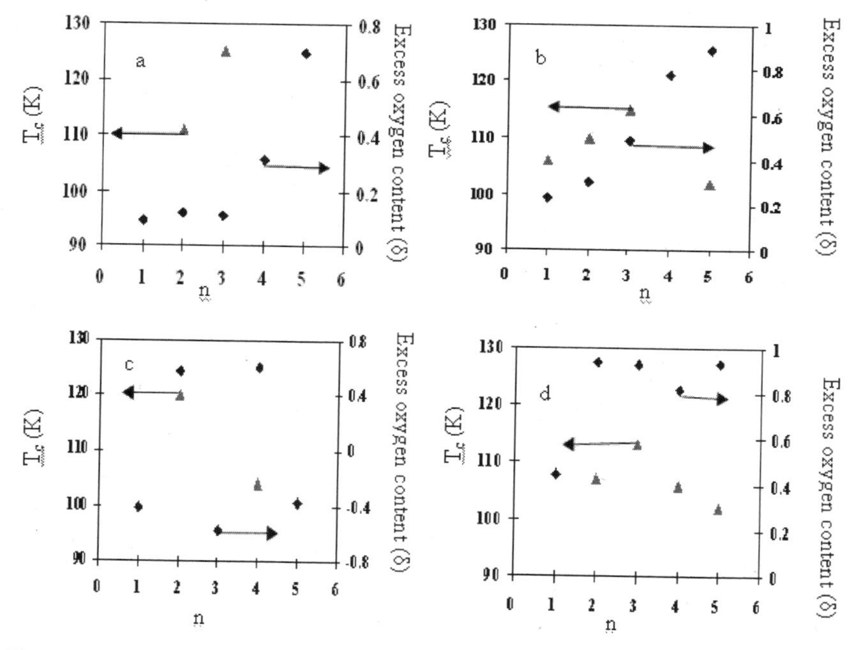

Figure 7.Excess oxygen content (δ), transition temperature (T_c) as a function of n sintered in capsule for (a) P=0.5 GPa with no flow of oxygen , (b) P= 0.5GPa with flow of oxygen (c) P=0.9 GPa with no flow of oxygen , (d) P=0.9 GPa with flow of oxygen

As already mentioned, the excess of oxygen content (δ) has increased with n for sample prepared in air, the same behavior was found for superconductor samples prepared in capsule after annealing with oxygen. However δ increases from 0.233, 0.446 and for n=1 to 0.890, 0.936 for n=5 under 0.5GPa and 0.9GPa respectively .While a random variation of δ was found for samples prepared under 0.9 GPa before the flow of oxygen .On other hand we found an enhancement of oxygen content for all the composition after annealing with oxygen. The explanation is that, if the substitution of $O_{(ions)}$ at the apical-oxygen sites change the charge concentration in the CuO_2 planes i.e. (adding holes) that improves the behavior of the system and convert it to superconductor. Such results were agreement with Chmarissem et.al. [11] .They believed that optimum annealing condition such as oxidation at different pressures together with varying of substitution ration in Hg side could induce superconductor behavior.

It is obvious that increasing δ causing a semi saturation in the oxygen content in the structure which tend to decrease in the amount of vacancy and the last cause decrease of T_c value [12]. These results were found for samples with n=3 where T_c= 125K, δ =0.111 after annealing with oxygen Tc=115K, δ =0.489, under 0.5 GPa, also we obtained the same behavior for samples n=2 under 0.5 GPa and 0.9 GPa.

It should be noticed that the δ values are directly associated with the average oxidation state of copper valence (ν).This value could be obtained from equation [13] below and their values are listed in the Table (I&II) for different n

$$\delta = \frac{(\nu - 2)n}{2} \quad (1)$$

Interestingly, our data are in accordance to this equation

CONCLUSIONS

It has been noticed that the preparation method plays a great role in optimizing good superconducting samples

X-ray fluorescent results have shown that the samples sintered in capsulation method are better than those prepared in air. The in capsulation keeps the quantity of the element compounds and the samples having high quality in spite of the fact that the sintering time has been reduced in comparison with the other method.

The superconducting T_c for multilayered high-T_c superconductors of Hg-based do not monotonically increase with the number of CuO_2 layer (n). The T_c increases with n up to n=3, but then starts dropping at n=4 in the $Hg_{0.8}Tl_{0.2}Ba_2Ca_{n-1}Cu_nO_{2n+2+\delta}$ system with all the methods of preparation. The T_c(n) dependence for Hg-based cuprate has a parabolic shape Iodometric titration results of $Hg_{0.8}Tl_{0.2}Ba_2Ca_{n-1}Cu_nO_{2n+2+\delta}$ show an enhancement of δ value with the increase of n.

REFERENCES

[1]M. Monteverde, C. Acha, M. Nunez-Regueiro, D. A. Pavlov, K. A. Lokshin, S. N. Putilin and E. Antipor: Euro phys Lett,**72**,458,(2005).
[2]J. Z. Liu, C. Chang, M. D. Lan, P.Klavins and P N.Shelton High-Tc Update-Mail version, 8, 3 ,(1994).
[3]P. Dai, B. C. Chakoumakes, G. F. Sun, K. W. Wong, Y. Xin, D. F. Lu, Phys C, 243, ,201 (1995).
[4]G. B. Peacock, A. Flecher, I. Gameson and P. P. Edwards, Phys C, Superconductivity: 30, 1, (1998).
[5]R. Giri, H. K. Singh, R. S. Tiwari, & O. N. Srivastova: Bull. Mater. Sci,24, 523,(2001) .
[6]R. Giri, G. D. Verma, R. S. Tiwari, and O. N. Srivastova: Cryst. Res. Technol,38,760,(2003).
[7]I. Hase, N. Hamada & Y. Tanaka: Phys, superconductivity C V.412 – 414,246,(2004).
[8]Zalesk, T. A. Kopec and T. K.: Phys. Rev. B, V.71, P.014519, (2005).
[9]Q. L. Xu, Tana, T. B. Chen, Zh: Supercon, Sci., Techno,**7**,828, (1994).
[10]K.A.Lokshin,D.A.Pavlov,M.L.Kovba,E.V.Antipov,I.G.Kuzemskaya,L.F.Kulikova, V.V.Davydov, I.V.Morozov and E.S.Itskevich, Phys C**:** 300, 71, (1998).
[11]O. Chmaissem, L.Wessels, and Z. Z. Sheng: Phys C, 230, 231, (1994).
[12]J.L.Macmanuus-Driscll,j.A.Alonso,P.C.Wanug,T.H.Geballe and J.C.Bravman , Phys C:232 ,288,(1994).
[13]I.Kirschoner., R.Laiho, P.Lokack, A.C.Bodi, M.Dimitrova-Lukacs,E. Lihderanta. and G. Zsolt, Z.Phys.B ,99,501,(1996).

ELECTRODEPOSITED Ag-STABILIZATION LAYER FOR HIGH TEMPERATURE SUPERCONDUCTING COATED CONDUCTORS

Raghu N. Bhattacharya and Jonathan Mann
National Renewable Energy Laboratory, 1617 Cole Boulevard, Golden CO 80401, USA

Yunfei Qiao
Superpower Inc., 450 Duane Avenue, Schenectady, NY 12304, USA

Yue Zhang and Venkat Selvamanickam
Department of Mechanical Engineering and Texas Center for Superconductivity, University of Houston, Houston, TX 77204, USA

ABSTRACT

We developed a non-aqueous based electrodepostion process of Ag-stabilization layer on YBCO superconductor tapes. The non-aqueous electroplating solution is non-reactive to the HTS layer thus does not detoriate the critical current capability of the superconductor layer when plated directly on the HTS tape. The superconducting current capabilities of these tapes were measured by non-contact magnetic measurements.

INTRODUCTION

The development of high-T_c superconductors has created the potential of economically feasible development of superconductor components and other devices in the power industry, including applications for power generation, storage, distribution power cables, transformers, and fault current interrupters/limiters. In addition, other benefits of high-temperature superconductors in the power industry include a factor of 3-10 increase of power-handling capacity, significant reduction in the size (i.e., footprint) and weight of electric power equipment, reduced environmental impact, greater safety, and increased capacity over conventional technology. While such potential benefits of high-temperature superconductors remain quite compelling, numerous technical challenges continue to exist in the production and commercialization of high-temperature superconductors on a large scale. Among the challenges associated with the commercialization of high-temperature superconductors, many exist around the fabrication of a superconducting tapes. A first generation of superconducting tape includes Bi(Pb)-Sr-Ca-Cu-O (BSCCO) high-temperature superconductor. This material is generally fabricated in form of discrete filaments, which are embedded in a matrix of noble metal, typically silver. Although such conductors may be made in extended lengths needed for implementation into the power industry (such as on the order of a kilometer), its widespread commercially feasibility is limited due to materials and manufacturing costs. Accordingly, a great deal of interest has been generated in the so-called second-generation HTS YBCO tapes that have superior commercial viability[1]. However, to date, numerous engineering and manufacturing challenges remain prior to full commercialization of such second generation-tapes. These tapes typically rely on a layered structure, generally including a flexible substrate that provides mechanical support, buffer layers overlying the substrate, an HTS YBCO layer overlying the buffer film, and an electrical stabilizer layer around the entire structure. In the following section we will discuss the importance and development of low-cost stabilizer layer.

High temperature superconducting (HTS) coated conductors have possibilities to be quenched due to several factors, e.g., local defects in the conductors, over-load operation, a failure in power supply and cooling system, etc. Quench produces excessive temperature or voltage in the coil winding, and overheating may cause the conductor to meltdown or high voltage to cause a dielectric break down[2-6]. Coated conductors that are made by deposition of thin HTS YBCO film on Hastelloy tapes or textured

Ni based substrates become highly resistive when they are quenched. Therefore to manufacture reliable and safe HTS coil, conducting layers such as Cu and Ag are necessary to attach to the HTS tapes to stabilize and protect the conductors from damage due to quenches. The stabilizer layers also serve as a protection layer against harsh environmental conditions. The stabilization layers are generally dense, thermally and electrically conductive, and bypass electrical current in case of failure of the superconducting layer or if the critical current of the superconducting layer is exceeded. It has been found that a capping Ag layer at least 1 micron in thickness is needed between the superconductor layer and the Cu-stabilizer layer to avoid interfacial reaction and reduction in the critical current capability of the superconductor layer. At present, the capping Ag-stabilization layer is fabricated by sputtering techniques. Sputtering techniques are suitable for large-area deposition; however, they requires expensive vacuum equipment and suffer from low material utilization. Non-vacuum electrodeposition techniques have the potential to prepare large-area uniform films using low-cost source materials and low-cost capital equipment. We are developing a non-aqueous based Ag-stabilization layer which is non-reactive to the HTS layer. While the content of silver in 2G HTS wire is small (< 2 μm), silver deposition is an important constituent of wire cost as well as production capacity.

EXPERIMENTAL

Electrodeposition of Ag on YBCO superconductor tape was performed from a bath containing $AgNO_3$ dissolved in organic solvent. A Fisher Scientific (FB300) power supply was used to electrodeposit Ag films. The electrodeposited Ag films were prepared by employing a two-electrode cell in which the counter electrode was Pt gauze and the working electrode (substrate) was YBCO tape.

Critical current measurements were conducted using non contact technique. A hall-probe-based measurement was used for the non contact technique with a spatial resolution of 1 mm.

RESULTS AND DISCUSSION

Electrodeposition uses electrolysis to deposit a coating of the desired form on conducting substrates from a solution ("bath") containing the ions of interest (e.g., Ag^+). In cathodic electrodeposition, when the potential of the substrate (electrode) is moved from its equilibrium value toward negative potentials, the cation will be reduced and metal film will be deposited on the substrates. In a solution containing Ag^+, when the potential is sufficiently negative, the deposition of Ag on electrode surface takes place, corresponding with the following Nernst equation.

$$Ag^+_{(aq)} + e^- \rightarrow Ag(s) \quad (1)$$

$$E = E^0_{Ag} + RT/F \ln [Ag^+] = 0.799 + 0.0591 \log [Ag^+] \quad (2)$$

where E is the electrode equilibrium potential with respect to the standard hydrogen electrode (SHE); and E^0_{Ag} is the standard electrode potentials of Ag. F is Faraday's constant equal to 96,485 Coulomb/mole.

The electrodeposited Ag thin film morphology and thickness were evaluated by the scanning electron microscopy (SEM). Figure 1(a) and 1(b) show the SEM of the electrodeposited Ag film deposited from an organic solvent on glass/Mo substrates. The surface morphology of the electrodeposited film shown in Fig.1a indicates the deposition of dense polycrystalline Ag film. The cross-sectional view of an electrodeposited Ag film deposited for 2 minutes is shown in Fig.2b. The SEM cross sectional image shows the film thickness is about 3.5 microns with a deposition rate of about 1.75 μm/min. The surface morphology of the bare YBCO tape and electrodeposited Ag film deposited on YBCO tape for 2 minutes are shown in Fig.2(a) and 2(b). The SEM images indicates that electrodeposition of Ag is strongly influenced by the surface topography of YBCO tape.

Figure 3(a) and 3(b) show I_c of two YBCO tapes with electrodeposited Ag-stabilization layer. The I_c of bare YBCO tape does not change after the electrodepostion of Ag thin film. The I_c improves slightly after annealing the tape with electrodeposited Ag. Figure 4 shows the results obtained from YBCO tape with sputtered Ag. Figure 5 shows the results obtained from the bare YBCO tape. The annealing process does not have any effect on the YBCO tape coated with sputtered Ag and also for bare YBCO tape. YBCO tapes with elctrodeposied Ag-stabilization layer, YBCO tape with sputtered Ag and bare YBCO tape were annealed together. The I_c improvement of the annealed electrodeposited Ag-coated YBCO tape is not clearly understood at this time, but could be due to the electrochemical reduction reactions which can change the stoichiometry/oxygen of the YBCO tape.

CONCLUSION

We developed a non-aqueous based electrodepostion process of Ag-stabilization layer on YBCO superconductor tapes. We demonstrated that direct Ag plating on YBCO tapes from the non-aqueous solvent does not destroy the superconducting YBCO layer. Overall, it is seen that the critical current of YBCO tape stays at the similar level after each step indicating no degradation in the tape quality after the electrodeposition and oxygenation processes.

ACKNOWLEDGEMENTS

The author thanks Bobby To for sacnning electron micrographs. This work has been authored by an employee of the Midwest Research Institute under contract number DE-AC36-08GO28308 with the U.S. Department of Energy. The United States Government retains and the publisher, by accepting the article for publication, acknowledges that the United States Government retains a non-exclusive, paid-up, irrevocable, worldwide license to publish or reproduce the published form of this work, or allow others to do so, for United States Government purposes.

REFERENCES

[1]V. Selvamanickam, Y. Chen, X. Xiong, Y.Y. Xie, M. Martchevski, A. Rar, Y. Qiao, R.M. Schmidt, A. Knoll, K.P. Lenseth, and C.S. Weber, *IEEE Trans. Appl. Supercond.* **19,** 3225 (2009)

[2]A. Ishiyama, H. Ueda, T. Ando, H. Naka, S. Bamba, and Y. Shiohara, *IEEE transactions on Applied Superconductivity,* **17**, 2 (2007).

[3]A. Ishiyama and M. Yanai *et al.*, *IEEE Trans. Appl. Supercond.*, **15**, 1659 (2005).

[4]Y. Iwasa, *Case Studies in Superconducting Magnets*. New York: Plenum Press, pp. 328–331 (1994).

[4]M. N. Wilson, *Superconducting Magnets*. Oxford: Clarendon Press, pp. 200–202 (1983).

[5]K. Tasaki, T. Yazawa, M. Ono and T. Kuriyama, *Journal of Physics: Conference Series, 7th European Conference on Applied Superconductivity,* **43,** 1047 (2006).

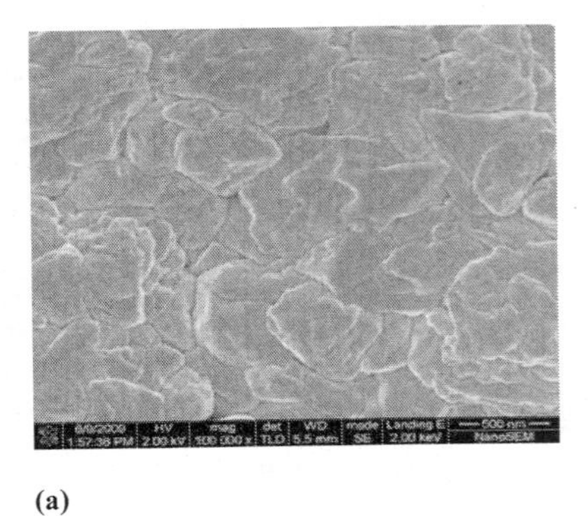

(a)

(b)

Figure 1. (a) SEM surface morphology of an electrodeposited Ag film on Mo/Glass; (b) SEM cross-section of a film deposited for 2 minutes on Mo/Glass.

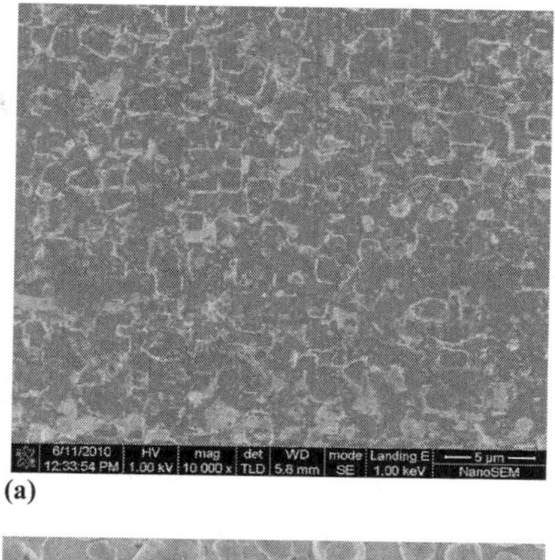

(a)

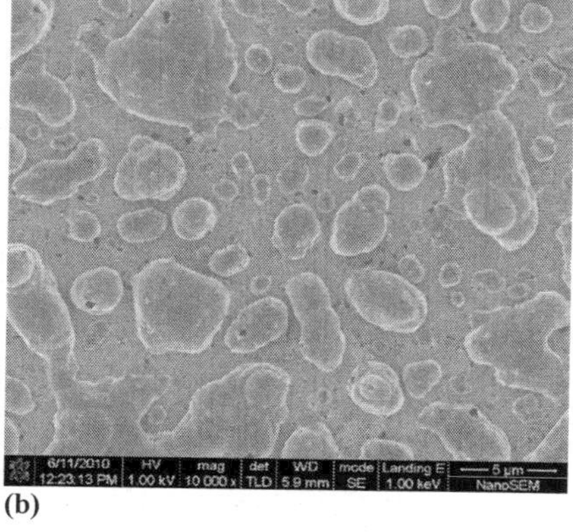

(b)

Figure 2. SEM surface morphology of (a) YBCO tape and (b) YBCO tape with electrodeposited Ag thin film.

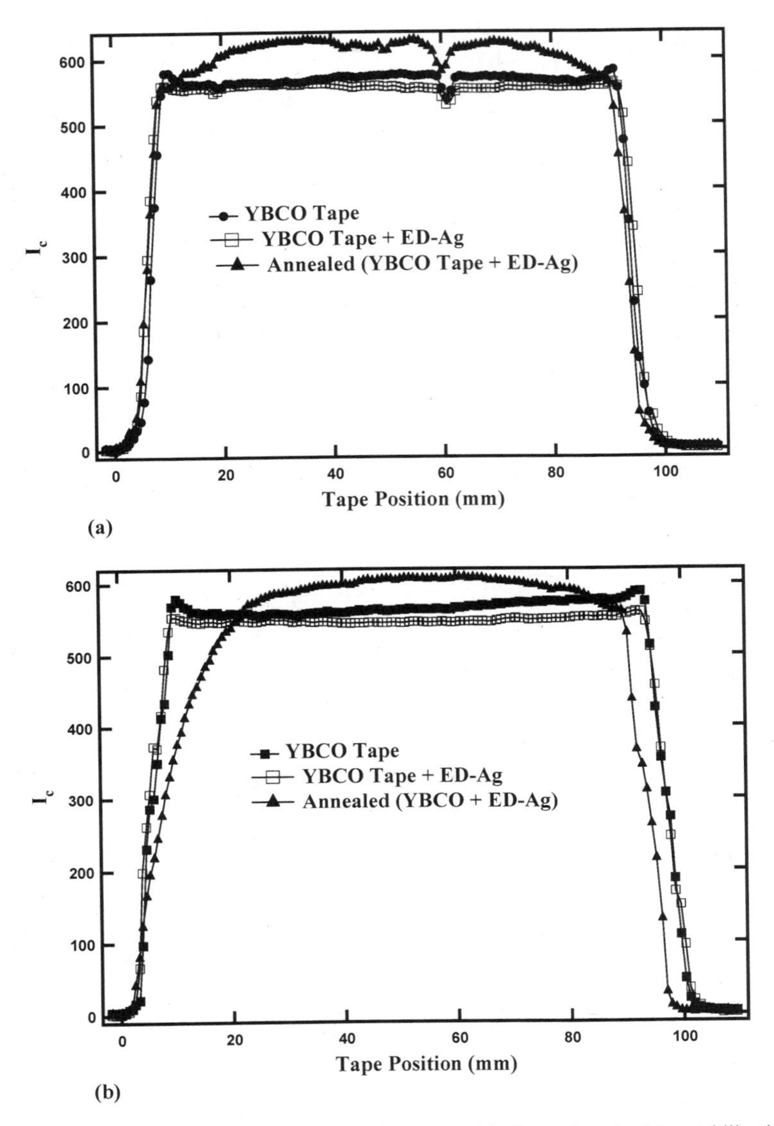

Figure 3(a) and (b). I_c of two 10 cm YBCO tapes with electrodeposited Ag-stabilization layer (before and after annealing). The I_c results are compare with bare YBCO tape

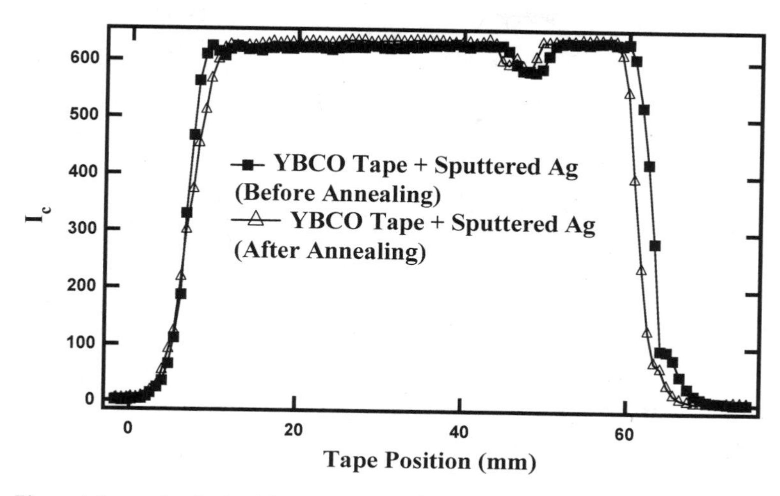

Figure 4. I_c results obtained from YBCO tape with sputtered Ag (before and after annealing).

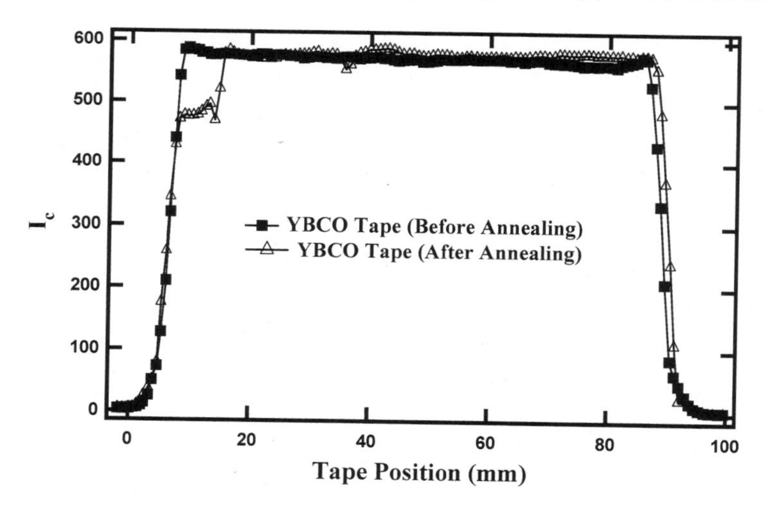

Figure 5. I_c results obtained from the bare YBCO tape (before and after annealing).

THE COMBINED INFLUENCE OF SiC AND RARE-EARTH OXIDES DOPING ON SUPERCONDUCTING PROPERTIES OF MgB_2 WIRES

Hui Fang, Brenden Wiggins, and Gan Liang

Department of Physics, Sam Houston State University, Huntsville, TX 77341-2267

ABSTRACT

Ti-sheathed MgB_2 wires doped with different amount of nanosized rare-earth oxide (Yb_2O_3, Gd_2O_3, and Dy_2O_3) and/or nanosized SiC were investigated. X-ray diffraction patterns suggested the existence of Mg_2Si phase due to the SiC addition, while no any phase related to rare-earth oxide could be detected. Strong enhancement of in-field current carrying capability was observed on 2.5 wt.% Yb_2O_3 doped sample annealed at 800 °C. Dual doping with rare-earth oxide and SiC does not enhance J_C due to the negative effect caused by SiC addition.

INTRODUCTION

Magnesium diboride, MgB_2 superconductor has been considered as a promising candidate for large-scale applications such as superconducting magnets at 20 K due to its relative high critical temperature in comparison with low-temperature superconductors and low raw material and process cost in comparison with high-temperature superconductors [1-4]. Currently, powder-in-tube (PIT) method is widely employed to fabricate long length MgB_2 superconducting wires/tapes for magnet winding [5-7]. In PIT process, reacted (*ex situ*) or non-reacted (*in situ*) Mg and B powders are packed into a metallic tube, followed by cold work, such as swaging, drawing, rolling, to make long length wires/tapes. The chosen metallic sheath has to be low-cost, chemically compatible with MgB_2 compound, and adequate for mechanical strength, which leaves Fe, Ni, Ti, and stainless steel (SS) ideal sheath metals [8].

Compared with the commercially available low-temperature superconductors, the main obstacle of MgB_2 for its practical applications in the presence of magnetic field is its lack of flux pinning [9]. Various approaches including chemical doping and irradiation have been employed to generate flux pinning centers such as nano-scale secondary non-superconducting phase, lattice defects, etc. Among all explored methods, adding nano-scale dopants has been proved a very effective way for introducing pinning centers into MgB_2 [10]. In the past seven years, a great deal of research effort has been made in searching for proper dopants and more than thirty different doping species have been investigated. These dopants include metal elements, borides, nitrides, oxides, carbon, carbides, organic compounds and other materials [11-14]. Currently, the most effective dopants are carbon and carbon compounds.

More recently, nano-scale rare-earth oxides such as Dy_2O_3, Y_2O_3, Eu_2O_3, Ho_2O_3, Pr_6O_{11}, have been emerging as effective dopants to enhance current carrying capability of MgB_2 in the presence of external magnetic field [15-19]. These nanosized rare-earth oxides react with B to form their respective borides without any significant substitution at the Mg/B site [20].

Despite extensive research on searching doping species, little attention has been paid to dual or even triple dopants, which may bring breakthrough towards the possible high field applications of MgB_2. In this research, we investigated the effects on Ti-sheathed MgB_2 wires doped with several nanosized rare-earth oxides Yb_2O_3, Gd_2O_3, and Dy_2O_3. We also reported the effect on Ti-sheathed MgB_2 wires dual doped with nanosized rare-earth oxides and SiC. Our selection of titanium as sheath material was based on the following facts: (1) the performance of Ti-sheathed wires is comparable to or even better than iron-sheathed wires [21]; (2) the magnetic measurement is much easier on Ti-sheathed wire than ferromagnetic metal sheathed wire.

EXPERIMENTAL

The powder-in-tube method was used to fabricate nanosized rear-earth oxides and SiC doped Ti-sheathed MgB_2 wires. The titanium tube used in this study had an outer diameter of 6 mm and a wall thickness of 1 mm. One end of a 10 cm long titanium tube was sealed by crimping. Then desired powder was filled into the tube gradually to ensure the tight pack. The remaining end of the tube was crimped and sealed afterwards. To make the desired powder, commercially available Mg powder (Alfa-Aesar, nominally 99.8% pure, -325 mesh), B powder (Alfa-Aesar, nominally 99.99%, amorphous phase, -325 mesh), Yb_2O_3 powder (MTI, nominally 99.9%, 50 nm), Gd_2O_3 powder (MTI, nominally 99.9%, 20 – 80 nm), Dy_2O_3 powder (MTI, nominally 99.9%, 40 nm), and SiC powder (Nanostructured & Amorphous Materials Inc., nominally 99%, amorphous phase, 15 nm) were stoichiometrically mixed. The powder mixture was then mechanically milled by a Spex-8000 high-energy ball mill for 100 minutes. Tungsten carbide balls and vial were used as milling medium and the mass ratio of ball to powder was 20:1.

The entire filling procedure was carried out in a high purity argon atmosphere. The powder-filled tube was rolled to a wire with a square cross-sectional area of about 1 mm by 1mm. A motorized groove rolling mill with 17 different grooves of sizes from 6 mm to 1 mm was used for the wire rolling. As-rolled wires were cut into 4 inch long segments and heat treated at desired temperatures for 30 minutes. A high purity argon gas flow was maintained throughout the heat treatment process to avoid the oxidation of titanium sheath.

The XRD patterns on the core materials of the wires were obtained by using a Rigaku D/Max Ultima II diffraction machine with Cu K_α radiation. To prepare XRD sample, core material of the wires was removed from the titanium sheath carefully using a scissor and a blade, followed by grinding. The temperature dependent magnetization, *M (T)*, was measured in zero-field-cooled (ZFC) mode by using a Magnetic Property Measurement System (MPMS) from Quantum Design. The applied field for *M (T)* measurement is 20 Oe. The hysteresis loops of magnetization, *M (H)*, up to 5 Tesla were measured using the same MPMS. The cross-sectional area of the MgB_2 wire for MPMS measurement was about 1 mm by 1 mm and the MgB_2 cores had cross-sectional area of about 0.55 mm by 0.55 mm. The magnetic Critical current density of the samples was calculated with the formula $J_c = 20\Delta M/[a(1-a/3b)]$ from Bean critical state model [22], where ΔM is the difference between the upper and lower branches of the hysteresis loops with unit emu/cm^3.

RESULTS AND DISCUSSIONS

MgB_2 wires doped with Yb_2O_3 and SiC

Figure 1 shows the XRD patterns of the core materials of Ti-sheathed MgB_2 wires doped with various amount of Yb_2O_3 and/or SiC and annealed at 800 °C. To our surprise, no any phase related to Yb_2O_3 could be observed on XRD patterns, which we believe is due to very small size or very small doping amount of Yb_2O_3. Meanwhile, the XRD patterns also reveal that with the increase of annealing temperature, the relative amount of MgO increases. For SiC doped wires, three different phases were identified: MgO, Mg_2Si, and the main phase MgB_2 for which the Miller indices are labeled in the figure. The MgO impurity phase was formed due to the trapped oxygen in the raw powder, and the formation of Mg_2Si should be due to the reaction between Mg and SiC. With the increase of SiC amount, Mg_2Si amount increases as well.

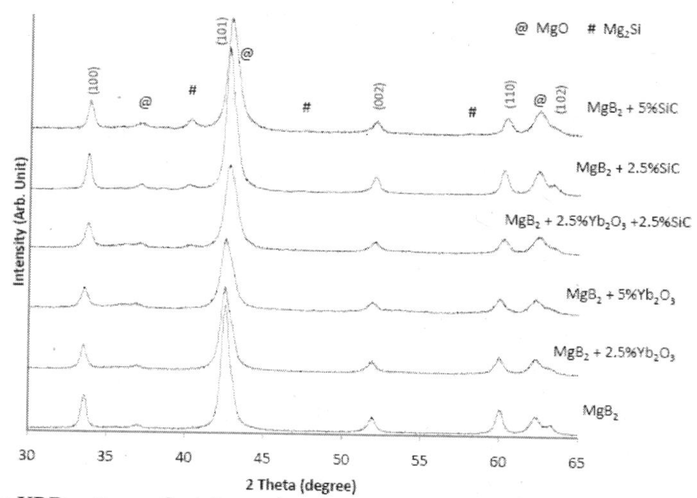

Figure 1. The XRD patterns of pristine and various amount Yb_2O_3 and/or SiC doped MgB_2 wires annealed at 800 °C.

Figure 2 shows the temperature dependent dc magnetization of Ti-sheathed MgB_2 wires doped with various amount of Yb_2O_3 and/or SiC and annealed at 800 °C. Addition of Yb_2O_3 only has slight effect on T_C of MgB_2 wire, while addition of SiC impacts T_C dramatically. 2.5 wt.% SiC doping lowers T_C from 36.2 K to 35.4 K, while 5 wt.% SiC drops T_C to about 33.6 K. This can be explained by the existence of Mg_2Si phase observed on XRD patterns. T_C of co-doped 2.5 wt.% Yb_2O_3 and 2.5 wt.% SiC is 35.2 K, only slightly lower than that of sample doped with SiC alone, which proved small effect on T_C brought by Yb_2O_3 doping.

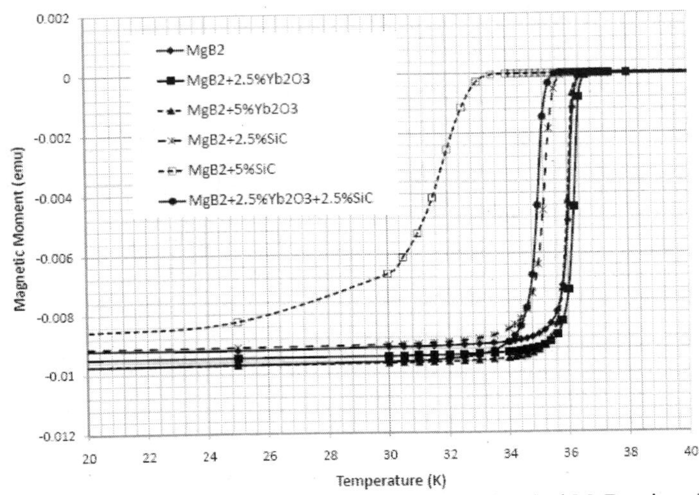

Figure 2. The temperature dependence dc magnetization of Ti-sheathed MgB_2 wires doped with various amount of Yb_2O_3 and/or SiC and annealed at 800 °C.

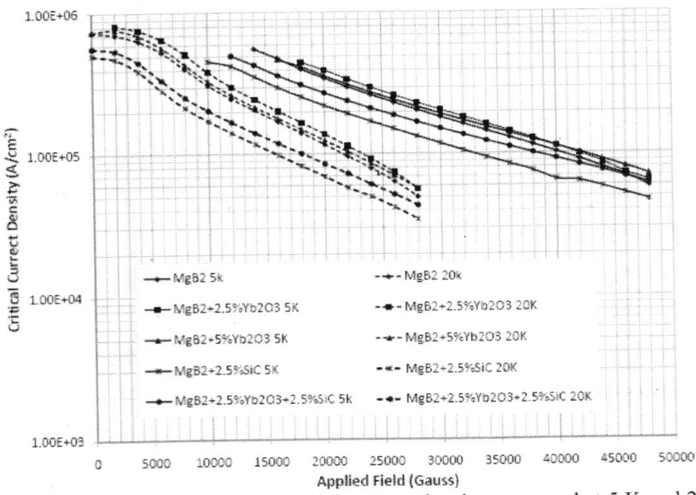

Figure 3. The field dependence magnetic critical current density measured at 5 K and 20 K of Ti-sheathed MgB_2 wires doped with various amount of Yb_2O_3 and/or SiC and annealed at 800 °C.

Figure 3 shows the field dependence of critical current density J_c measured at 5 K and 20 K of

Ti-sheathed MgB_2 wires doped with various amount of Yb_2O_3 and/or SiC and annealed at 800 °C. The addition of 2.5 wt.% Yb_2O_3 enhances the current carrying capability of MgB_2 wires at almost all measured field. When increasing the doping amount of Yb_2O_3 to 5 wt.%, J_C values are slightly higher than that of pristine MgB_2 wires, which we believe is due to the decrease of the net superconducting phase amount per unit volume. The addition of SiC generates the negative effect on J_C, which is the same as our previous results [23]. 5 wt.% of SiC addition causes dramatically drop of J_C, which is not shown in the figure. Due to the negative effect on J_C of SiC addition, the J_C of wires dual doped with 2.5 wt.% Yb_2O_3 and 2.5 wt.% SiC is smaller than that of the pristine MgB_2 wire.

MgB_2 wires doped with Gd_2O_3 and SiC

Figure 4 shows the temperature dependent dc magnetization of Ti-sheathed MgB_2 wires doped with various amount of Gd_2O_3 and/or SiC and annealed at 800 °C. Addition of Gd_2O_3 also only has slight effect on T_C of MgB_2 wire, decreasing T_C of pristine MgB_2 wire from 36.4 K to 36.2 K. There is almost no difference on T_C of 2.5 wt.% Gd_2O_3 doped wire and 5 wt.% Gd_2O_3 doped wire. T_C of co-doped 2.5 wt.% Gd_2O_3 and 2.5 wt.% SiC is 35.4 K.

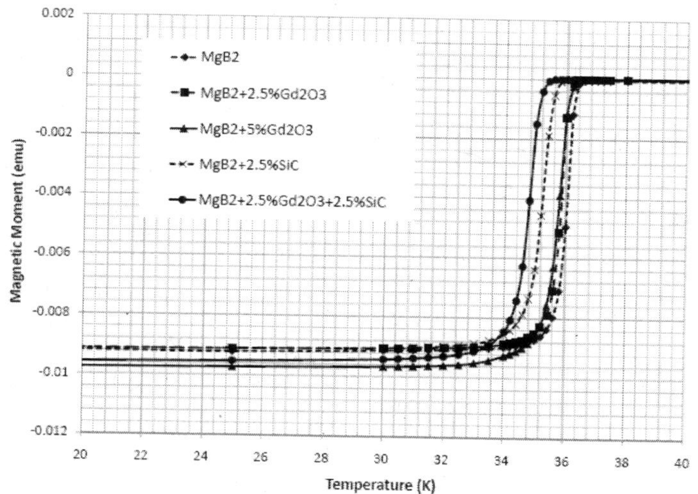

Figure 4. The temperature dependence dc magnetization of Ti-sheathed MgB_2 wires doped with various amount Gd_2O_3 and/or SiC and annealed at 800 °C.

Figure 5 shows the field dependence of critical current density J_c measured at 5 K and 20 K of Ti-sheathed MgB_2 wires doped with various amount of Gd_2O_3 and/or SiC and annealed at 800 °C. The addition of 2.5 wt.% and 5 wt.% Gd_2O_3 shows almost no influence on the current carrying capability of MgB_2 wires at almost all measured field. The negligible difference on J_C of 2.5 wt.% SiC doped wire and 2.5 wt.% SiC and 2.5 wt.% Gd_2O_3 co-doped wire also indicates

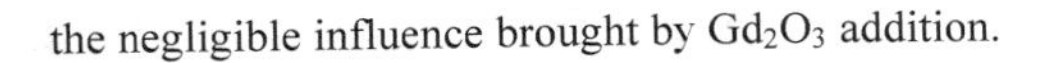

the negligible influence brought by Gd_2O_3 addition.

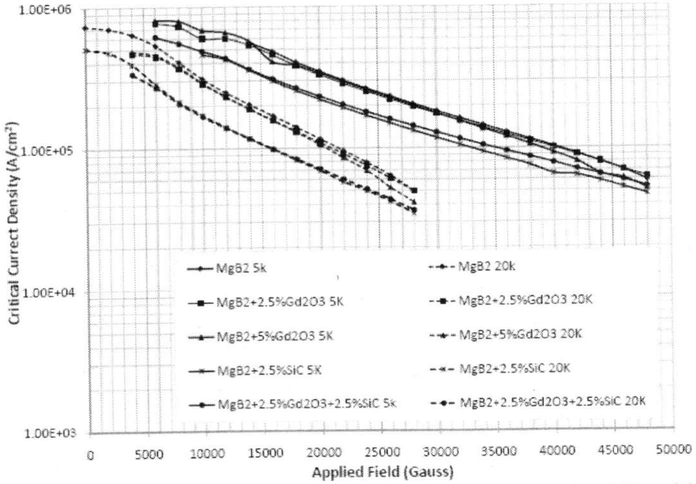

Figure 5. The field dependence magnetic critical current density measured at 5 K and 20 K of Ti-sheathed MgB_2 wires doped with various amount of Gd_2O_3 and/or SiC and annealed at 800 °C.

MgB_2 wires doped with Dy_2O_3 and SiC

Figure 6 shows the temperature dependent dc magnetization of Ti-sheathed MgB_2 wires doped with various amount of Dy_2O_3 and/or SiC and annealed at 800 °C. Addition of Dy_2O_3 decreases the T_C of pristine MgB_2 wire from 36.4 K to 36.0 K, while there is almost no difference on T_C of 2.5 wt.% Dy_2O_3 doped wire and 5 wt.% Dy_2O_3 doped wire. Due to the dramatic decrease of T_C on SiC doping, the T_C of 2.5 wt.% Dy_2O_3 and 2.5 wt.% SiC co-doped wire drops to 34.0 K. Therefore, addition of Dy_2O_3 has negative effect on T_C of MgB_2 wire.

Figure 7 shows the field dependence of critical current density J_c measured at 5 K and 20 K of Ti-sheathed MgB_2 wires doped with various amount of Dy_2O_3 and/or SiC and annealed at 800 °C. The addition of Dy_2O_3 shows negative effect on the current carrying capability of MgB_2 wires at almost all measured field, while with the increase of Dy_2O_3 amount from 2.5 wt.% to 5 wt.%, the J_C value of the wire increase slightly, which is still much lower than that of pristine wire. The addition of Dy_2O_3 shows the comparable negative effect to the addition of SiC on MgB_2 wires. Consequently, the current carrying capability of co-doped wire with 2.5 wt.% Dy_2O_3 and 2.5 wt.% SiC shows the worst value. The different results we obtained in comparison with that reported by other research groups, we believe, is due to the different process we applied and/or different raw materials we used. More systematically work is needed to clarify the effect on superconducting properties of MgB_2 wires doped with nanosized rare-earth oxides.

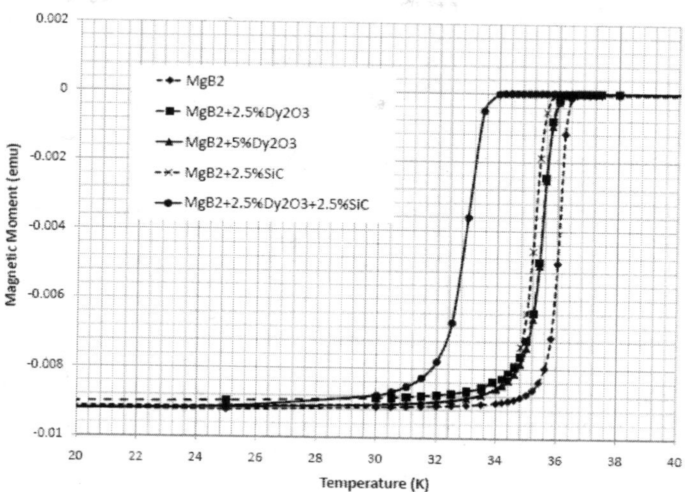

Figure 6. The temperature dependence dc magnetization of Ti-sheathed MgB_2 wires doped with various amount of Dy_2O_3 and/or SiC and annealed at 800 °C.

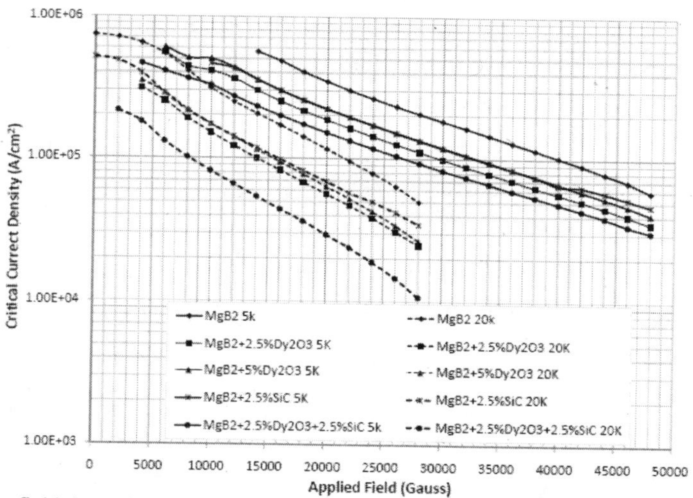

Figure 7. The field dependence magnetic critical current density measured at 5 K and 20 K of Ti-sheathed MgB_2 wires doped with various amount of Dy_2O_3 and/or SiC and annealed at 800 °C.

CONCLUSION

In this research, we studied the nanosized rare-earth oxides and nanosized SiC doping effects on superconducting properties of Ti-sheathed MgB_2 wires. The addition of Yb_2O_3 slightly alters the critical temperature, and enhances the critical current density in the presence of external magnetic field at 2.5 wt.% doping amount. Addition of nanosized Gd_2O_3 shows only negligible influence on superconducting properties of MgB_2 wires, while addition of nanosized Dy_2O_3 provides negative effects on both critical temperature and critical current density. The addition of SiC causes the dramatic decrease of T_C and J_C of MgB_2 wires. Consequently, the T_C and J_C of dual doped MgB_2 wire with rare-earth oxides and SiC are lower than that of non-doped MgB_2 wires.

ACKNOWLEDGEMENT

This work was supported by the National Science Foundation under Grant No. CHE-0718482, an award from Research Corporation for Science Advancement and Enhancement Grants for Research (EGR) of Sam Houston State University.

REFERENCES

[1] J. Nagamatsu, N. Nakagawa, T. Muranaka, Y. Zenitani, and J. Akrimitsu, "Superconductivity at 39 K in magnesium diboride", Nature, vol. 410, 2001, pp. 63-64.

[2] D. C. larbalestier, L. D. Cooley, M. O. Riikel, A. A. Polyanskii, J. Jiang, S. Patnaik, X. Y. Cai, D. M. Feldmann, A. Gurevich, A. A. Squitieri, M. T. Naus, C. B. Eom, E. E. Hellstrom, R. J. Cava, K. A. Regan, N. Rogado, M. A. Hayward, T. He, J. S. Slusky, P. Khalifah, K. Inumaru, and M. Hass, " Strongly linked current flow in polycrystalline forms of the superconductor MgB_2 ," *Nature*, vol. 410, pp. 186–189, Mar. 2001

[3] S. Jin, H. Mavoori, C. Bower, and R. B. van Dover, "High critical currents in iron-clad superconducting MgB_2 wires", Nature, vol. 411, 2001, pp. 563-565.

[4] H. Fang, S. Padmanabhan, Y. X. Zhou, and K. Salama, "High critical current density in iron-clad MgB2 tapes", Appl. Phys. Lett., vol. 82, 2003, pp. 4113-4115.

[5] H. Kumakura, A. Matsumoto, H. fuji, and K.Togano, "High transport critical current density obtained for powder-in-tube processed MgB_2 tapes and wires using stainless steel and Cu-Ni tubes", Appl. Phys. Lett. vol. 79, 2001, pp. 2435-2437.

[6] G. Grasso, A. Malagoli, C. Ferdeghini, S. Roncallo, V. Braccini, and A. S. Siri, " Large transport currents in unsintered MgB_2 superconducting tapes", Appl. Phys. Lett., vol. 79, 2001, pp. 230-232.

[7] H. Suo, C. Beneduce, M. Dhalle, N. Musolino, J. Y. Genoud, and R. Flukiger, "Large transport critical currents in dense Fe- and Ni-clad mgB_2 superconducting tapes", Appl. Phys. Lett., vol. 79, 2001, pp. 3116-3118..

[8] A.V. Pan, S. Zhou, and S.X. Dou, "Iron-sheath influence on the superconductivity of MgB_2 core in wires and tapes," Superconductor Science and Technology **17**, S410 (2004).

[9] K. Vinod, R.G. Abhilash Kumar, and U. Syamaprasad, "Prospects for MgB_2 superconductors for magnet application," Superconductor Science and Technology **20**, pp. R1-R13 (2007).

[10] S. X. Dou, A. V. Pan, S. Zhou, M. Ionescu, H. K. Liu, and P. R. Munroe, "Substitution-induced pinning in MgB_2 superconductor doped with SiC nano-particles", Supercond. Sci. Technol., vol. 15, 2002, pp. 1587-1591.

[11] Y. Yamada, M. Nakatsuka, Y. Kato, K. Tachikawa, and H. Kumakura, "Superconducting properties of in-situ PIT MgB_2 tapes", Advances in Cryogenic Engineering, vol 52, 2006, pp. 631.

[12] S. X. Dou, S. Soltanian, J. Horvat, X. L. Wang, S. H. zhou, M. Lonescu, H. K. Liu, P. Munroe, and M. Tomsic, "Enhancement of the critical current density and flux pinning of MgB_2 superconductor by nanoparticle SiC doping", Appl. Phys. Lett., vol. 81, 2002, pp. 3419-3421.

[13] W. Gruner, M. Herrmann, A. Nilsson, H Herrmann, W. HaBler, and B. Holzapfer, "Reactive nanostructured carbon as an effective doping agent for MgB_2", Superconductor Science and Technology, 20 (2007) 601.

[14] R. Zeng, L. Lu, and S. X. Dou, "Significant enhancement of the superconducting properties of MgB_2 by polyvinyl alcohol additives", Supercond. Sci. Technol. vol. 21, 2008, pp. 085003.

[15] N. Ojha, V.K. Malik, C. Bernhard, and G.D. Varma, "The effect of Pr_6O_{11} doping on superconducting properties of MgB_2," Phys. Status Solidi A 207, No. 1, pp 175-182 (2010).

[16] C. Cheng, and Y. Zhao, "Enhancement of critical current density of MgB_2 by doping Ho_2O_3," Applied Physics Letters 89, 252501 (2006).

[17] J. Wang, Y. Bugoslavsky, A. Berennov, L. Cowey, A.D. Caplin, L.F.Cohen, J.L. MacManus Driscoll, L.D. Cooley, X. Song, and D.C. Larbalestier, "High critical current density and improved irreversibility field in bulk MgB_2 made by a scaleable, nanoparticle addition route," Applied Physics Letters 81, pp 2026-2028, (2002).

[18] S.K. Chen, M. Wei, and J.L. MacManus-Driscoll, "Strong pinning enhancement in MgB_2 using very small Dy_2O_3 additions," Applied Physics Letters 88, 192512 (2006).

[19] N. Ojha, G.D. Varma, H.K. Singh, and V.P.S. Awana, "Effect of rare-earth doping on the superconducting properties of MgB_2," Journal of Applied Physics 105, 07E315 (2009).

[20] N. Varghese, K. Vinod, M.K. Chattopadhyay, S.B. Roy, and U.Syamaprasad, "Effect of combined addition of nano-SiC and nano-Ho_2O_3 on the in-field critical current density of MgB_2 superconductor," Journal of Applied Physics 107, 013907 (2010).

[21] G. Liang, H. Fang, M. Hanna, F. Yen, B. Lv, M. Alessandrini, S. Keith, C. Hoyt, Z. Tang, and K. Salama, "Development of Ti-sheathed MgB_2 wires with high critical current density", Supercond. Sci. Technol. vol. 19, 2006, pp. 1146-1151.

[22] Bean, C. P., "Magnetization of high-field superconductors," Reviews of Modern Physics **36**, pp. 31-39 (1964).

[23] G. Liang, H. Fang, Z.P. Luo, C. Hoyt, F. Yen, S. Guchhait, B. Lv, and J.T. Markert, "Negative effects of crystalline-SiC doping on the critical current density in Ti-sheathed $MgB_2(SiC)_y$ superconducting wires," Superconductor Science and Technology 20, pp 697-703 (2007).

FABRICATION OF GdBCO COATED CONDUCTORS ON CLAD-TYPE TEXTURED METAL SUBSTRATES FOR HTS CABLES

Kazuya Ohmatsu, Kenji Abiru, Yoshihiro Honda, Yuki Shingai, Masaya Konishi
Sumitomo Electric Industries, Ltd.
1-1-3, Shimaya, Konohana-ku, Osaka, 554-0024 Japan

ABSTRACT

We have been fabricating $Gd_1Ba_2Cu_3O_x$ (GdBCO) coated conductors on clad-type textured metal substrates by using PLD for HTS power cables. In this project, high Ic and homogeneous characteristics of GdBCO tapes are required for fabricating a 66 kV-5 kA class, 3-in-One HTS model cable system. In order to produce this HTS cable, it is important for the conductor to establish stable and high-throughput manufacturing process. Therefore, we have newly installed a 300 W laser equipment and introduced electron beam (EB) evaporation method for buffer layer to optimize surface morphology. As a result, the maximum I_c for a short sample achieved 500 A/cm level at 77 K. In combination with this improved process, long length GdBCO tapes with high I_c properties have been successfully fabricated.

INTRODUCTION

R & D efforts of HTS conductors have been made for several applications such as power cables, high field magnets, transformers, etc. In Japan, a national project to develop materials & power application of coated conductors has been started since 2008 [1], and we have been fabricating PLD-GdBCO tapes for developing a 15 m long, 5 kA, 66 kV class 3-in-One HTS model cable system as a part of this project [2, 3]. In order to construct this HTS cable, GdBCO tapes with high I_c and stable production process are required. We applied a wide tape approach using 30 mm width substrates to obtain the large production throughput, and successfully obtained uniform superconducting characteristics across the tape width [4, 5]. Our coated conductors are composed of a clad-type textured metal substrate, a multi-buffer layer of $CeO_2/YSZ/CeO_2$, GdBCO superconducting layer, Ag protection layer, and Cu stabilizing layer. The clad-type metal substrate, which consists of a thin bi-axially textured Ni layer on a non-crystalline stainless tape, has less magnetism and higher mechanical strength than typical Ni-alloy textured substrates. As I_c performances of 200 ~ 300 A/cm for 100 m length are confirmed, GdBCO tapes with the clad-type textured metal substrate become suitable for AC applications, especially for HTS cable with its low AC losses.

However, long GdBCO tapes with high and uniform I_c properties such as 400 A/cm are necessary to construct the 66 kV-5 kA HTS cable. The deposition process of buffer and superconducting layers on the clad-type substrate are key technologies for this purpose. Therefore, we have newly installed a 300 W high power excimer laser for the PLD equipment to obtain thick GdBCO layer. In addition, we have improved surface conditions of the buffer layers to enhance the critical current density (J_c) by using electron beam (EB) evaporation method.

In this paper, we investigated the effect of laser power on I_c values and stable manufacturing process of GdBCO. The I_c values of the GdBCO layers deposited by the 300 W laser and a previous 200 W laser were discussed. In addition, the influences of surface morphologies and cube textured characteristics of buffer layers were also discussed in combination with I_c variations along the long length tapes.

EXPERIMENTAL

The clad-type textured metal substrates were prepared by using the conventional cold-rolling and heat treatment processes, in which a clad technique with a thin bi-axially textured layer on a thick non-crystalline tape was included. The typical value of in-plain alignment, a full width at half maximum (FWHM) of XRD scan for Ni(100), were 4.6 degree. The multi-buffer layers of CeO_2/YSZ/CeO_2 were deposited on 30 mm wide clad-type textured metal substrate by a reel to reel RF sputtering or EB evaporation method. The first CeO_2 layer deposited on the substrate plays the role of a seed layer. The second YSZ layer prevents Ni inter-diffusion and reaction to superconducting layer. The third CeO_2 layer acts as the relaxation of lattice-mismatch between the YSZ and the superconducting layer. The buffer layer showed around 6 degree of in-plain alignment for CeO_2(111) pole across the 30 mm tape width. In addition, we optimized the EB evaporation process for the seed layer. The surface morphologies of buffer layers were observed by a scanning electron microscope (SEM).

The GdBCO layer was fabricated on the CeO_2/YSZ/CeO_2 buffer layers by a reel to reel PLD system with a 300 W Kr-F excimer laser. The pulse energy and the repetition rate were 1 J and 300 Hz, respectively. The deposition was repeated for several times with tape moving speed of 24 m/h under 10 Pa of the O_2 partial pressure. In this study, we compared the I_c properties of the GdBCO layers deposited by the 300 W laser and a previous 200 W laser. After the deposition of GdBCO, Ag was over-coated by DC sputtering as a protection layer and Cu was deposited by the conventional electro-plating method. Furthermore, we investigated the relationship between I_c properties and cube textured ratio of seed layer. Table 1 summarizes the architecture of GdBCO coated conductor on clad-type textured metal substrate.

Table 1. Architecture of GdBCO coated conductors on clad-type textured metal substrates

Function	Process	Material	Thickness (micron)
Textured substrate	Clad-type metal	Ni/SUS	120
Buffer layers			
Seed layer	RF sputtering, EB	CeO_2	0.1
Barrier layer	RF sputtering,	YSZ	0.2
Cap layer	RF sputtering,	CeO_2	0.1
Superconducting layer	PLD	GdBCO	2~3
Protection layer	DC sputtering	Ag	2~8
Stabilizer	Electro-plating	Cu	20

RESULTS AND DISCUSSION

Improvement of surface morphology for buffer layers

Fig. 1 (a) shows a SEM image of a surface morphology of the 140 nm CeO_2 seed layer deposited by the RF sputtering method. A number of small cracks with the sub-micron scale were observed on the seed layer. This surface condition was considered to degrade the flatness of cap layer and crystal orientation as shown in Fig. 1 (b). These small cracks could not be disappeared by decreasing the thickness of CeO_2 below 100 nm. Therefore, we have used the EB evaporation to make a crack-free seed layer. Although the process window to obtain the cube textured seed layer was very narrow, we found the growth mechanism of the seed layer on the clad-type textured metal substrate.

Fig. 2 (a) shows a SEM surface image of seed layer deposited by the EB method. After optimizing the deposition process, the surface became significantly flat. Moreover, numerous precipitates or a grain boundary on the cap layer were almost disappeared as shown in Fig 2 (b). The crystallinity of the third CeO_2 layer, cube textured ratio defined by the orientation ratio of $I(200)/[I(200)+I(111)]$, was analyzed by XRD. The value showed highly uniform crystallinity over 95 %.

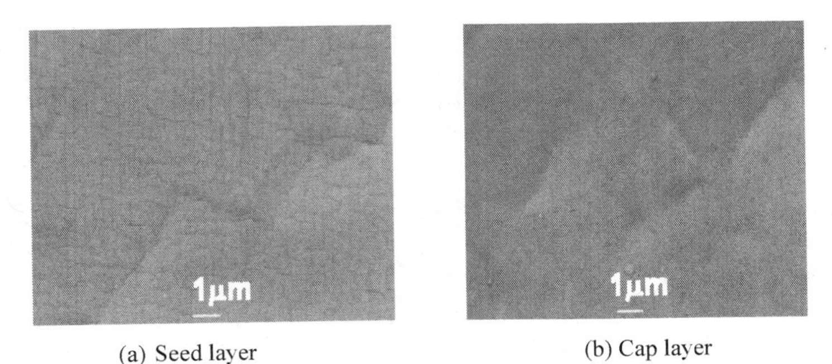

(a) Seed layer (b) Cap layer

Fig. 1 SEM image of seed and cap layers; both layers were deposited by RF sputtering

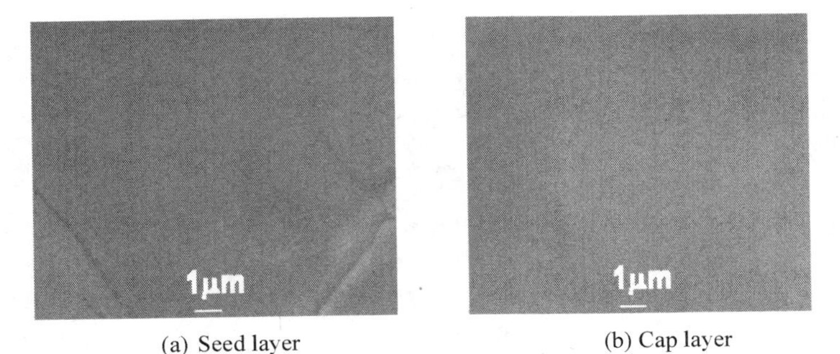

(a) Seed layer (b) Cap layer

Fig. 2 SEM image of seed and cap layers; seed layer was deposited by EB evaporation

Enhancement of I_c properties of GdBCO layers

We first observed the effect of laser power on I_c properties of GdBCO layers. Fig. 3 shows the I_c distributions of GdBCO layer across the 30 mm wide tape, which were deposited by the previous 200 W laser and the new 300 W laser. The frequencies of each laser were 180 Hz and 300 Hz, respectively. The total deposition time was fixed to be equal between the samples.

In Fig. 3, the I_c distribution of the GdBCO layer grown by the new 300 W laser was about 1.5 times larger than that of the layer deposited by the previous 200 W laser. This increased I_c was simply caused by the effect of thick GdBCO with 2.8 micron, which was 1.3 times thicker than the previous GdBCO layer with 2.1 micron. In Fig. 4, the 300 W laser obviously enlarged the plasma plume on GdBCO target. In our PLD equipment, the plasma plume has been optimized to cover the 30 mm wide substrates completely in combination with the line beam technique. Thus, the production rate and yield with high I_c value for GdBCO layer was significantly improved by using the new 300 W laser.

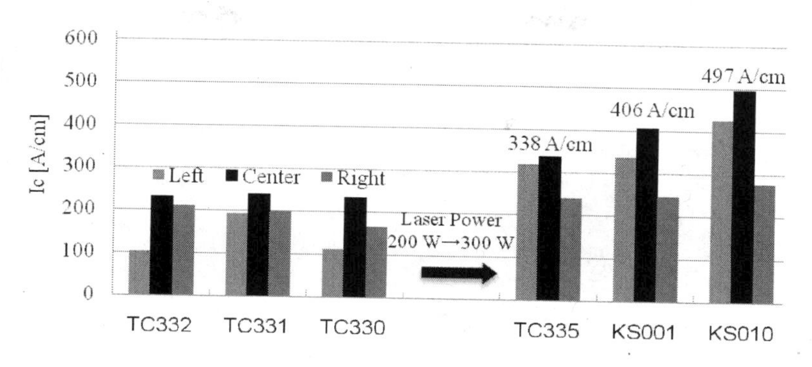

Fig. 3 I_c distributions of GdBCO layer across the 30 mm wide tape. TC330~332 were made by previous 200 W laser and TC335, KS001, KS010 were made by new 300 W laser.

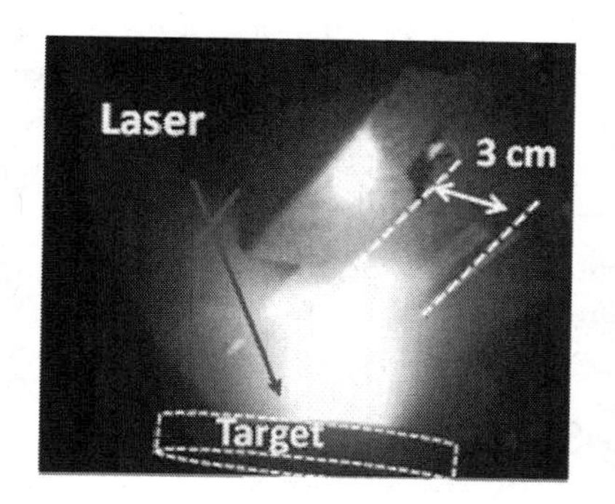

(a) Plume by previous 200 W laser

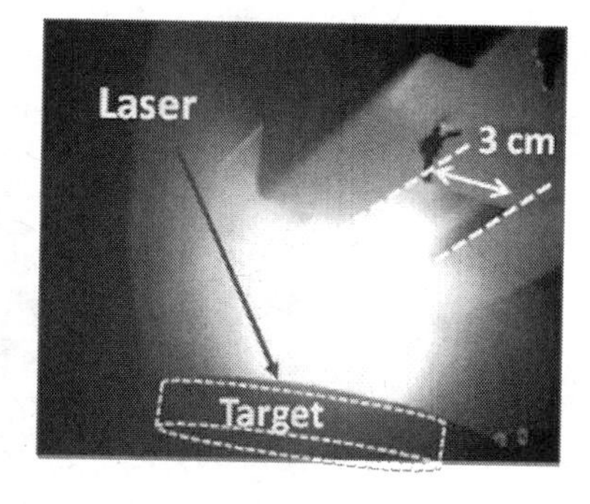

(b) Plume by newly installed 300 W laser

Fig. 4 Plasma plume on GdBCO target by 200 W and 300 W lasers during the deposition

The I_c property of 2.1 micron GdBCO films (TC330 ~ TC332) tend to be saturated to 250 A/cm. Although the I_c values 2.8 micron GdBCO films (TC335, KS001) were improved to 350 ~ 400 A/cm after the installation of the 300 W laser, the I_c value tend to be saturated. Fig 3 also indicates the I_c value of the 2.8 micron GdBCO tape (KS010) goes up to 500A/cm level fabricated on the buffer layer by EB evaporation. Therefore, I_c property has been greatly improved by applying thick GdBCO layers on the improved crack free buffer layer. The liner relationship between I_c values and GdBCO thickness up to 1.5 micron was also observed for this tape (KS010) ,and high J_c ~ 3 MA/cm^2 was confirmed within this area. Above 1.5 micron, J_c was degraded 2 MA/cm^2 level and I_c becomes close to 500A/cm at 3 micron. This phenomenon seems to be the growth of a-axis grain among the c-axis grains. To obtain the higher J_c in thick GdBCO layer over 2 micron is future challenge in our PLD process.

Long length performances of GdBCO conductors

By using the improved manufacturing processes of buffer layer and superconducting layer, we have routinely fabricated high Ic exceeding 400 A/cm, 30 mm wide, 100 m long GdBCO tapes. A 30 mm wide tape deposited by Ag protection layer was cut into 6 tapes in 4 mm width by a continuous slitter machine. After the each 4 mm tapes were plated electrically with 20 micron Cu stabilizer, I_c was measured for every 1.5 m by adopting a continuous four-probe technique.

Fig. 5 shows a longitudinal I_c distribution of 4 mm wide GdBCO tapes fabricated as a first trial. The minimum and maximum I_c were 178 A (445 A/cm) and 190 A (475 A/cm), respectively. The I_c property consistently increases from start to end. We found that this Ic distribution was strongly influenced by the cube textured ratio of buffer layer and that it could be solved by optimizing the deposition temperature of RF sputtering and EB evaporation. Based on these results, we started long length production of GdBCO tapes with high I_c over 400 A/cm for aiming at the construction of the 66 kV-5 kA class, 3-in-One HTS model cable in next year.

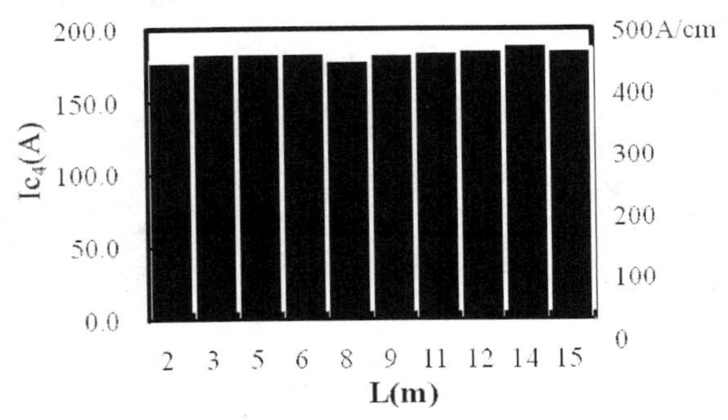

Fig.5 Longitudinal I_c distribution of a 4 mm wide GdBCO tape

CONCLUSION

The manufacturing process of GdBCO tapes on the clad-type textured metal substrates were successfully improved by the newly installed 300 W excimer laser for the PLD process and EB evaporation method for the buffer layers. The maximum I_c has achieved 500 A/cm level (J_c=1.8 MA/cm^2) at 77 K under the self-field, and the production process in 30 mm width has constructed. We have also demonstrated a 15 m long 4 mm wide GdBCO tapes with maximum I_c of 190 A. The improvement of the stable manufacturing process for long length GdBCO tapes with high I_c over 400

A/cm has been in progress to construct the 66 kV-5 kA class, 3-in-One HTS model cable system.

ACKNOWLEDGMENT

This work was supported in part by the New Energy and Industrial Technology Development Organization (NEDO) as the Project for Development of Materials & Power Application of Coated Conductors, M-PACC.

REFERENCES

[1] Y. Shiohara, N. Fujiwara, H. Hayashi, S. Nagaya, T. Izumi, M. Yoshizumi, "Japanese efforts on coated conductor processing and its power applications: New 5 year project for materials and power applications of coated conductors (M-PACC)," Physica C, vol. 469, 15 Oct. 2009, pp. 863-867.

[2] T. Masuda, H. Yumura, M. Watanabe, "Recent progress of HTS cable project", Physica C, vol. 468, 15 Sep. 2008, pp. 2014-2017.

[3] T. Minamino, M. Ohya, H. Yumura, T. Masuda, T. Nagaishi, Y. Shingai, X. Wang, H. Ueda, A. Ishiyama, N. Fujiwara, "Design and evaluation of 66 kV class RE-123 superconducting cable", Physica C, to be published.

[4] T. Nagaishi, Y. Shingai, M. Konishi, T. Taneda, H. Ota, G. Honda, T. Kato, K. Ohmatsu, "Development of REBCO coated conductors on textured metallic substrates", Physica C, vol. 469, 15 Oct. 2009, pp. 1311-1315.

[5] T. Yamaguchi, H. Ota, K. Ohki, M. Konishi, K. Ohmatsu, "Development of buffer layers on 30 mm wide textured metal substrates for REBCO coated conductors", Physica C, to be published.

CHARACTERISTICS OF SUPERCONDUCTING YBCO PHASE FORMATION THROUGH AUTO COMBUSTION CITRATE-NITRATE SOL-GEL

Hassan Sheikh[a,*], Ebrahim Paimozd[a], Ali Sharbati[b], Sedigheh Sheikh[c]
[a]Materials Engineering Department, Malek Ashtar University of Technology, Shahin Shahr, Iran, P. O. Box 83145/115
[b] Electroceramics Research Center, Malek Ashtar University of Technology, Shahin Shahr, Isfahan, Iran
[c] Department of Physics, Isfahan University of Technology, Isfahan 84156-83111, Iran

ABSTRACT

Fine $YBa_2Cu_3O_{7-\delta}$ (YBCO) powder was synthesized through a sol-gel auto-combustion technique, in which nitrates were used as an oxidizing agent, and citric acid as an organic fuel/chelating agent. X-ray diffraction (XRD), scanning electron microscopy (SEM), energy dispersive spectrometer (EDS), Fourier transform infrared spectroscopy (FTIR), and thermogravimetric and differential thermogravimetric analyses (TGA/DTG) were employed to characterize the superconducting phase formation. The results show that the formation of $YBa_2Cu_3O_{7-\delta}$ phase is completed at 750-770 °C. Also the SEM photograph at a high magnification illustrates that the preferred mechanism of crystal growth is a spiral type.

1. INTRODUCTION

The ceramic of $YBa_2Cu_3O_{7-\delta}$ has many applications in scientific and engineering fields which are increasing [1,2]. The conventional method of preparing $YBa_2Cu_3O_{7-\delta}$ powder is the solid-state reaction of oxides and carbonates. Despite of simplicity, this process is time consuming and energy-intensive [3, 4]. Recently, a variety of wet-chemical synthesis methods such as coprecipitation [5] and sol-gel [6] techniques have also been reported to be effective in generating fine and homogenous powders of ceramic oxides. Relatively complex schedules and low production rates are the common problems of wet-chemical methods. Fortunately the drawbacks of these methods could be partially eliminated by the combustion synthesis method. Several researchers have utilized a sol-gel auto-combustion technique to synthesize many inorganic materials such as ferrites [7, 8], piezoelectric [9] and garnet [10] powders, many of which are technologically important. This is a novel and unique technique which combines the chemical sol-gel process and combustion synthesis process. The advantages of this technique are simple preparation, high energy efficiency, low processing cost, a resulting submicron particle size with high homogeneity at atomic level, and possibility to reduce the calcination temperature to limit thermal agglomeration of the particles. It involves a highly exothermic reaction between an oxidizer (such as a mixture of metal nitrates) and a fuel (such as urea, glycine, and citric acid) which causes spontaneous combustion. The auto-combustion of citrate-nitrate gel forms as-burnt powder (ash) converted to desired powder during a calcination process. This technique produces fine and homogenous powders after calcination of the burnt ash at appropriate temperatures [11, 12]. It is especially interesting that the obtained fine $YBa_2Cu_3O_{7-\delta}$ powder from this technique can be used for fabrication of the second generating high-T_c superconducting tapes using rolling process, which makes these fine grains connect to each other directly forming a melting texture [13]. Therefore, this production could induce a revolution in the production of superconducting tapes and take great effect on the extensive application of the high temperature superconducting materials [12].

* **Corresponding author**: Tel:+98-9133358098; Fax:+98-3125228530
Email address: sheikh_scientific@yahoo.com

The kind of synthesis process has an effect on the structural aspects and the desirable phase formation. In previous works, the formation of the superconducting phase using citrate-nitrate sol-gel technique has not been investigated in detail. Therefore, the study of characteristics of this technique for synthesizing the$YBa_2Cu_3O_{7-\delta}$ g phase is the prime objective of the present work.

2. EXPERIMENTAL PROCEDURES

The synthesis of $YBa_2Cu_3O_{7-\delta}$ superconductor was carried out according to the procedure shown in Fig. 1.Yittrium nitrate (Y $(NO_3)_2.5H_2O$), barium nitrate (Ba $(NO_3)_2$), copper nitrate (Cu $(NO_3)_2.3H_2O$), and citric acid ($C_6H_8O_7.H_2O$) were used as raw materials. An aqueous solution was prepared by dissolving of metal nitrates in distilled water at the cationic mole ratio of Y:Ba:Cu=1:2:3 at temperature of 50 °C for 1 h under continuous stirring using magnetic agitator. An appropriate amount of citric acid was added to the as-prepared metal nitrates solution with stirring. The molar ratio of metal nitrates to citric acid was 1. After 2h, small amount of ammonia was added to adjust the pH value to about 7. The neutralized dark-blue solution was stirred continuously at 80 °C to transform it into a gel. The dried gel was obtained by heating the gel at 100 °C and then was fired at temperature of 250 °C. Finally, the as-burnt powder was calcined at 890 °C for 12 h in air.

TGA and DTG were used to study the decomposition behavior of the as-burnt gel, which was examined with the heating rate of 10 °C/min on the Pyris Diamond SII instrument. The phase identification of the as-burnt and calcined powders was performed using SIFERT 2000 (3003 TT) X-ray diffractometer with CuKα radiation (2θ range of 10-90°).The morphology and the mechanism of crystal growth was investigated by SEM. The ions concentration of the obtained powder was characterized by EDS (XL-SERIES). FTIR spectra for the gel precursor, the as-burnt powder, and the calcined powder were recorded on a Perkin Eaelmer (Spectrum One) spectrometer in the range of 450-4000 cm^{-1}.

3. RESULTS AND DISCUSSION

The clear dried gel prepared in the present study is a result of citric acid and pH role in citrate-nitrate sol-gel technique. Citric acid (≡H_3Cit) is a tribasic acid that can be dissociated in the aqueous solution to give H_2Cit^-, $HCit^{2-}$, and Cit^{3-} ions depending upon the pH of the solution. At low pH, barium nitrate has relatively low solubility which can be avoided by increasing the pH of the initial citrate-nitrate solution up to value close to 7 by addition of ammonia. At this pH, Cit^{3-} ions are the predominant species that interact more strongly with barium ions to form a stable $BaCit^-$ complex than H_2Cit^- ions. The high stabilization of barium ions with Cit^{2-} ions reduces the concentration of free barium ions in the solution, thus being able to lead to the formation of a clear gel with no precipitates upon the evaporation [13, 14].

FTIR spectra of the dried gel, as-burnt and calcined powders were examined to investigate the chemical and structural changes that take place during the combustion and calcination processes. Fig. 2 shows the FTIR spectra in the range of 450-4000 cm^{-1}. It is clearly seen from the figure that the dried gel has several absorption bands at about 3150, 1580 and 1385 cm^{-1} corresponding to the OH group, carboxyl group and NO_3^-, respectively. The appearance of the characteristic bands of NO_3^- indicates that the NO_3^- ion exists as a group in the structure of the citrate gel during the gelation of the mixed solution formed from nitrates and citric acid (Fig. 2a). The consecutively peaks in the ranges 3100-4000 cm^{-1} and 2850-3100 cm^{-1} can be attributed to OH groups and C-H and/or CH_2 stretching vibrations, respectively. The weak peaks observed in the range of 2200-2400 cm^{-1} are assigned to CO and N_2O bands. The O-C=O and C-O-H/C-O-C bands have absorption peaks in the range of 1000-1200 cm^{-1}.Finally, the NO_3^- and –CO-OH groups are the responsible for the absorption peaks observed in the range of 825-950 cm^{-1}.

As can be seen in Fig. 2b for the as-burnt powder, two characteristic bands arising from ionic carbonate (CO_3^-) are shown in the spectrum near 858 and 1427 cm^{-1}. The peaks related to the bonding of metal and oxygen appear at 693 and 611 cm^{-1} .These peaks imply that the as-burnt powder may contain metal oxides and barium carbonate. This will be verified by XRD analysis. As mentioned for the dried gel, the absorption in the range of 3100-4000 cm^{-1} is related to O-H vibration bands. Two weak peaks of CO_2 and CO appear at 2447 and 2171 cm^{-1}, respectively. The CH_3 or CH_2 group has an absorption peak in the range of 2800-3000 cm^{-1}. The peak at 1620cm^{-1} can be assigned to COO^- group. The disappearance of the characteristic bands of carboxyl group and a large part of $NO3^-$ ion on the spectrum of the as-burnt powder reveals that the organic groups and NO_3^- ion take part in the reaction during combustion. The auto-ignition of the gel occurs by a thermally induced oxidation-reduction between the citrate and nitrate ions, which can be written according to Pederson's reaction model [15]:

$$9M(NO_3)_X - 2.5x(C_6H_8O_7) - 15xCO_2 - (9/2)xN_2 - 10xH_2O - 9MO_{X/2} \quad (1)$$

Wherein the citrate ion and the NO_3^- ion act as a reductant agent and an oxidant agent, respectively. Since NO_3^- ion provides an in situ oxidizing environment for the decomposition of the organic components, the rate of the oxidation reaction relatively increases, resulting in a self-propagating combustion of the dried gel. The Barium oxide reacts with CO_2 gas as follows:

$$BaO + CO_2 \rightarrow BaCO_3 \quad (2)$$

Also, Fig. 2b shows the presence of nitro group at 1620 cm^{-1} demonstrating the small value of this group after the combustion reaction.
FTIR spectrum of the calcined powder has been shown in Fig. 2c. According to this figure, it is noteworthy that the absorption peaks of CO_3^{2-} have been deleted. Only a band at 1377 cm^{-1} is the characteristic of the stretching vibration of the CO_3^{2-} . Therefore, the calcination at 890 °C leads to a pure orthorhombic phase. This Result agrees well with XRD analysis of the calcined powder.

The XRD patterns of the as-burnt and calcined powders have been shown in Fig. 3. The characteristics peaks of Y_2O_3, $BaCO_3$, CuO, and Ba $(NO_3)_2$ are detected in the XRD pattern of the as-burnt powder. This reveals that the dried gel has been converted to metal oxides, barium carbonate and amorphous phases after the combustion reactions. The XRD pattern of the calcined powder shows the formation of $YBa_2Cu_3O_{7-\delta}$ phase. The superconducting phase has an orthorhombic provskite-like structure.

TGA/DTG results for the as-burnt powder are given in Fig. 4. The first weight loss is seen in the temperature range of 230-300 °C. It can be attributed to a combustion reaction among the residual unreacted materials of firing stage. The increase in the weight at 300-340 °C is originated from the reaction between CO_2 and BaO according to reaction 4.The residual of barium nitrate decomposes to BaO and NO_2 at 590 °C. The increase in the weight at 650 °C is due to the formation of oxides. The superconducting phase formation carries out at 750 °C which can be seen from a peak at the same temperature in the DTG curve. The reaction occurred at this temperature can be presented in the following state:

$$Y_2O_3 + 4BaCO_3 + CuO - 2YBa_2Cu_3O_{7-\delta} + 4CO \quad (3)$$

As can be seen in Fig. 2b for the as-burnt powder, two characteristic bands arising from ionic carbonate (CO_3^-) are shown in the spectrum near 858 and 1427 cm^{-1}. The peaks related to the bonding of metal and oxygen appear at 693 and 611 cm^{-1} .These peaks imply that the as-burnt powder may contain metal oxides and barium carbonate. This will be verified by XRD analysis. As mentioned for the dried gel, the absorption in the range of 3100-4000 cm^{-1} is related to O-H vibration bands. Two weak peaks of CO_2 and CO appear at 2447 and 2171 cm^{-1}, respectively. The CH_3 or CH_2 group has an absorption peak in the range of 2800-3000 cm^{-1}. The peak at 1620cm^{-1} can be assigned to COO^- group. The disappearance of the characteristic bands of carboxyl group and a large part of $NO3^-$ ion on the spectrum of the as-burnt powder reveals that the organic groups and NO_3^- ion take part in the reaction during combustion. The auto-ignition of the gel occurs by a thermally induced oxidation-reduction between the citrate and nitrate ions, which can be written according to Pederson's reaction model [15]:

$$9M(NO_3)_X - 2.5x(C_6H_8O_7) - 15xCO_2 - (9/2)xN_2 - 10xH_2O - 9MO_{X/2} \tag{1}$$

Wherein the citrate ion and the NO_3^- ion act as a reductant agent and an oxidant agent, respectively. Since NO_3^- ion provides an in situ oxidizing environment for the decomposition of the organic components, the rate of the oxidation reaction relatively increases, resulting in a self-propagating combustion of the dried gel. The Barium oxide reacts with CO_2 gas as follows:

$$BaO + CO_2 \rightarrow BaCO_3 \tag{2}$$

Also, Fig. 2b shows the presence of nitro group at 1620 cm^{-1} demonstrating the small value of this group after the combustion reaction.
FTIR spectrum of the calcined powder has been shown in Fig. 2c. According to this figure, it is noteworthy that the absorption peaks of CO_3^{2-} have been deleted. Only a band at 1377 cm^{-1} is the characteristic of the stretching vibration of the CO_3^{2-} . Therefore, the calcination at 890 °C leads to a pure orthorhombic phase. This Result agrees well with XRD analysis of the calcined powder.

The XRD patterns of the as-burnt and calcined powders have been shown in Fig. 3. The characteristics peaks of Y_2O_3, $BaCO_3$, CuO, and Ba $(NO_3)_2$ are detected in the XRD pattern of the as-burnt powder. This reveals that the dried gel has been converted to metal oxides, barium carbonate and amorphous phases after the combustion reactions. The XRD pattern of the calcined powder shows the formation of $YBa_2Cu_3O_{7-\delta}$ phase. The superconducting phase has an orthorhombic provskite-like structure.

TGA/DTG results for the as-burnt powder are given in Fig. 4. The first weight loss is seen in the temperature range of 230-300 °C. It can be attributed to a combustion reaction among the residual unreacted materials of firing stage. The increase in the weight at 300-340 °C is originated from the reaction between CO_2 and BaO according to reaction 4.The residual of barium nitrate decomposes to BaO and NO_2 at 590 °C. The increase in the weight at 650 °C is due to the formation of oxides. The superconducting phase formation carries out at 750 °C which can be seen from a peak at the same temperature in the DTG curve. The reaction occurred at this temperature can be presented in the following state:

$$Y_2O_3 - 4BaCO_3 - CuO - 2YBa_2Cu_3O_{7-\delta} - 4CO \tag{3}$$

REFERENCES

[1] Soares, E.R., Fuller, J.D., Morozick, P.J. & Alvarez, R.L., Applications of high-temperature-superconducting filters and cryoelectronics for satellite communication. IEEE Trans. Micro. Theo. Tech., 2000, 48, 1190-1198.

[2] Lee, B.W., Kang, J.S., Park, K.B., Kim, H.M. & Oh, I.S., Optimized current path pattern of YBCO films for resistive superconducting fault current limiter. IEEE Trans. App. Super. , 2005, 15, 2118-2121.

[3] Sozeri, H., Özkan , H. & Ghazanfari, Properties of YBCO superconductors prepared by ammonium nitrate melt and solid-state reaction methods. N., J. Alloys Compd., 2007, 428, 1-7

[4] Bolzan, A.A., Millar, G.J., Bhargava, A., Mackinnon, I.D.R. & Fredericks, P.M., A spectroscopic comparison of YBCO superconductors synthesised by solid-state and co-precipitation methods. Mat. Lett., 1996, 28 , 27-32.

[5] Barboux, P., Campion, I., Daghish, S., Livage, J., Genicon, J.L. & Sulpice, A., Tournier , R., Synthesis of $YBa_2Cu_3O_{6+x}$ from coprecipitated hydroxides. J. Non-Cryst. Solids, 1992, 147&148, 704-710.

[6] Yang, W., Chang, A. & Yang, B., Preparation of Barium Strontium Titanate Ceramic by Sol-Gel Method and Microwave Sintering. J. Mater. Syn. Proces., 2002, 10, 303-309.

[7] Yue, Z., Li, L., Zhang, H. & Gui, Z., Preparation and characterization of NiCuZn ferrite nanocrystalline powders by auto-combustion of nitrate–citrate gels. Mater. Sci. Eng. B, 1999, 64 , 68-72.

[8] Xu, G., Ma, H., Zhong, M., Zhou, J., Yue, Y. & He, Z., Influence of pH on characteristics of $BaFe_{12}O_{19}$ powder prepared by sol–gel auto-combustion. J. Magn. Magn. Mater. 2006,301 , 383-388.

[9] Chakrabarti, N. & Maiti, H.S., Chemical synthesis of PZT powder by auto-combustion of citrate-nitrate gel. Mater. Lett., 1997, 30 , 169-173.

[10] Hosseini Vajargah, S., Madah Hosseini, H.R. & Nemati, Z.A., Synthesis of nanocrystalline yttrium iron garnets by sol–gel combustion process: The influence of pH of precursor solution. Mater. Sci. Eng. B, 1999, 129 , 211-215.

[11] Peng, C.H., Preparation and characterization of YBa2Cu3O7−x superconductor by means of a novel method combining sol–gel and combustion synthesis techniques. J. Mater. Sci.,2004, 39 , 4057-4061.

[12] Xu, X.L., Guo, J.D., Wang, Y.Z. & Sozzi, A., Nanocrystalline $YBa_2Cu_4O_8$ *c*-oriented films prepared by Nd:YAG laser deposition. Physica C, 2002, 371 , 129-132.

[13] Kakihana, M., Invited review “sol-gel” preparation of high temperature superconducting oxides. J. Sol-Gel Sci. Tech., 1996, 6 , 7-55.

[14] Ghobeiti Hasab, M., Ebrahimi, S.A. & Badiei, A., Effect of different fuels on the strontium hexaferrite nanopowder synthesized by a surfactant-assisted sol–gel auto-combustion method. J. Non-Cryst. Solids, 2007, 353, 814-816.

[15] Pederson, L.R., Maupin, G.D., Weber, W.J., McReady, D.J. & R.W. Stephens, Combustion synthesis of $YBa_2Cu_3O_{7-x}$: glycine/metal nitrate method. Mater. Lett., 1991, 10 , 437-443.

[16] Luo, T., Zhang, Y., Li, X., Lin, L., Zhang, Y. & Feng, Q., Studies of synthesizing behaviors and superconductivity of sol-gel YBa2Cu3O7−x samples in flowing oxygen atmosphere. Front. Phys. China, 2008, 3, 55-60.

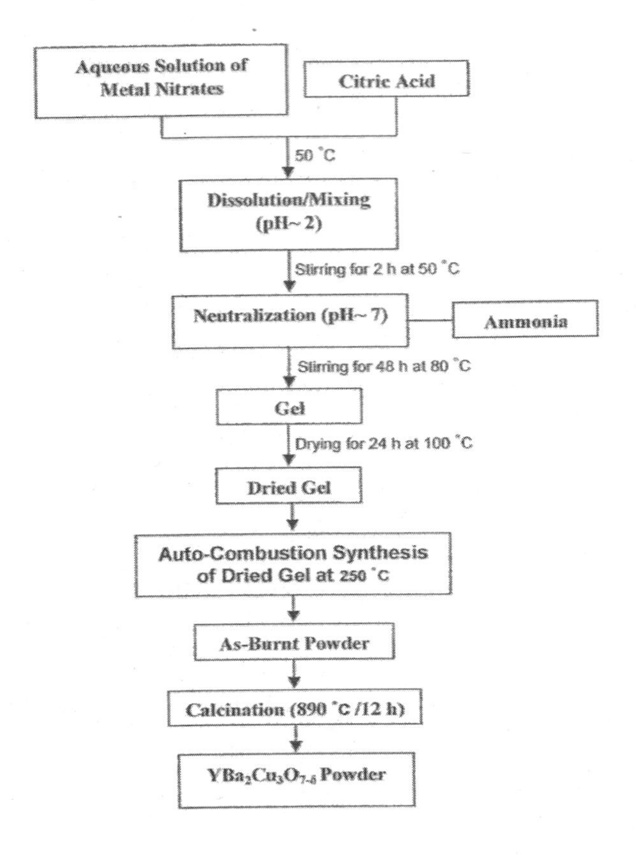

Figure 1: Scheme of the steps involved in the citrate-nitrate sol-gel process used for the synthesis of the superconducting $YBa_2Cu_3O_{7-\delta}$ phase.

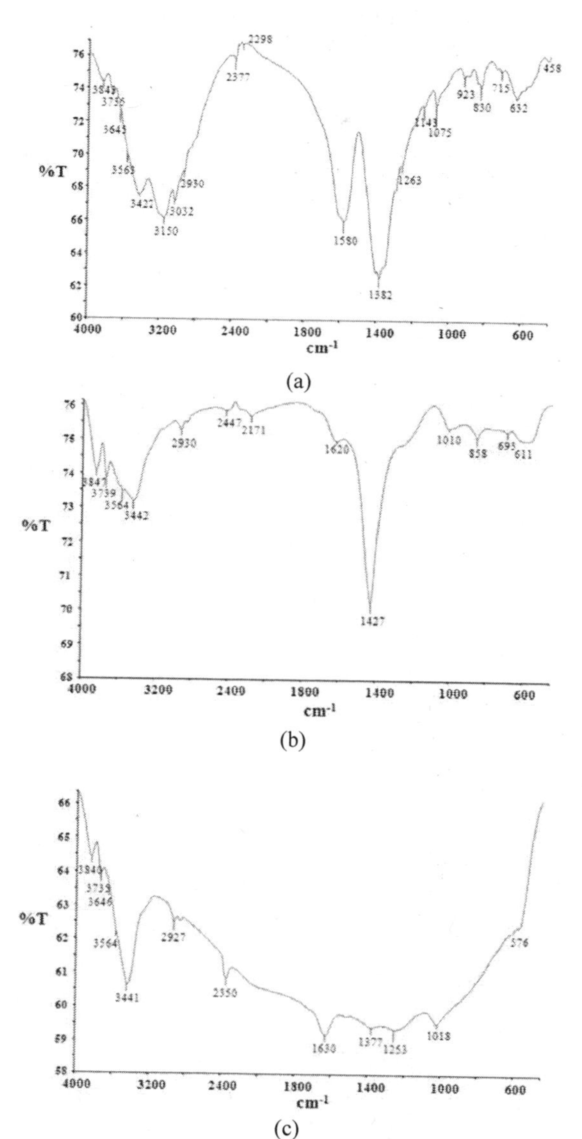

Figure 2: FTIR spectra of (a) the dried gel (b) the as-burnt powder (c) the calcined powder.

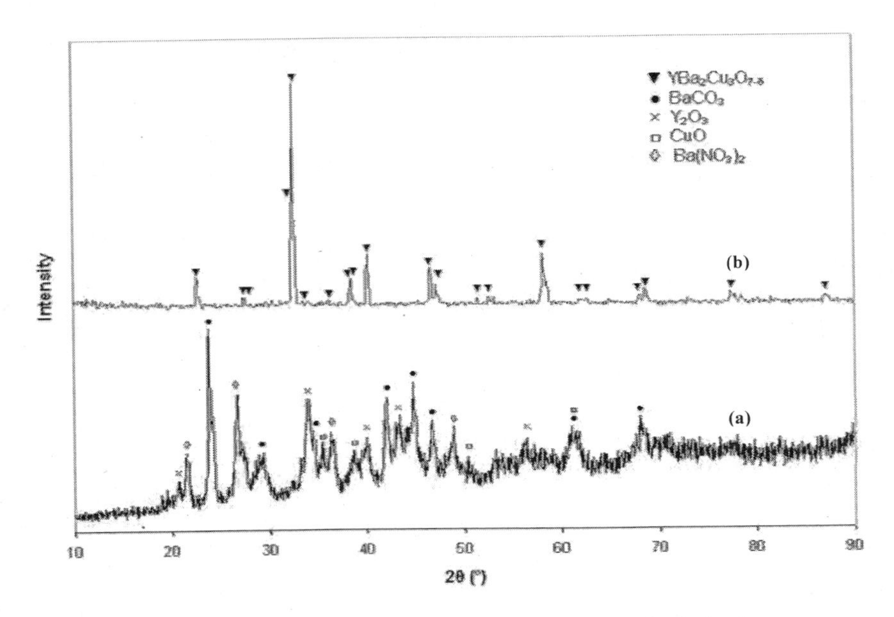

Figure 3: XRD patterns for (a) the as-burnt (b) the calcined powders.

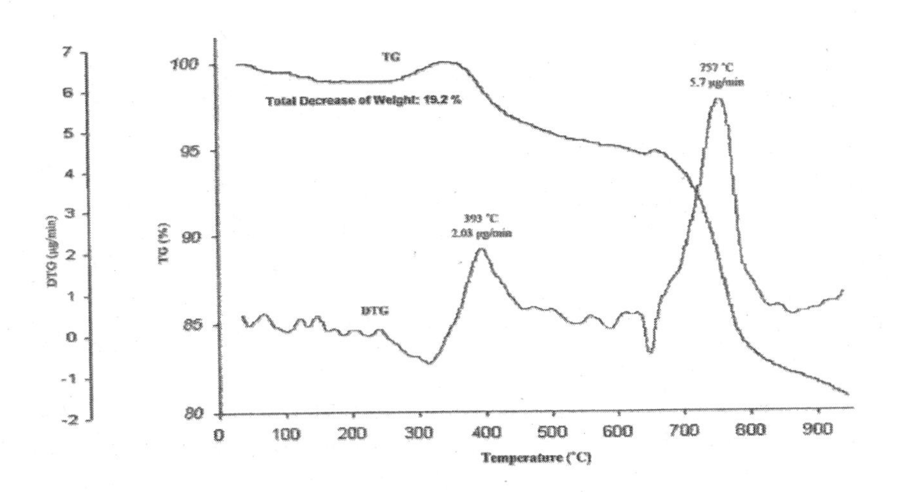

Figure 4: TG/DTG curves of the as-burnt powder.

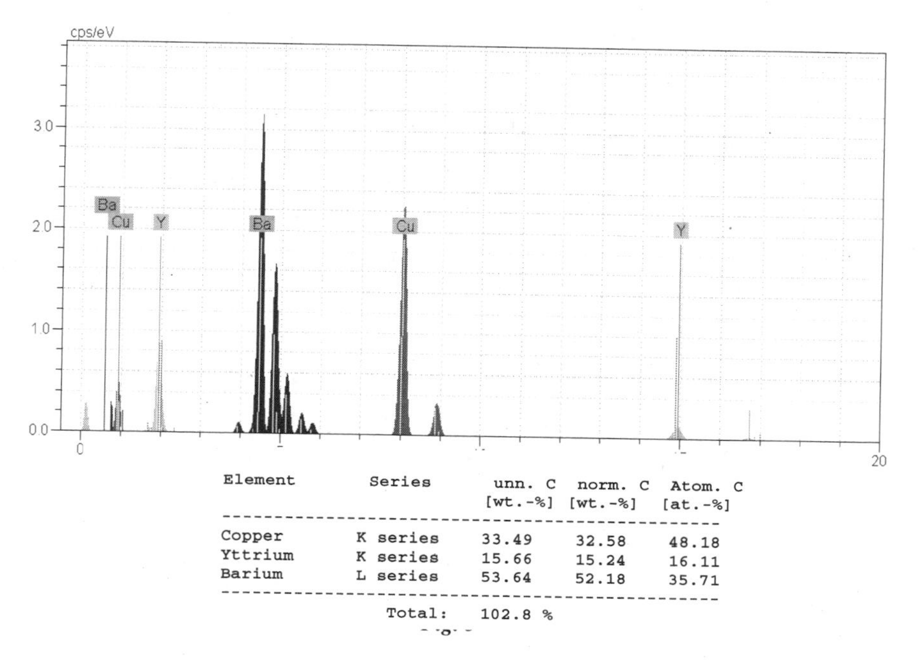

Element	Series	unn. C [wt.-%]	norm. C [wt.-%]	Atom. C [at.-%]
Copper	K series	33.49	32.58	48.18
Yttrium	K series	15.66	15.24	16.11
Barium	L series	53.64	52.18	35.71
	Total:	102.8 %		

Figure 5: EDS analysis graph of the superconducting $YBa_2Cu_3O_{7-\delta}$ phase.

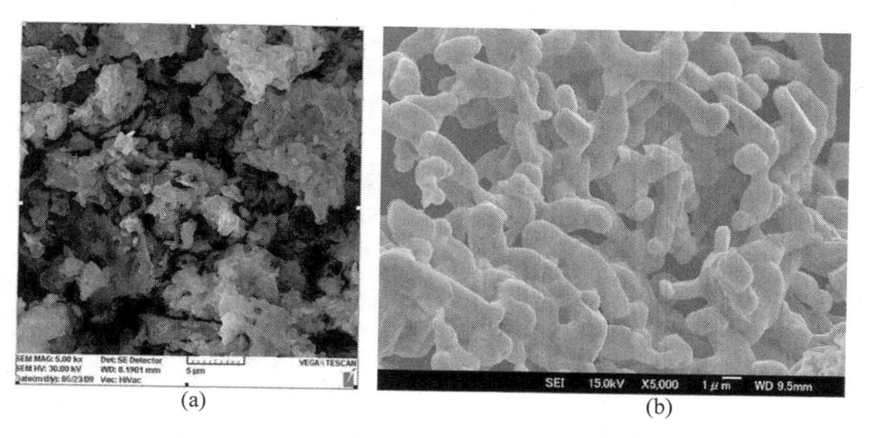

(a) (b)

Figure 6: SEM photographs of (a) the as-burnt (b) calcined powders.

Figure 7: SEM photograph of a calcined particle showing screw dislocations.

CHEMICAL INTERACTIONS OF THE $Ba_2YCu_3O_{6+x}$ SUPERCONDUCTOR WITH COATED CONDUCTOR BUFFER LAYERS

[1]W. Wong-Ng, [1]Z. Yang, [1]G. Liu, [1]Q. Huang, [1]L.P. Cook, [1]S. Diwanji, [1]C. Lucas, [2]M-H. Jang, and [3]J.A. Kaduk
[1]NIST, Gaithersburg, MD 20899.
[2]Yonsei University, Seoul, Republic of Korea, 120-740.
[3]Illinois Institute of Technology, Chicago IL 60616

ABSTRACT

Reactions between the high T_c superconductor $Ba_2YCu_3O_{6+x}$ and the potential buffer layer materials $SrTiO_3$, Gd_3NbO_7, $LaMnO_3$, and CeO_2 have been investigated using x-ray powder diffraction and other analytical techniques. Reaction products include $(Ba,Sr)_3YTi_2O_x$, $(Ba,Sr)_2YCu_3O_{6+x}$, $(Sr,Br)TiO_3$, $(Sr,Ba)_2TiO_4$, CuO_x, $(Gd_xY_{3-x})NbO_7$, $(Y,Gd)_2Cu_2O_5$, $Ba_2(Gd_xY_{1-x})NbO_6$, $(Gd,Y)_2CuO_4$, $Ba_2(Gd_xY_{1-x})Cu_3O_{7-x}$, $Ba_{2-x}(La_{1+x-y}Y_y)Cu_3O_{6+z}$, $Ba(Y,La)_2CuO_5$, $(La,Y)MnO_3$, $(La,Y)Mn_2O_5$, and $Ba(Ce,Y)O_{3-x}$, and most of these have been characterized structurally. The reaction of $Ba_2YCu_3O_{6+x}$ and CeO_2 is diffusion-limited, and a kinetic model was derived.

INTRODUCTION

High-temperature superconductors have demonstrated potential for meeting the technological needs associated with more efficient utilization of electrical distribution grids. In particular, there has been an accelerated effort within the high T_c community on research and development of coated conductors for wire/tape applications [1-7]. These conductors are based on $Ba_2YCu_3O_x$ (Y-213) and $Ba_2RCu_3O_x$ (R=lanthanides) as the principal superconducting materials. They can be deposited on flexible metallic tapes, and the resulting materials show excellent current-carrying capability.

Specially prepared substrates form the basis for coated conductor fabrication. The two state-of-the-art technologies for producing biaxially-textured substrates are commonly known as Ion Beam Assisted Deposition (IBAD) [1-2], and Rolling Assisted Bi-axially Textured Substrate (RABiTS) [3-7]. Typically, the architecture of a RABiTS film includes a number of buffer layers of different materials deposited on a biaxially textured metallic substrate. The general functions of these multilayers are to provide a physical/chemical barrier to substrate oxidation and substrate reaction with the superconductor layer, and to provide texture for crystallographic alignment [8]. Examples of the lower cap layer over the substrate include CeO_2, $SrTiO_3$, Gd_3NbO_7, $LaMnO_3$, and $SrRuO_3$. Often there may be chemical reactions at the interface between $Ba_2YCu_3O_{6+x}$ and the buffer layers, which may partially mitigate the benefits of buffer layers, including the promotion of epitaxial growth of Y-213. Understanding of interfacial reactions of $Ba_2YCu_3O_{6+x}$ with the buffer layers will provide information about how to avoid and/or control the formation of second phases. Phase formation data will further allow better interpretation of the results of TEM analysis of coated conductor interfaces.

We have studied the interfacial reactions of $Ba_2YCu_3O_{6+x}$ with four substrate materials, namely, CeO_2 [9-11], Gd_3NbO_7 [12], $LaMnO_3$ [13], and $SrTiO_3$ [14]. This paper reviews the studies of the multi-component systems representing the interaction of Y-213 with these substrates, particularly the phases formed at the interface. Phase formation and structural studies of the non-binary systems $Ba_2YCu_3O_{6+x}$-$LaMnO_3$, $Ba_2YCu_3O_{6+x}$-Gd_3NbO_7, $Ba_2YCu_3O_{6+x}$-CeO_2, and $Ba_2YCu_3O_{6+x}$-$SrTiO_3$ were conducted using x-ray diffraction and differential thermal analysis/thermal gravimetric analysis (DTA/TGA).

EXPERIMENTAL[1]

A 10 g master batch of $Ba_2YCu_3O_{6+x}$ was prepared by heating a mixture of BaO, Y_2O_3 and CuO under purified air (CO_2- and H_2O-scrubbed). The BaO starting material was produced from $BaCO_3$ (99.99 % purity, metals basis) by vacuum calcination in a vertical tube furnace. The following heating schedule was used: room temperature to 1300 °C in 20 h; isothermal at 1300 °C for 10 h; 1300 °C to room temperature in 20 h. To prepare $Ba_2YCu_3O_{6+x}$, stoichiometric amounts of BaO, Y_2O_3 and CuO were weighed out in a glove-box, well mixed and calcined in an atmospherically controlled high temperature furnace, first at 850 °C, then at 930 °C repeatedly with intermediate grindings for about two weeks. Ten grams of single-phase master batches of the buffer materials $LaMnO_3$, Gd_3NbO_7, CeO_2, and $SrTiO_3$ were also prepared. The details of the preparation for $LaMnO_3$, Gd_3NbO_7, CeO_2 were described earlier [9-13].

Subsequently, nine samples with different ratios of $Ba_2YCu_3O_{6+x}$ to buffer material were prepared using these master batches of $Ba_2YCu_3O_{6+x}$ and the buffer materials. The ratios prepared were 10:90, 20:80, 30:70, 40:60, 50:50, 60:40, 70:30, 80:20, and 90:10. Pelletized samples were placed inside individual MgO crucibles for annealing in a horizontal box-type controlled-atmosphere furnace. Transfer from the glove-box to the box furnace and vice versa was achieved via a second transfer vessel and an interlock system attached to the furnace. Samples were annealed at 810 °C at p_{O2} = 100 Pa_2, or at 950 °C in purified air repeatedly until no further changes were observed in x-ray diffraction patterns.

X-ray powder diffraction was used to identify the phases synthesized, to confirm phase purity, and to determine phase relationships. A computer-controlled automated Philips diffractometer equipped with a θ-compensation slit and CuK_α radiation was operated at 45 kV and 40 mA. The radiation was detected by a scintillation counter and a solid-state amplifier. All x-ray patterns were measured using a hermetic cell designed for air-sensitive materials [15]. The commercially-supplied Siemens software package and the reference x-ray diffraction patterns of the Powder Diffraction File (PDF) [16] were used for performing phase identification.

Structures of selected member of the product phases were studied using the x-ray Rietveld refinement technique [17]. These samples were mounted as acetone slurries in zero-background quartz holders. A Bruker D8 Advanced Diffractometer equipped with a VANTEC-1 position-sensitive detector was used to measure the powder patterns (CuK_α radiation, 40 kV, 40 mA) [17].

Simultaneous differential thermal analysis and thermogravimetric analysis (DTA/TGA) were used to study thermal events. Most experiments utilized mainly the DTA signal; the TGA signal was useful primarily in following oxygen gain/loss associated with the CuO_x component. DTA/TGA experiments were performed using an electronically upgraded Mettler TA-1 system

[1] Certain commercial equipment, instruments, or materials are identified in this paper in order to specify the experimental procedure adequately. Such identification is not intended to imply recommendation or endorsement by the National Institute of Standards and Technology, nor is it intended to imply that the materials or equipment identified are necessarily the best available for the purpose.

fitted with an Anatech digital control and readout system. The DTA/TGA apparatus was calibrated against the α/β quartz transition (571 °C) and the melting point of NaCl (801 °C), and temperatures reported in this study have a standard uncertainty of ± 5 °C. Event temperatures were determined as the intersection of the baseline with the extrapolated linear portion of the rising DTA peak. Oxygen partial pressure during DTA/TGA was controlled using a previously analyzed Ar/O_2 mixture. During the experiments, gas was continuously flowed through the sample region at a rate of 150 mℓ/min, and the oxygen pressure at the outlet of the DTA/TGA system was periodically checked with a zirconia sensor.

DISCUSSION

The system BYC+ $SrTiO_3$

The study of the interactions of BYC with $SrTiO_3$ have been completed at 810 °C under 0.1% O_2. We first prepared master batches of BYC and $SrTiO_3$. Then we prepared 9 mixtures with different proportions from these two master batches. From the X-ray study of these 9 samples, we determined the products of reactions. Typical examples of x-ray patterns are shown in Fig. 1. We found that all samples contained $(Ba,Sr)_3YTi_2O_x$, BYC phase, $BaTiO_3$,

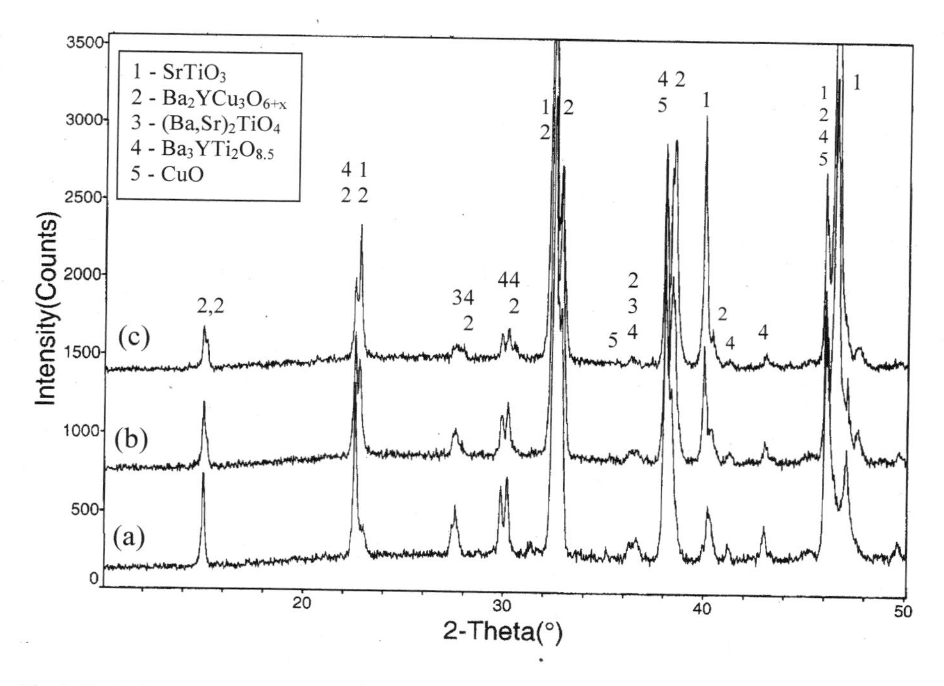

Fig. 1. Typical x-ray patterns showing the presence of small amounts of the phases formed as a result of interaction of $Ba_2YCu_3O_{6+z}$ with $SrTiO_3$ for mixtures with the ratio of (a) 2:8, (b) 6:4, and (c) 8:2. Different amounts of the reaction products are shown as a result of different composition ratios.

Ba_2TiO_4 phase and CuO_x, with different amounts depending on the mixture ratio. The melting temperatures of these compositions ranged from 940 to 970 °C under 0.1% O_2. Therefore, melting is not likely to be a problem for BYC films on $SrTiO_3$ substrates at processing temperatures below 900 °C.

Based on a Ba-Sr-Y-Cu-O quaternary diagram that we published earlier, we found that Sr substitutes Ba to a certain degree in a large number of phases in this system. Therefore the phases are written as solid solutions between Ba and Sr. The reactants and products can be represented in the following unbalanced equation:

$$Ba_2YCu_3O_{6+x} + SrTiO_3 \rightarrow (Ba,Sr)_3YTi_2O_x + (Ba,Sr)_2YCu_3O_{6+x} + (Sr,Ba)TiO_3 + (Sr,Ba)_2TiO_4 + CuO_x$$

The 50-50 (Ba,Sr)-BYC solid solution is structurally related to that of the well known BYC phase. We have also studied the crystal chemistry of the lanthanide substituted analogs. The structure of the larger lanthanide compounds (from Pr to Eu) was found to be tetragonal, while that of the smaller size cation compounds are orthorhombic. Convergence of *a* and *b* parameters as a function of ionic radius [18] is demonstrated in Fig. 2.

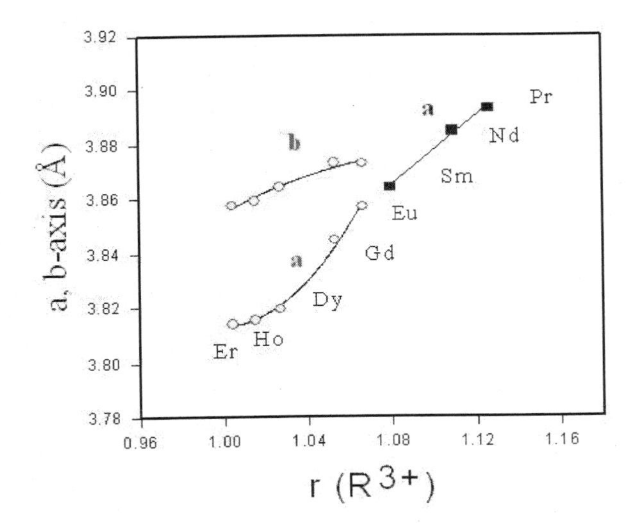

Fig. 2. Plots of a, and b axis of the $(BaSr)_2RCu_3O_{6+x}$ members showing the phase transition from orthorhombic to tetragonal structures between R=Gd and Eu. Convergence of *a* and *b* parameters as a function of ionic radii is demonstrated.

The $(Ba,Sr)_3YTi_2O_x$ phases are hexagonal *P6₃/mmc*. The structure can be considered as related to the 12-layer hexagonal-pervoskite family (Fig. 3). We have also determined the lanthanide series of compounds, $Ba_3RTi_2O_x$ (R=lanthanides). The structure can be viewed as stacking of cubic 'c' and hexagonal 'h' layers (BaO_3) as well as ordered, oxygen deficient pseudocubic c' [BaO_2] layers in the sequence of $(c'cchcc)_2$ along the *c*-axis. The structure units consist of face-sharing TiO_6 octahedral layers, RO_6 layers and TiO_4 tetrahedral layers.

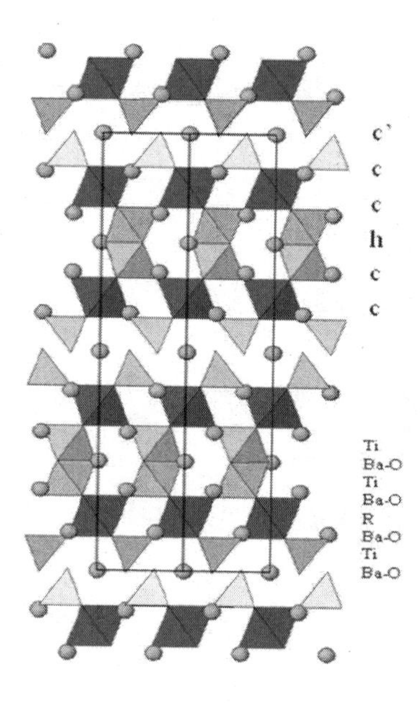

Fig. 3. Perspective view of the $Ba_3RTi_2O_{8.5}$ structure along the [110] direction, showing the stacking of face-sharing TiO_6 octahedral layers, RO_6 layers and TiO_4 tetrahedral layers [18].

The System BYC+ Gd_3NbO_7

X-ray patterns showed that there is a maximum of 5 phases present in each sample. Gd and Y in most cases form solid solutions. DTA temperature of the lowest melting event is around 938 °C in air and 850 °C at 0.1% O_2. Fig 4 gives a representation of the progression of phases from one end to the other, with BYC to Gd_3NbO_7 ratio from 1:9 to 9:1. The compositions can be considered as lying in two 5-phase volumes and a common three-phase boundary (actually a three phase volume due to the presence of solid solutions) between the two five phase volumes.

Compositions on the left of the three-phase boundary consist of the 3 common phases plus $(Y,Gd)_3NbO_7$ and $Ba_2(Gd,Y)NbO_6$; the compositions on the right of the three-phase boundary consist of the 3 common phases plus $Ba_2(Y,Gd)Cu_3O_x$ and the green phase,

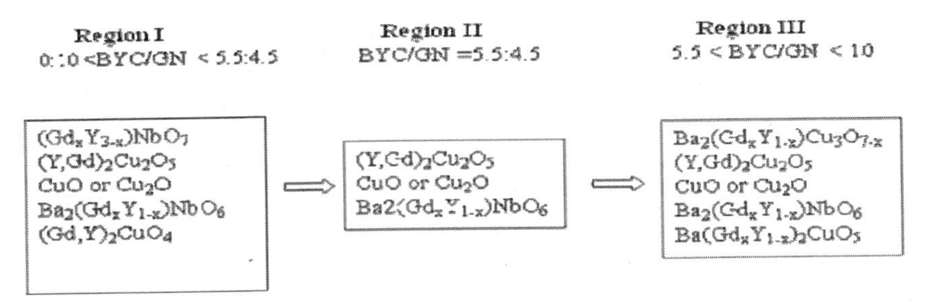

Fig. 4. A schematic representation of three regions of phase assemblages as a function of the compositional ratio of $Ba_2YCu_3O_{6+y}$ (BYC)/Gd_3NbO_7 (GN).

$Ba(Y,Gd)_2CuO_5$. The crossing of the boundary takes place between 5:5 and 6:4 composition ratio between the two end members.

The structure of $(Y,Gd)_3NbO_7$ consists of alternate stacking of distorted NbO_6 octahedra with GdO_8 polygon, forming two dimensional slabs. The additional GdO_7 polyhedra which lie in between these layers are omitted for clarity (Fig. 5). $Ba_2(Gd,Y)NbO_6$ forms a complete solid solution between Gd and Y at high temperature. The NbO_6 octahedra alternates with the $(Gd,Y)O_6$ octahedra in the cubic structure.

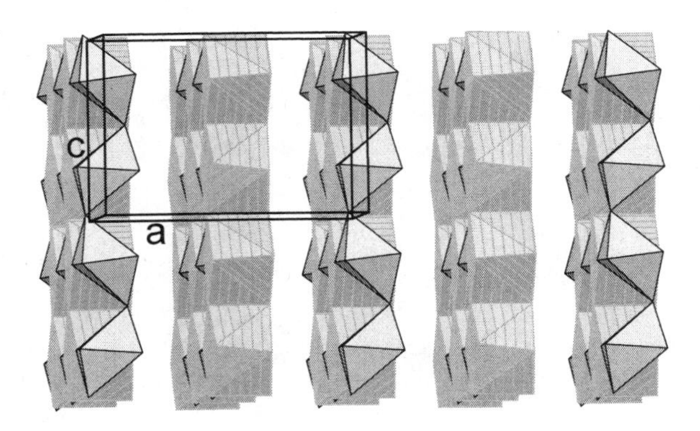

Fig. 5. Crystal structure for Gd_3NbO_7 (*C222₁*) showing the partial layered feature. The alternate stacking of distorted NbO_6 octahedra (plain pattern) and $(Gd,Y)O_8$ polyhedra (ruled pattern) are illustrated. The $(Gd,Y)O_7$ polyhedra are omitted for clarity.

The System BYC+ $LaMnO_3$

The interaction between BYC and $LaMnO_3$ have been investigated by determining the phases formed in the non-binary join of the Y213-$LaMnO_3$ system. Based on the x-ray diffraction patterns (Fig. 6) of the 9 samples of different ratios of Y213 and $LaMnO_3$ prepared at 950 °C in purified air and at 810 °C in 100 Pa oxygen, the unbalanced equation representing the chemical reactions can be written as:

$$Ba_2YCu_3O_{6+z}+LaMnO_3 \rightarrow Ba_{2-x}(La_{1+x-y}Y_y)Cu_3O_{6+z}+Ba(Y,La)_2CuO_5+(La,Y)MnO_3+(La,Y)Mn_2O_5$$

In this equation, general formulas which are written with '(La, Y) or (Y, La)' are used to represent possible solid solution formation between La, and Y. We found that La and Y indeed form solid solution in the $Ba(Y,La)_2CuO_5$ phase to a small extent in air and in 0.1% O_2. But there is no evidence of solid solution formation in the $(La,Y)Mn_2O_5$ phase under the present synthesis conditions. The structure of the $LaMn_2O_5$ phase consists of infinite chains of $Mn^{4+}O_6$ edge-sharing octahedra, linked by $Mn^{3+}O_5$ and YO_6 units. The Mn ions adopt 3+ and 4+ valence states.

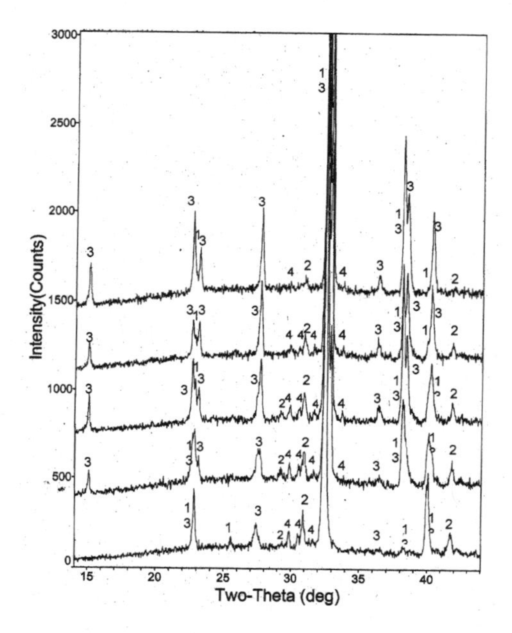

Fig. 6. XRD patterns representing the reaction of $Ba_2YCu_3O_{6+z}$ with $LaMnO_3$ in different mole ratios of 10:90, 30:70, 50:50, 70:30, and 90:10 (810 °C, p_{O2} = 100 Pa). The labeling in the figure is as follows: 1- $(La,Y)MnO_3$, 2-$(La,Y)Mn_2O_5$, 3- $Ba_{2-x}(La_{1+x-y}Y_y)Cu_3O_{6+z}$, 4-$Ba(Y,La)_2CuO_5$.

The System BYC+CeO_2

Since CeO_2 is an important buffer material, we have conducted more detailed work on this system. We described their interactions in terms of both phase equilibria and kinetics. The equilibria along the $Ba_2YCu_3O_{6+x}$-CeO_2 non-binary join are shown in the context of the BaO-Y_2O_3-CuO_x-CeO_2 framework in Fig. 7. This join passes through two 4-phase regions, namely: $Ba_2YCu_3O_{6+x}$-$BaCeO_3$-BaY_2CuO_5-Cu_2O and $BaCeO_3$-BaY_2CuO_5-Cu_2O-CeO_2. These two volumes are mutually consistent. These two tetrahedra share a common plane defined by BaY_2CuO_5, Cu_2O and $BaCeO_3$. The green phase and $BaCeO_3$ are part of the products of reaction. As a composition vector passes through the two tetrahedra, only three phases are observed at the boundary. This phase boundary exists at the mole ratio of $Ba_2YCu_3O_{6+x}$: CeO_2 = 40 : 60. Cerium is known to possess various oxidation states. Under the present conditions, CeO_2 is the stable form.

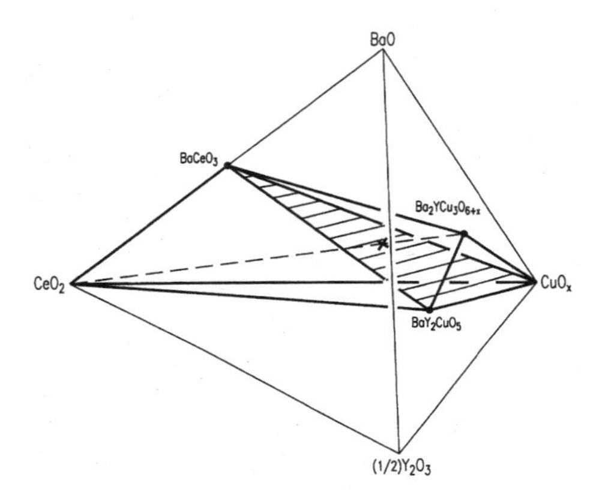

Fig. 7. The BaO-Y_2O_3-CuO_x-CeO_2 tetrahedron showing the two sub-volumes $Ba_2YCu_3O_{6+x}$ - $Ba(Ce,Y)O_{3-x}$ - BaY_2CuO_5 - Cu_2O and $BaCeO_3$-BaY_2CuO_5- Cu_2O-CeO_2, within which the compositions of the $Ba_2YCu_3O_{6+x}$-CeO_2 join lie. "$BaCeO_3$" is a solid solution with the formula $Ba(Ce_{1-z}Y_z)O_{3-x}$. The "x" indicates the intersection of the $Ba_2YCu_3O_{6+x}$-CeO_2 join with the $BaCeO_3$ – BaY_2CuO_5 - CuO_x plane.

We conducted a preliminary study to obtain the appropriate temperature range for heat treatments: 790 °C, 810 °C, and 830 °C. For each sample, each subsequent heat treatment was in general doubled in time. X-ray diffraction was used to obtain intensity of selected reflections of these samples. For example, we used the (031) reflection for Y213, (213) reflection for $BaCeO_3$, and (131) reflection for the green phase. The intensity values are assumed to be proportional to the amount of a particular phase, or the thickness of the reaction layer. TEM study was carried out to obtain the microstructure, including the thickness of the reaction layer. A kinetics model was then derived to understand the progress of reaction.

The x-ray patterns of the sample heat-treated at 810 °C at various times is shown in Fig. 8. The development of peaks related to $Ba(Ce,Y)O_3$, the green phase and CuO phases are shown. As the amount of reaction time increases, we clearly see the decrease of the peak intensity of the reactant BYC phase, the increase of intensity of the products $Ba(Ce,Y)O_3$ phase and the green phase, BaY_2CuO_5.

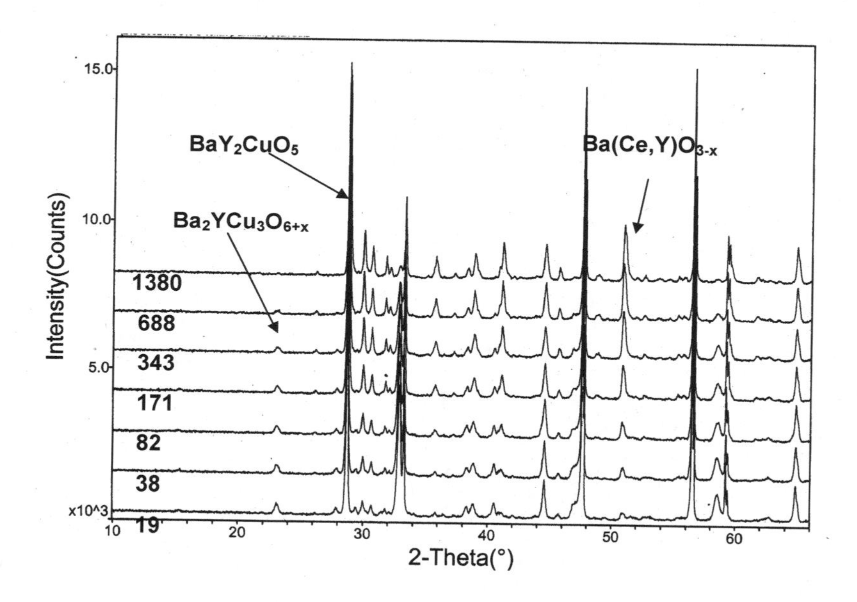

Fig. 8. Sequential x-ray diffraction patterns of $Ba_2YCu_3O_{6+x}/CeO_2$ pellet heat-treated at 810 °C for increasing cumulative time (minutes).

To study the kinetics of reaction, we deposited BYC films on CeO_2 pellets using our NIST laser deposition facility. The optical images of the CeO_2 pellet on the left, and the one with BYC deposited on the right (Fig. 9). The diameter of the pellets is about 10 mm. They both have smooth surfaces. In most ceramic reactions such as this, there is a reaction interface between reacting phases. The overall process involves transport of material to the interface, reaction at the interface, and sometimes transport of reaction products away from the interface [19]. For a reaction in which a compound is formed as the *planar* reaction layer, and the rate of product formation is controlled by diffusion through the product layer, then the reaction kinetics obey the parabolic rate rule, $x = Kt^{1/2}$ (where x is thickness of the reaction layer, t is time, and K is a constant related to the rate constant).

We indeed found in all samples studied, that a parabolic rate law is obeyed. The plots of intensity vs. $t^{1/2}$ for the experiments at 810 °C are shown in Fig. 10. For the products $Ba(Ce,Y)O_3$ and green phase, the shape of the curves are found to be rather similar. There is the straight-line portion of these curves at a relatively short reaction time period, and then the change of intensity decreases. While the intensity of product increases, that of the reactant, namely the BYC, decreases. The intensity vs. $t^{1/2}$ plots for the experiments at 830 °C are similar to those of

Fig. 9. Optical images of uncoated CeO_2 (left) and $Ba_2YCu_3O_{6+x}/CeO_2$ (right) (scale indicator = 10 mm).

810 °C. A linear portion of the curves for the products at short reaction times is found, and then the slope becomes flat. If we compare the phase formation feature of $Ba(Ce,Y)O_3$ at the three different temperatures (Fig. 11), we observe that at higher temperature, 830 °C, the formation of $Ba(Ce,Y)O_3$ was completed earlier, i.e. the reaction progressed faster, as expected. The reaction to form $Ba(Ce,Y)O_3$ was nearly complete at $t^{1/2}$ of about 13.5 $min^{1/2}$. At 810 °C the corresponding time parameter was longer at $t^{1/2}$ = 23 $min^{1/2}$. At 790 °C, the reaction was still occurring at $t^{1/2}$ = 60 $min^{1/2}$.

The bright-field image of a cross-section of a sample heat-treated at 830 °C for 123 minutes was used to obtain the thickness of the reaction layer of reaction. Electron energy loss spectroscopy/energy dispersive spectroscopy (EELS/EDS) measurements in TEM confirmed existence of a monophasic reaction layer about 0.4 μm thick, comprised of $Ba(Ce,Y)O_{3-x}$. A mixture of BYC, green phase, and CuO was also observed in the immediate vicinity of $Ba(Ce_{1-z}Y_z)O_{3-x}$, whereas the outer part of the film consisted primarily of unreacted Y-213. Using a thickness of 0.40 μm, the parabolic equation gives a value of 4.7 x 10^{-3} $\mu m/s^{1/2}$ for K at 830 °C. The activation energy for the reaction was derived using the Arrhenius Equation [20]:

$$K_1/K_2 = [Ae^{(-Eact/RT_1)}/Ae^{(-Eact/RT_2)}] \qquad (1)$$

where K_1 and K_2 are constants relating to the reaction rate at temperatures T_1 and T_2 (Kelvin), respectively. A is an empirical pre-exponential factor, E_{act} is the activation energy, and R is the gas constant. K_1 and K_2 were obtained from the parabolic law $x = Kt^{1/2}$, as discussed above. From the experiments at 790 °C (1063 K) and 830 °C (1103 K), K_1 = 0.0016 $\mu m/s^{1/2}$ and K_2 = 0.0047 $\mu m/s^{1/2}$, and E_{act} = 2.67 x 10^5 J/mol. This activation energy is similar in magnitude to activation energies reported for reactions between other multicomponent ceramic oxides [21,

22]. The fact that a monophasic product layer of $Ba(Ce_{1-z}Y_z)O_{3-x}$ was formed adjacent to the CeO_2 suggests that the reaction occurred primarily by diffusion of Ba into the CeO_2. This process left behind products of green phase and CuO.

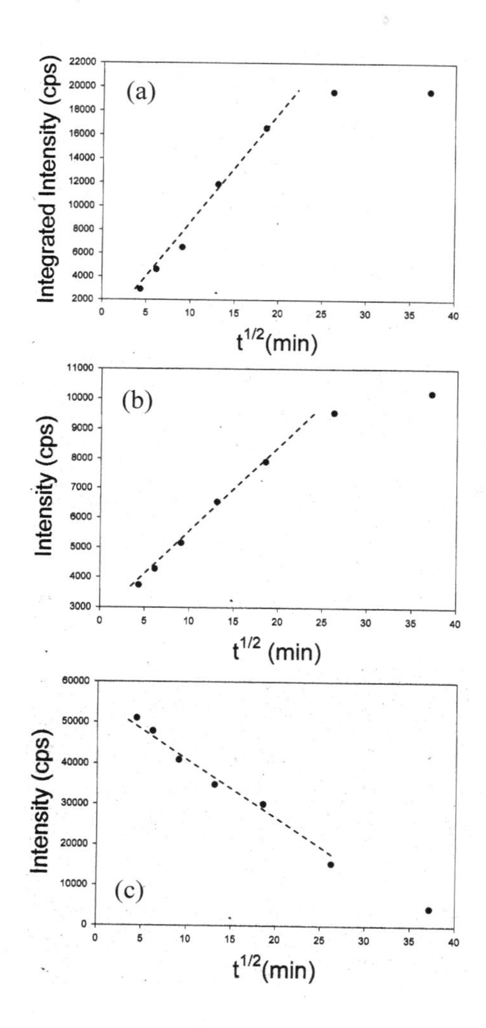

Fig. 10. Plots of integrated intensity vs. cumulative (time, minutes)$^{1/2}$ (a) $Ba(Ce_{1-z}Y_z)O_{3-x}$, (reflection 213), (b)BaY_2CuO_5 (reflection 131), and (c) $Ba_2YCu_3O_{6+x}$ (reflection 031) for samples heat-treated at 810 °C. 'cps' indicates counts per second.

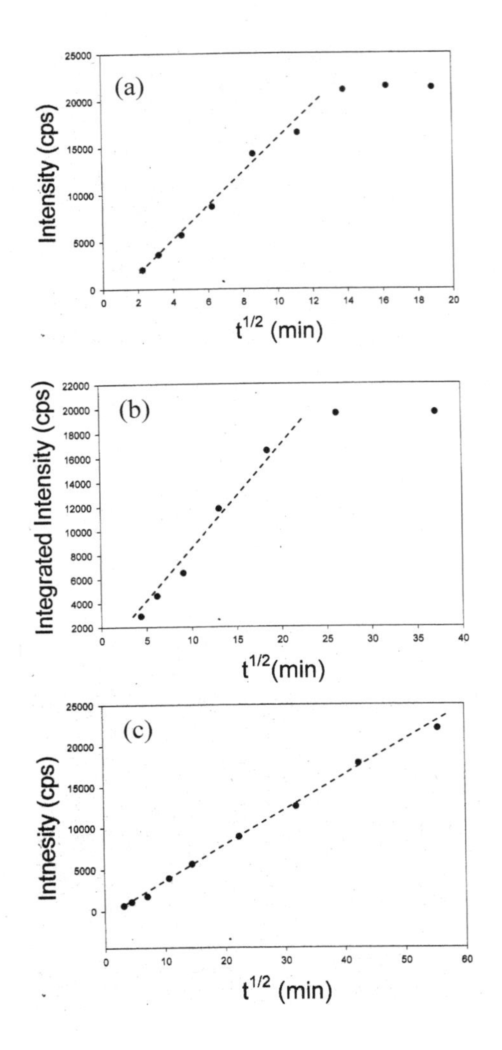

Fig. 11. Plots of integrated intensity vs. cumulative (time, minutes)$^{1/2}$ of $Ba(Ce_{1-z}Y_z)O_{3-x}$ (reflection 213) at three different temperatures, (a) 830 °C, (b) 810 °C, and (c) 790 °C. ‘cps’ indicates counts per second.

SUMMARY

As a summary, we have studied the reactions of BYC with $SrTiO_3$, Gd_3NbO_7, and $LaMnO_3$ in terms of crystal chemistry and crystallography of the phases formed. With CeO_2 we also studied kinetics of reactions. The reaction falls in two 4-phase fields. We followed the kinetics of reaction by monitoring the product phases $Ba(Ce_{1-z}Y_z)O_{3-x}$ and BaY_2CuO_5 as a function of time and temperature. The reaction kinetics obey a simple parabolic rate law, characteristic of diffusion-limited processes. The $Ba_2YCu_3O_{6+x}$ /CeO_2 reaction is limited by solid-state diffusion, and the reaction kinetics obey the parabolic rule, $x = Kt^{1/2}$ (where x is thickness of the reaction layer, *t* is time, and K is a constant related to the rate constant); K was determined to be 1.6×10^{-3} $\mu m/s^{1/2}$ at 790 °C and 4.7×10^{-3} $\mu m/s^{1/2}$ at 830 °C. The activation energy for the reaction was determined to be $E_{act} = 2.67 \times 10^5$ J/mol using the Arrhenius Equation.

REFERENCES

[1] P. N. Arendt, S.R. Foltyn, L. Civale, R.F. DePaula, P.C. Dowden, J.R. Groves, T.G. Holesinger, Q.X. Jia, S. Kreiskott, L. Stan, I. Usov, H. Wang, and J.Y. Coulter, *Physica C* 412-414 (2004) 795-800.

[2] S. R. Foltyn, E.J. Peterson, J.Y. Coulter, P.N. Arendt, Q.X. Jia, P.C. Dowden, M.P. Maley, X.D. Wu, D.E. Peterson, *J. Mater. Res.* 12 (1997) 2941-2946.

[3] A.P. Malozemoff, S. Annavarapu, L. Fritzemeier, Q. Li, V. Prunier, M. Rupich, C. Thieme, W. Zhang, A. Goyal A., Paranthaman M., and Lee D.F., Supercond. Sci. Technol. **13** 473-476 (2000).

[4] M. Paranthaman, C. Park, X. Cui, A. Goyal, D.F. Lee, P.M. Martin, T.G. Chirayil, D.T. Verebelyi, D.P. Norton, D.K. Christen, and D.M. Kroeger, J. Mater. Res **15** (12) 2647-2652 (2000).

[5] A. Goyal, D.F. Le, F.A. List, E.D. Specht, R. Feenstra, M. Paranthaman, X. Cui, S.W. Lu, P.M. Martin, D.M. Kroeger, D.K. Christen, B.W. Kang, D.P. Norton, C. Park, D.T. Verebelyi, J.R. Thompson, R.K. Williams, T. Aytug, and C. Cantoni, Physica C **357** 903-913 (2001).

[6] T. Aytug, A. Goyal, N. Rutter, M. Paranthaman, J.R. Thompson, H.Y. Zhai, and D.K. Christen, J. Mater. Res. **18** [4] 872-877 (2003).

[7] M. Paranthaman, A. Goyal, F.A. List, E.D. Specht, D.F. Lee, P.M. Martin, Q. He, D.K. Christen, D.P. Norton, J.D. Budai, D.M. Kroeger, Physica C **275** 266-272 (1997).

[8] M. Rupich, American Superconductor Corporation, Devens, MA 01434, private communication.

[9] W. Wong-Ng, L.P. Cook, P. Schenck, I. Levin, Z. Yang, Q. Huang, and J. Frank, Proceedings of the PACRIM meeting, sponsored by ACerS, Maui, Hawaii, September, 2005; Ceramic Transactions series **191** 83-98 (2006).

[10] W. Wong-Ng, Z. Yang, L.P. Cook, Q. Huang, and J. Frank, "Chemical Interaction Between $Ba_2YCu_3O_{6+x}$ and CeO_2 at p_{O2} = 100 Pa," Solid States Sci, **7** 1333-1343 (2005).

[11] L. P. Cook, W. Wong-Ng, P. Schenck, Z. Yang, I. Levin, and J. Frank, "Kinetics Studies of the Interfacial Reactions of the $Ba_2YCu_3O_{6+x}$ Superconductor with CeO_2 Buffer Systems," J. Electronic Mater. **36**(10), 1293-1298 (2007)

[12] W. Wong-Ng, Z. Yang, J. A. Kaduk, L.P. Cook, and M. Paranthaman, ", J. Solid State Chem. **470**(5-6) 345-351 (2010).

[13] G. Liu, W. Wong-Ng, J. A. Kaduk, and L.P. Cook, "Interactions of $Ba_2YCu_3O_{6+y}$ with Substrate Layer $LaMnO_3$ in Coated Conductors", Physica C, **470** (5-6) 341-351 (2010).
[14] W. Wong-Ng, Y. Zhi, J. A. Kaduk, L. P. Cook, S. Diwanji, and C. Lucas, "Interactions of $Ba_2YCu_3O_{6+x}$ with $SrTiO_3$ Substrate," to be submitted to J. Solid State Chem. (2010).
[15] Ritter J.J., Powd. Diffr. **3** 30-31 (1988).
[16] PDF, Powder Diffraction File, produced by International Centre for Diffraction Data, 12 Campus Blvd., Newtown Squares, PA. 19073-3273, USA.
[17] Larson, A.C., von Dreele, R.B., *General Structure Analysis System (GSAS)*, Los Alamos National Laboratory Report LAUR 86-748, Los Alamos, USA (2004).
[18] R.D. Shannon, Acta Crystallogr. **A32**, 751-767 (1976).
[19] Kingery W.D., Bowen H.K., and Uhlmann D.R., *Introduction to Ceramics*, 2nd Ed., pp. 381-447, John Wiley & Sons, New York, USA, 1976.
[20] Laidler, K. J., *Chemical Kinetics*,Third Edition, Benjamin-Cummings, (1997).
[21] Shiue J-T. and Fang T-T., J. Mater. Res. 18(11) 2594-2599 (2003).
[22] Xu J., Zhu X. H. and Meng Z. V., IEEE Trans. Components and Packaging Technol. 22(1) 11-16 (1999).

CHEMICAL TAILORING OF ELECTRONIC DOPING IN $Y_{1-x}Gd_xBa_{1.9}Sr_{0.1}Cu_3O_{7-\delta}$ HIGH T_C SUPERCONDUCTORS

M.M. Abbas,* M.N. Makadsi,* and E.K. Al-Shakarchi**
* Physics department, College of Science, Baghdad University, Baghdad, Iraq
** Physics Department, College of Science, Al-Nahrain University, Baghdad, Iraq

ABSTRACT

Solid state thermodynamical reaction method involving mixing, calcinations and sintering were used to prepare high temperature superconductor with a nominal composition $Y_{1-x}Gd_x Ba_{1.9}Sr_{0.1}Cu_3O_{7-\delta}$ for Gd ($0.1 \leq x \leq 0.3$). The effect of the substitution of Gd for Y sites and Sr for Ba sites on sintering time, annealing temperature and oxygen content of the superconductor has been investigated to obtain the optimum conditions for the formation and stabilization of the superconducting samples. The critical temperature (T_C) of the samples prepared with different conditions were 120.5 K and 118 K for x=0.1. While increasing x =0.2 and 0.3 led to decrease T_C to 86.5 K and 81 K respectively, X-ray diffraction analysis showed single phase with orthorhombic structure for all samples.

INTRODUCTION

Since the discovery of the high temperature superconductor (HTSc) $YBa_2Cu_3O_{7-\delta}$ (YBCO) by Chu *et al.* [1] there has been a flurry of activity in the preparation and study of these materials. The YBCO compound contains CuO planes and chains. The unit cell of $YBa_2Cu_3O_{7-\delta}$ consists of 3-orthorhombic elementary perovskite unit cells. The middle one has ytterium Y atom at its center the upper and the bottom ones have each one Barium atom Ba at the center. The Y-atom is sandwiched by two CuO_2 planes while the upper plane and bottom plane are of CuO planes. The stoichiometry of this compound needs 9 oxygen atoms (O)-while the crystallographic examination shows very near to 7-O-atoms, thus it is referred as oxygen deficient perovskite. Hence the basic structure of these compounds has been considered as distorted.

The HTSc property is essentially related to the oxygen deficient multi-layered perovskite structure. The electronic properties of copper oxide superconductors are strongly affected by subtle points of the structure namely the vacant sites and vortices, also it should be mentioned that the configurations of Cu-O chains and CuO_2 planes seem to play an important role in superconducting mechanism [2] as well as the O deficiency and the related distortion. Hence almost all the researches have been concerned with substitution of some isovalent atoms.

Critical temperature T_c of the YBCO compounds depends on the chemical composition; cations substitutions or structural deformation and oxygen content [3-4]. Maple and his group have examined the normal and HTSc state properties of the rare earth (RE) element of $YBa_2Cu_3O_{7-\delta}$ and they found that this compound has considerable sensitivity to oxygen content, heat treatment and the rate of cooling after sintering [5]. Basically, YBCO releases oxygen with increasing temperature and absorb oxygen when temperature decreases, which can be called as an oxygen respiration. Moreover, the amount of the released oxygen is different for samples with different rare earth element [6].

Buchner *et al.* have studied the substitution of RE like Gd in $Y_{1-x}RE_xBa_2Cu_3O_{7-\delta}$ system. They reported that the RE ionic radius scales linearly not only with the lattice parameters, but also with the superconducting transition temperature [7]. It has been poited that some rare earth substitution such as Gd in YBCO system, the transition temperature remains either the same or moves to lower values [8]. However the partial substitution of Sr instead of Ba, in the $YBa_2Cu_3O_{7-\delta}$ superconductor results in a chemical pressure which induces the shortening of the interactomic distances and the contraction of the unit cell. These structural modifications induce an increase in the T_C for the majority of the superconducting cuprates [9].

Our research concerns in finding which are the suitable processing variables that would be an alternative route for tailoring the electronic doping with the oxygen content by partial substitution Gd for Y and Sr for Ba of the series $Y_{1-x}Gd_x Ba_{1.9}Sr_{0.1}Cu_3O_{7-\delta}$ systems with various concentration x=0.1, 0.2 and 0.3.

EXPERIMENTAL DETAILS

The samples of the system $Y_{1-x}Gd_xBa_{1.9}Sr_{0.1}Cu_3O_{7-\delta}$ with concentration x= 0.1, 0.2, and 0.3 were prepared by conventional solid –state reaction route. The appropriate weights of the starting materials, Y_2O_3, Gd_2O_3, $BaCo_3$, CuO and $SrNO_3$ were taken with precised values. Then the powders were well mixed using agate mortar. A sufficient quantity of 2-propane was used to homogenize the mixture. The mixture was ground to a fine powder and then calcined in air at (850-950) °C for (15-30) h and followed by an intermediate re-grinding. Further grinding followed to ensure the finest powder and then pressed into disc-shaped pellets with 12 mm in diameter and 2mm thickness using manually hydraulic press type (Specac) under different pressures around 7.5 ton/cm^2 for 2 min.

In order to investigate the effective factors, which act on the superconducting properties series of pellets were prepared with different conditions.

First series were sintered at 950°C for 5 h in air then 24 h in an oxygen environment and followed by subsequent slow air cooling to room temperature (process 1) figure (1-a).

Second series were sintered at 950°C for 5 h in air then 24 h in an oxygen environment and also followed by subsequent slow air cooling but to 550°C and hold at this temperature for 20 h with flow of oxygen after that they were cooled in ambient atmosphere to room temperature (process 2) figure (1-b).

Third series were sintered at 950°C for 10 h in air then cooled to 550°C and held at this temperature for 20 h under oxygen environment then air-cooled to room temperature (process 3) figure (1-c). Flow of pure oxygen has been used over the pellets, at a rate of 0.9 l/min in a tube furnace. The rate of heating was 1°C/min, while for cooling was 0.5°C/min. figure 1 shows the details of the three processes.

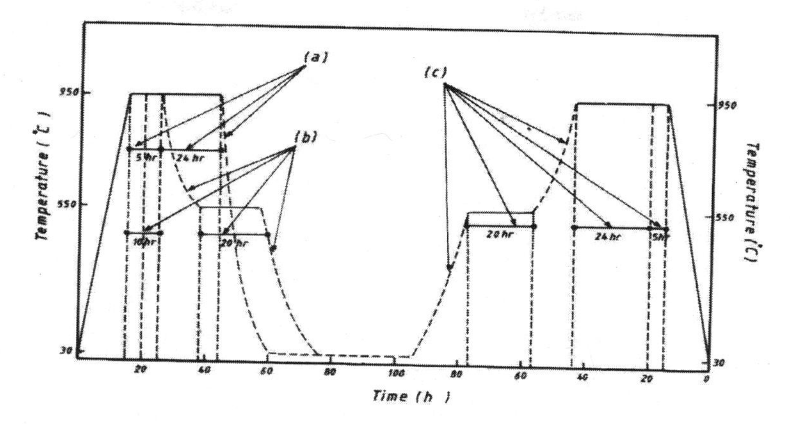

Figure1. A schematic diagram of the sintering processes for $Y_{1-x}Gd_x Ba_{1.9}Sr_{0.1}Cu_3O_{7-\delta}$.

Idometric titration was used to assess the (δ) value in the oxygen content of the samples.

The structure of the prepared samples was obtained by using X-ray diffractometer type Philips with the Cu-kα radiation. A computer program was used to calculate the lattice parameters, based on Cohen's least square method.

The resistivity measurements were performed by the standard four-probe method.

RESULTS AND DISCUSION

The particular, various conditions of the preparation procedure of the samples, are highlighted in Table1.The samples were investigated to determine the optimum treatment for producing high temperature superconductor. The temperature dependence of the electrical resistivity of $Y_{1-x}Gd_xBa_{1.9}Sr_{0.1}Cu_3O_{7-\delta}$ substituted with Gd concentrations (x= 0.1, 0.2 and 0.3) and prepared by different processes have been studied, the results are shown in figure 2. We have found that process 1 gave superconductor sample where T_C=120.5 K for x = 0.1 the others are semiconductors. Processes 2 gave negative results this elucidate that decreasing sintering time causes reduction of oxygen content and thus produce incomplete oxidation in the sample .. Whereas, process 3 decreased T_C to 118 K, the reason certainly caused by the mechanism of the oxygen redistribution in the CuO plane due to oxygen saturation such that it lowered the vacant O-sites necessary for optimum electronic transition, which tend to decrease T_C value.

Table 1. Values of T_C for different preparing conditions of $Y_{1-x}Gd_xBa_{1.9}Sr_{0.1}Cu_3O_{7-\delta}$.
* Annealing at 550°C for 20 h.

No. of Sample	Substituting Concentration	Time of Sintering h at 950°C with		Cooling with		T_C (K)
		Air Furnace	O_2 Atmosphere	Air Furnace	O_2 Atmosphere	
	Gd= 0.1					
(1)		5	24	×		120.5
(2)*		5	24		×	semi.
(3)*			10		×	118
(4)*			10	×		semi.
	Gd= 0.2					
(1)		5	24	×		86.5
(2)*		5	24		×	semi.
(3)			24	×		semi.
(4)			18	×		insu.
	Gd= 0.3					
(1)		5	24	×		81
(2)*		5	24		×	semi.
(3)			24	×		semi.
(4)			30	×		semi.

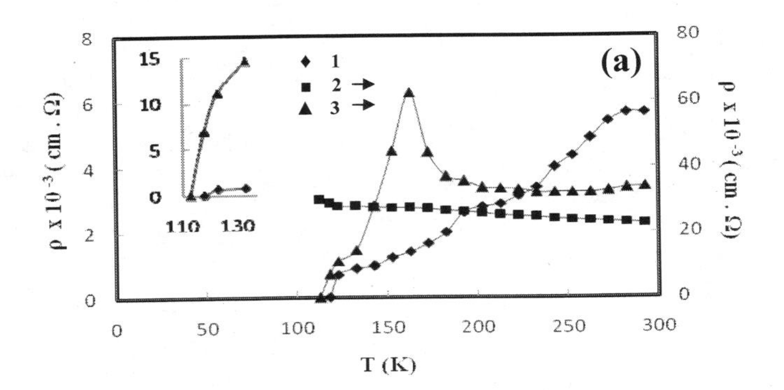

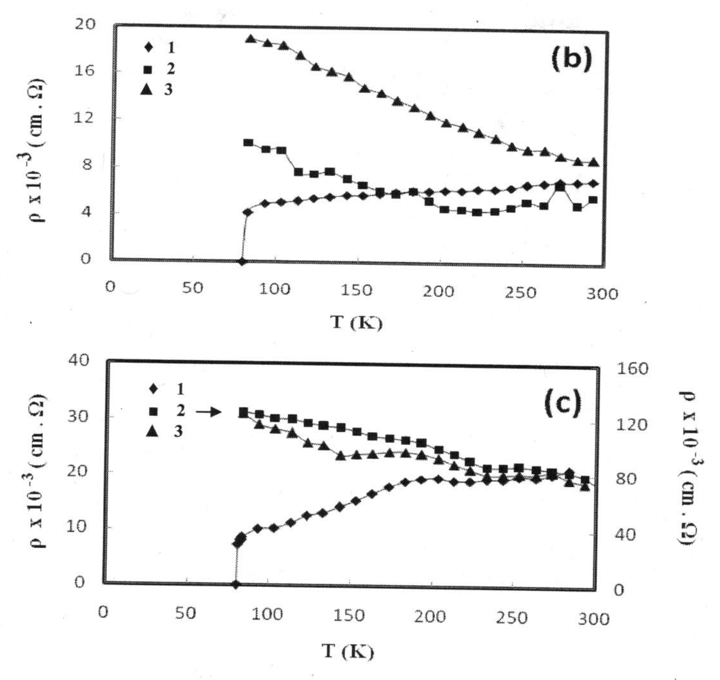

Figure 2. Temperature dependence of resistivity with different x. (a) x= 0.1; (b) x= 0.2 and (c) x=0.3 for $Y_{1-x}Gd_xBa_{1.9}Sr_{0.1}Cu_3O_{7-\delta}$ prepared by different processes. Symbols denoted the number of samples.

The X-ray characterization indicates that the samples with the substitution concentration x=0.1, 0.2 and 0.3 prepared by process 1 showed single phase with orthorhombic structure as shown in figure 3. They showed a small shift in 2θ with increasing x, similar behavior was pointed out by Wakim [10]. The Intensity that represented by the peaks decreased with increasing the substitution except at x = 0.2, this may need further work for the future.

The lattice parameters a, b and c of the unit cells and their volumes of these samples are presented in Table 2 and were found to vary unsystematically with increasing x, this results are similar to Ganepathi [11] . It should be emphasized that as the doping concentration increases the cell volumes are decreasing due to the both oxygen vacancies and cations ordered arrangements at the same time [12], besides that the substitution of Sr causes a shrinkage of the unit cell resulting in decreasing c-axis [10,13]. Thus a lattice distortion took place and the structure unit cell preserved the type of $YBa_2Cu_3O_{7-\delta}$ perovskite.

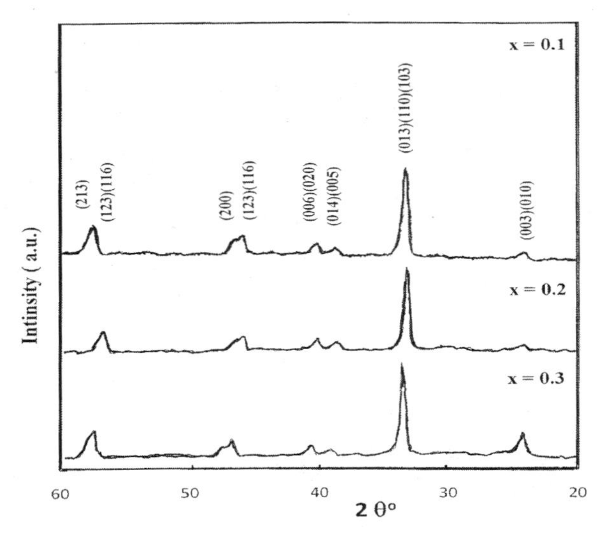

Figure 3. X-ray diffraction patterns with different x for $Y_{1-x}Gd_x Ba_{1.9}Sr_{0.1}Cu_3O_{7-\delta}$.

Table 2. Values of lattice parameter, oxygen content δ and T_C for the samples for different composition of $Y_{1-x}Gd_xBa_{1.9}Sr_{0.1}Cu_3O_{7-\delta}$.

x	a (Å)	b (Å)	c (Å)	V $(Å)^3$	δ	T_C (K)
0.1	3.8412	3.8751	11.7004	174.16	6.84	120.5
0.2	3.8241	3.8777	11.706	173.58	6.69	86.5
0.3	3.8310	3.8838	11.6417	173.214	6.60	81

In general increasing Gd concentrations decreases T_C as shown from Table 1. All the samples reveal a superconductor behavior as prescribed in figure 4 for samples prepared by processes 1.The critical temperature, T_C, is found to be 120.5 K, 86.5 K and 81 K for x=0.1, 0.2 and 0.3 respectively.

However, in addition of the volume contraction with increasing Gd concentration, see table (2), there was also a decreasing of T_C, and the oxygen content, this could be interpreted on bases that superconductivity is associated with the ordering of Y, Gd and Ba atoms. In certain ordering, as for x =0.1, offers a venue to even higher T_C [14]. But increasing x decreased T_C due to high deformation of the unit cell produced from the surrounding atoms that led to redistribution of charge in the lattice which changed the hole concentration in the CuO plane; and or affect the high vacancy concentration on the bridging apical oxygen[15]. Furthermore, substituting with Sr causes local distortion of the lattice in the neighborhood of the Sr site and the introduction of additional oxygen vacancies [16]. The interesting point in the sample of x =0.1 that has been shown in figure 4, is that, it exhibited two stages of transition the 1st is at about 180 K if it is extrapolated, such sample is called type II superconductor .

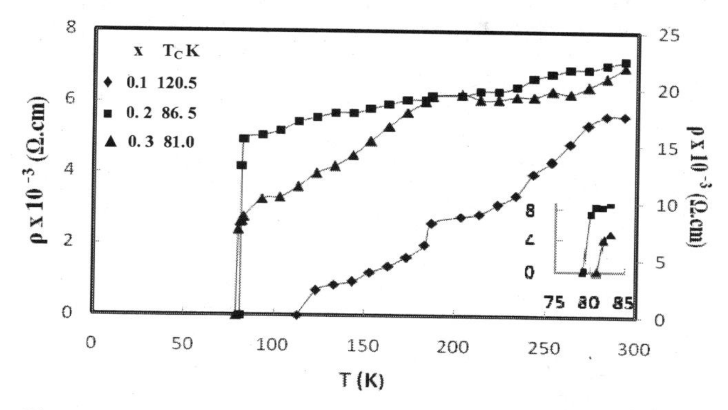

Figure 4. Temperature dependence of resistivity with different x prepared by process 1 for $Y_{1-x}Gd_x Ba_{1.9}Sr_{0.1}Cu_3O_{7-\delta}$.

CONCLUSIONS

It has become clear that the physical properties of high Tc superconductor $Y_{1-x}Gd_x Ba_{1.9}Sr_{0.1}Cu_3O_{7-\delta}$ for Gd ($0.1\leq x\leq 0.3$) depend greatly not only on the elemental composition but also on the details of the preparation method.

Optimum sintering conditions for $Y_{1-x}Gd_xBa_{1.9}Sr_{0.1}Cu_3O_{7-\delta}$ was found to be at 950°C for 29 h (5 h without oxygen and 24 h with 0.9 l/min flow of oxygen).

The substitution of Gd for Y in the compound $Y_{1-x}Gd_xBa_{1.9}Sr_{0.1}Cu_3O_{7-\delta}$ with (x=0.1, 0.2 and 0.3) exhibited maximum values of T_C 120.5, 86.5 and 81 K respectively. Our results also showed that sintering at 950°C for 10 h with consequence annealing at 550°C for 20 h when x=0.1 decreased T_C to 118 K.

The XRD showed single phase with orthorhombic structure for the samples sintered at 950°C for 5 h in air then 24 h in an oxygen environment and followed by subsequent slow air cooling to room temperature.

REFERENCES

[1] M.K.Wu, J. R. Ashburn, C. J. Torng, P. M. Hor, R. L. Meng, L. Gao, Z. J. Huang, Y. Q. Wang and C. W. Chu, Superconductivity at 93K in a new Mixed –Phase YBaCuO compound system at an ambient Pressure, *Phys. Rev. Lett*, 58, 9, 908-910,(1987).

[2] G. Alecu. Crystal Structures of Some High- Temperature Superconductors, *Romanian Reports in Physics*, 56, 3, 404 – 412, (2004) .

[3] J.Tahir-Kheli, W.A. Goddard . The Chiral Plaquette Polaron Paradigm for high Temperature Cuprate Superconductors, *Chemical Physics letter,* 472,153-165, (2009).

[4]A. Tavana and M. Akhavan, How T_c can go above 100 K in the YBCO family. *Eur. Phys. J.* B (2009) DOI: 10.1140/epjb/e2009-00396-7.
[5]M. B. Maple, High T_C Oxide Superconductors, *Proce. Mat. Res. Soc.* 14, 1, 20, (1989).
[6]F.Ya-Shan, S. Hong-Zhang, I.I.Zhi-Hui,Y. De-Lin and H.Xing, Influence of the rare earth element substitution on oxygen adsorption-desorption properties of YBCO. *Phys. China* 3, 368-370,(2006).
[7]B. Buchner. U. Calleib, H. D. Jostarndt. W. Schlabitz and D. Whohlleben, Correlation of Spectroscopic and Superconducting Properties of $REBa_2Cu_3O_{7-y}$ with the Rare Earth Ionic Radius , *Solid State Comm.* , 73, 357- 5,(1990).
[8]A. El- Ali, K. A. Azez, I. A. Al- Omari, J. Shobaki, M. K. Hasan, B. A. Albiss, Kh. Khasawnieh, Kh. Al- ziq and A. F. Salem, The Paramagnetic Contribution in Magneti-zation Behavior of $Y_{1-x}Gd_xBa_2Cu_3O_7$, *Physica B*.321,.1-4,320-323,(2002) .
[9]E.Gilioli, F.Licci, A. Prodi , A.Gauzzi and M.Marezio. Chemical Tailoring of Electronic Doping in Y $Sr_2\,Cu_3\,O_{7-\delta}$ Superconductor, *International Journal of Modern Physics B* 17,.4,5 &6, 685-689, (2003).
[10]N. N.Wakim, Study the Effect of Gd Substitution for Y in High Temperature YBaCuO Superconductor, Ms. C. Thesis, Baghdad University College of Science, (1994).
[11]L. Ganapathi, L. Ganapathia, Ashok Kumara and J. Narayan, Characteristics of high-Tc $La_xY_{1-x}Ba_2Cu_3O_{7-\delta}$ Superconductors, *Physica C* 167, 669-(5-6), (1990).
[12]X. T. Xu, J. K. Liang, S. S. Xie,G. C. Che, X. Y. Shao, Z. G. Duan and C. G. Cui,
Crystal Structure and Superconductivity of BaGdCuO System, *Solid State Commu*, 63, 649- 7, (1987).
[13]A.Prodi, A. Gauzzi, E. Gilioli, F. Licci, M. Marezio and F. Bolzoni. Correlation Between Local Oxygen Disorder and Electronic Properties in Superconducting $RESR_2Cu_3O_{6+x}$(RE=Y,Yb), *International J. of Modern Physics B* 17, 4,5 and 6, 873-878, (2003) .
[14]P.H.Hor, R.L.Meng,Y.Q.Wang, L.Gao, Z.J.Huang, J .Bechtold, K.Forster,and C.W.Chu, *Phys. Rev. Lett.* 58, 18, 1891-1894, (1987).
[15]J.L. Mac Manus- Driscoll, J.A.Alonso,P.C.Wang, T.H.Geballe and J.C.Bravman, Studies of Structural Disorder in $ReBa_2Cu_3O_{7-x}$ Thin Films as a Fucction of Rare Erth Ionic Radius and Film Deposition Conditions, *Physica C*, 232, 288-308,(1994).
[16]G. W. Crabtree, J. W. Downey, B. K. Flandermeyer, J. D. Jorgensen, T. E. Klippert, D. S. Kupperman, W. K. Kwok, D. J. Lam, A. W. Mitchell, A. G. McKale, M. V. Nevitt, L. J. Nowicki, A. P. Paulikas, R. B. Poeppel, S. J. Rothman, J. L. Roubort, J. P. Singh, C. H. Sowers, A. Umezawa, B. W. Veal, and J. E. Baker, Fabrication, Mechanical Properties, Heat Capacity, Oxygen Diffusion, and the Effect of Alkali Earth Ion Substitution on High T_C Superconductors, *Advanced Ceramic Materials* 2, 444-3B (1987).

PROCESSING –PROPERTY RELATIONS FOR $Y_{1-x}Gd_xBa_2Cu_3O_{7-\delta}$ HIGH T_C SUPERCONDUCTORS

Matti N. Makadsi,* Emad K. Al-Shakarchi,** and Muna M. Abbas*
* Physics Department, College of Science, Baghdad University, Baghdad, Iraq
** Physics Department, College of Science, Al-Nahrain University, Baghdad, Iraq

ABSTRACT

Solid state thermochemical reaction method involving mixing, calcining and sintering, has used to prepare high temperature superconductor with a nominal composition $Y_{1-x}Gd_xBa_2Cu_3O_{7-\delta}$ for Gd ($0 \leq x \leq 0.3$). The samples were studied to investigate the effect of preparational conditions and Gd substitution at Y sites on the superconducting properties. The critical temperature T_C of the samples were 86, 93, and 80 K for x= 0.0, 0.1, and 0.3 respectively. The X-ray diffraction analysis showed single phase with orthorhombic structure for all samples.

INTRODUCTION

Extensive investigations have been carried on in the past two decades and a halve to increase T_C of the high temperature superconductor (HTSc) compounds since first found in $YBa_2Cu_3O_{7-\delta}$ (YBCO) system of a perovskite unit cell, that was the breakthrough in this field i.e. $T_C > 77K$ [1,2]. The unit cell of $YBa_2Cu_3O_{7-\delta}$ consists of 3-orthorhombic elementary perovskite unit cells. The middle one has Yttrium Y atom at its center the upper and the bottom ones have each one Barium atom Ba at the center. The Y-atom is sandwiched by two CuO_2 planes while the upper plane and bottom plane are of CuO planes. The stoichiometry of this compound needs nine O-atoms while the crystallographic examination shows very near to seven O-atoms, thus it is referred as oxygen deficient perovskite. The advent of this breakthrough spawned a flurry of research directed toward these materials to modify the chemical composition by partial isotopic replacement of the elements in order to raise the transition temperature and to explore the electronic mechanism that lead to the high temperature superconductor transition T_C.

The planes in the perovskite unit cell which is the finger print of these materials play the major role in generating superconductivity while the chains act as electron reservoirs, these sites can be filled or emptied either by changing the oxygen atoms or by others type of doping [3,4]. It is claimed that the T_C value of hole – doped superconductors is on the average few times higher than that of the electron – doped superconductors [5].

It has been pointed out that the superconductor properties of $LnBa_2Cu_3O_{7-\delta}$ with larger ion radius lanthanide was improved by high temperature annealing [6], such that the sintering temperature 950°C was found more practical in processing its superconductivity, and this could be due to the increasing of contact areas (necking) between grains during the sintering process, in other words, decreases the porosity that led to higher densities and thus large grains has been observed [7].

Yu-Shan *et al.* [8] reported that YBCO compounds released oxygen with increasing the temperature, and the released oxygen can be absorbed back into the samples when the temperature decreases. The oxygen partial pressure of the atmosphere is important parameter in the solid – state reaction of the superconductors to stabilize the phases, especially on the Y-compounds superconducting. The control of charge carriers and oxygen content has a crucial effect on the superconducting properties [9].

Thus the structure and electrical development of $YBa_2Cu_3O_{7-\delta}$ compound depends on the preparing condition and partial substitution, therefore we thought, by carrying this work, that it is useful to know how these materials are affected by the processing variables together with the partial substitution of Y by Gd to promote T_C and to find the subtle and delicate balance between composition, structure and superconducting properties.

EXPERIMENTAL DETAILS

The samples of the system $Y_{1-x}Gd_xBa_2Cu_3O_{7-\delta}$ with concentration x=0.0, 0.1, 0.2, and 0.3 were prepared by conventional solid – state reaction route. The appropriate weights of the starting materials, Y_2O_3, Gd_2O_3, $BaCo_3$, and CuO were accurately taken. Then the powders were well mixed using agate mortar. A sufficient quantity of 2-propane was used to homogenize the mixture. The mixture was ground to a fine powder and then calcined in air at (850-950)°C for (15-30) h with repeated intermediate re-grinding. The powder obtained examined by XRD then ground again and pressed into disc-shaped pellets with 12 mm in diameter and (2-3) mm thickness, using a manually hydraulic press type (Specac) under different pressures (7-10) ton/cm^2 for (1-2) min.

In order to investigate the effective factors, which act on the superconducting properties, series of pellets were prepared with different conditions as shown in figure 1.

The first series; is sintered at 950°C for 5 h in air then for 24 h in an oxygen environment then we carried a subsequent slow air cooling to room temperature (process 1).

Second series: sintering at 950°C for 5 h in air then for 24 h in an oxygen environment and followed by slow air cooling to 550°C, it is hold at this temperature for 20 h with flow of oxygen and also followed by subsequent slow air cooling to room temperature (process 2).

Third series; sintered at 950°C for 10 h then air cooled to 550°C and hold at this temperature for 20 h under oxygen environment then air -cooled to room temperature (process 3).

It should be pointed out that the flow of pure oxygen over the pellets was at a rate of (0.7-0.9) l/min., the rate of heating was 1°C /min., while the rate of cooling was 0.5°C/min in a tube furnace.

Idometric titration was used to access the oxygen deficiency δ in the samples.

The structure of the prepared samples was examined by using X-ray diffractometer type Philips with the Cu-K_α radiation. A computer program was used to calculate the lattice parameters, based on Cohen's least square method.

The resistivity measurements were performed by the standard four-probe method.

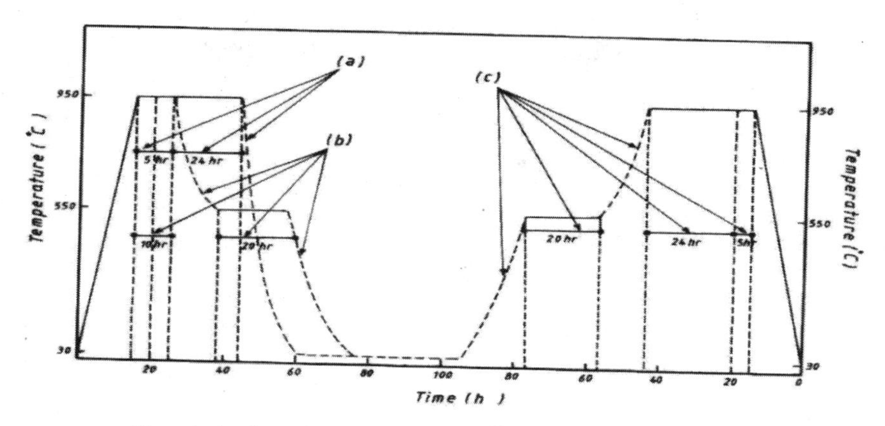

Figure1. A schematic diagram of the sintering processes for $Y_{1-x}Gd_x Ba_2Cu_3O_{7-\delta}$.

RESULTS AND DISCTION

This work showed that high calcination temperatures in the range (850-950)°C for a long period of time (15-30) h are necessary to produce an active powder with higher densities and having reduced particle size. We have observed that the first step of calcination gave significant inhomogeneous reaction especially when the sample was large amount of powder. The powder near the surface quickly acquired black color but the rest of the powder was green.

Intermediate milling and minimizing the thickness of the powder bed was performed in order to speed the process and produce chemically homogeneous powder.

It has been found that the best value of pressure on the powder to form a pellet is about 7.5 ton/cm^2 for (1-2) min., which produces samples with the highest T_C value. It should be pointed out that a pressure lowers than 7 ton/cm^2 is not efficient and produces brittle pellets and yields sample failure. On the other hand using pressure in the range (9-10) ton/cm^2 for (1-3) min. produces cracks inside the pellets during the sintering process, which render them to be broken easily during the investigations.

In particular, treatment at 950°C for various times was carefully examined, to determine the optimum sintering duration required for producing optimum HTSc T_C of $YBa_2Cu_3O_7$ sample.

It was interesting to note that, sintering for 5 h at 950°C without oxygen environment did not give useful modification for the preparation process, while when followed by 24 h sintering time at the same temperature with oxygen environment gave good results. This procedure decreased the sample failure and improved the superconducting behavior; furthermore the samples became quite stable in air. This is probably because samples sintered at 950°C in air are denser than those sintered at the same temperature in oxygen [10]. As the density of the samples became higher the open porosity closes and the distance between the grains decreases as well as decreasing the number and size of included pores, which limit the diffusion of oxygen in the matrix [11, 12]. Ling [13] found that the shrinkage increased with the sintering time during the first 2 h, reaching a saturated value after (4-6) h. However the shrinkage has stopped after 3 h at 950°C in air, while under flow of oxygen the shrinkage was still occurring. Moreover the oxygen treatment played an important role in determining the appropriate value of oxygen diffusion coefficient which is of critical importance in the optimizing the fully dense polycrystalline samples [10]. Glowaki *et al.* [14] showed that the temperature at which the rate of oxygen absorption was a maximum and is independent of the powder particle size, and the time constant for oxygen diffusion depends upon sample grain size. Various conditions in the preparation procedure for the samples are highlighted in Table 1.

Table 1. Values of oxygen content and T_C for different preparing conditions of $YBa_2Cu_3O_{7-\delta}$.

No. of Sample	Sintering Time at 950 °C (h)		Flow of O_2 (l/min)	Cooling Rate (°C/min)	Oxygen Content	T_C (K)
	Air Furnace	O_2 Atmosphere				
1	5	18	0.9	0.5	6.13	Semi
2		24	0.9	0.5	7.00	86.0
3		30	0.9	0.5	7.11	Semi
4		24	0.9	1.0	7.13	Semi
5		24	0.7	0.5	6.89	80.5
6		24	1-1.5	0.5	7.17	Semi

Figure 2 shows the resistivity vs. temperature for all $YBa_2Cu_3O_{7-\delta}$ samples prepared by process 1 but with different preparing conditions. The optimum preparing condition was for sample 2 which gave us the best superconductor behavior at T_C =86 K the related data are listed in Table 1 as shown, however T_C decreased to 80.5 K as the flow of oxygen decreased to 0.7 l/min.

Furthermore, when the sintering time was less than 24 h sample 1 exhibited incomplete zero-resistance because decreasing sintering time causes incomplete oxidation and this causes reduction of oxygen content. Based on the above observations, the parameters used to prepare sample 2 were the optimum for obtaining the highest T_C for the HTCs $YBa_2Cu_3O_7$ compound. Superconducting behavior and T_C were very sensitive to the preparing condition as we mentioned before. As shown in figure 2, sample 1 exhibited metallic - like conductivity and the transition to superconducting state has a kink up then a moderate drop albeit not complete. Sample 5 is belonging to $YBa_2Cu_3O_{6.89}$ prepared by process 1 but with low oxygen flow 0.7 l/min, as we notice there is a steep step like decreasing the resistivity, leading to T_C equal 80.5 K. Sample 1 has a sharp drop at 230 K but it didn't complete to zero albeit has metallic like behavior.

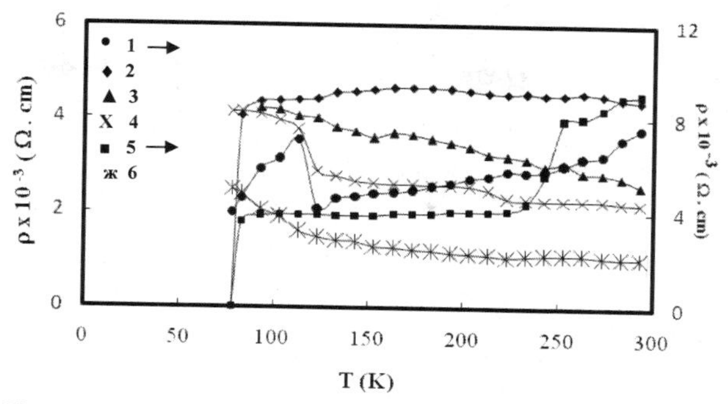

Figure 2. Temperature dependence of resistivity for $YBa_2Cu_3O_{7-\delta}$ prepared different conditions.

Table 2. Values of T_C for $Y_{1-x}Gd_xBa_2Cu_3O_{7-\delta}$ prepared at different conditions.
*Annealing at 550°C for 20 h.

No. of Sample	Time of Sintering (h) at 950° C with		Cooling with		T_C (K)	Oxygen Content
	Air Furnace	O_2 Atmosphere	Air Furnace	O_2 Atmosphere		
Gd concentration = 0.1						
(1)	5	24	X		93	6.69
(2)*		24		X	Semi	
(3)*		24	X		Semi	
(4)*		10		X	Semi	
(5)*		10	X	X	Semi	
Gd concentration = 0.2						
(1)	5	24	X		-	
(2)		30	X		Semi	
(3)*		24		X	Insulator	
Gd concentration = 0.3						
(1)	5	24	X		80.5	6.68
(2)*		24		X	Semi	
(3)		30	X		Semi	
(4)		24	X		Semi	

The Structural and electrical development of $Y_{1-x}Gd_xBa_2Cu_3O_{7-\delta}$ compound depend on the preparing conditions therefore it is useful to clarify the effect of time, temperature and flow of oxygen on T_C, such data are listed in Table 2 and shown in figures (3-5) .

Temperature dependence of the electrical resistivity of $Y_{1-x}Gd_xBa_2Cu_3O_{7-\delta}$ prepared by process1 with x=0.1, 0.2 and 0.3 have been studied. As shown in figures. (3-5) respectively, it is found that for $Y_{1-x}Gd_xBa_2Cu_3O_{7-\delta}$ samples with x=0.2 there was a decrease in the resistivity with low $T_C < 77$ K. While for the other superconductor compositions under study we observed sudden transitions to zero resistivity, such that T_C is found for $Y_{0.9}Gd_{0.1}Ba_2Cu_3O_{6.69}$ and $Y_{0.7}Gd_{0.3}Ba_2Cu_3O_{6.68}$ compound prepared by process 1 to be 93 K and 80.5 K respectively. Whereas other samples prepared by processes 2 and 3 were not superconducting as shown in figures (1 and 3), because annealing for 20 h after sintering at high temperature for long time didn't lead to superconducting phase.

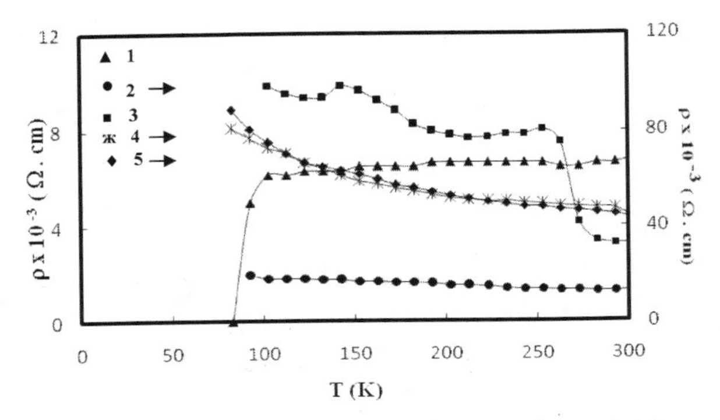

Figure 3. Temperature dependence of resistivity for $Y_{0.9}Gd_{0.1}Ba_2Cu_3O_{7-\delta}$ prepared at different conditions. Symbols denoted the number of samples.

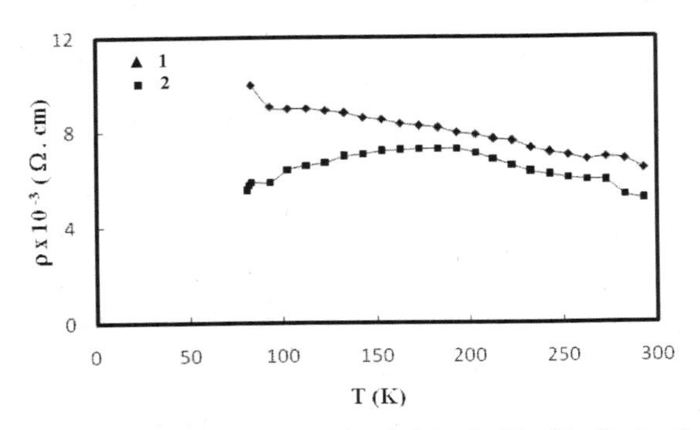

Figure 4. Temperature dependence of resistivity for $Y_{0.8}Gd_{0.2}Ba_2Cu_3O_{7-\delta}$ prepared at different conditions. Symbols denoted the number of samples.

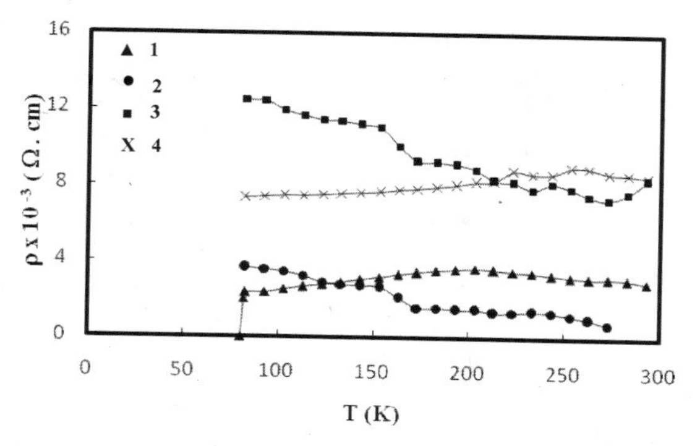

Figure 5. Temperature dependence of resistivity for $Y_{0.7}Gd_{0.3}Ba_2Cu_3O_{7-\delta}$ prepared at different conditions. Symbols denoted the number of samples.

The investigation of XRD pattern of calcinated powder showed a small amount of the desired phase of $YBa_2Cu_3O_7$ system formed in this process. The ordinary phase $BaCO_3$ was observed in the XRD pattern with the small amount as in figure (6-a).

XRD pattern of sintered samples revealed an orthorhombic phase of $YBa_2Cu_3O_7$. The positions of the peaks were in general agreement with that of the XRD patterns for calcinated powder. The sintering by process 1 gave polycrystalline with relatively large grain size. This estimation comes from the XRD sample 2 as shown in figure (6-b).

Excellent agreement is obtained between the measured peak position and intensities for this phase with those calculated on the basis of the crystal structure as determined by [15]. The highest peak is in fact shared by the reflection's 103, 013 and 110 planes and the intent of splitting on this peak is then a measure of the difference between the values of a and b.

Additional diffraction peaks presented in the spectra are associated with the compound $BaCuO_2$. The lattice constants deduced for $YBa_2Cu_3O_7$ are a = 3.8394 Å, b = 3.8895 Å and c = 11.6942 Å. Similar results observed by Elam *et al.* [16] with respect to all peak positions and intensities.

It is obvious from the diffraction spectra that increasing the sintering time up to 29 h enhances the relative amounts and quality of $YBa_2Cu_3O_7$ phase. In other words, the growths of the grains with the progress of the sintering is due to the usual theory of the growth of grains on account of smaller ones through grain boundaries, which in turn reduce the vacancy diffusions and the point defects or planer defects.

Figures (6-e and f) represents the X-ray profile of sample 3 and 4 respectively, and is showing the orthorhombic structure with small grain size as referred from the small intensity of the peaks this could be due to that, in the case of increasing sintering time caused the destruction of the intergrain junction;

the reason lies probably in the changes of the microstructure during the sintering, this treatment led to recrystallization processes causing in interruption of intergrain contacts [17].

While rapid cooling may cause pores trapped inside grains, small grain size pores on two and three grains junction and large pores are due to poor packing.

Trapped pores tend to be large and about the same size, indeed the pores are due to packing defects and tend to be inside large grains. The pores on boundaries are much smaller and are mostly at triple junctions. Slow cooling is necessary to form a phase with a well-characterized orthorhombic distortion, the same result was found by [18]. The lattice constants evaluated from 2θ of the major peaks are also listed in Table 3.

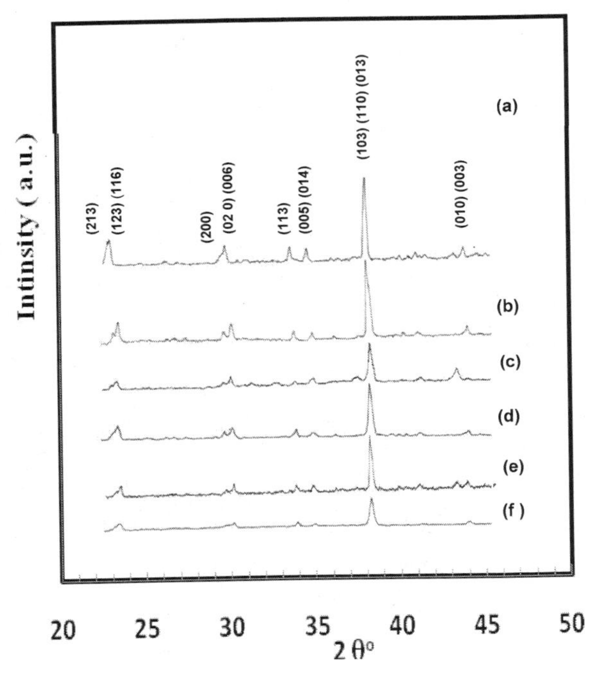

Figure 6. XRD pattern of $YBa_2Cu_3O_{7-\delta}$□ compound. (a) calcining powder ; (b) optimum sample; (c) sample sintering time < 24 h; (d) sample with flow of Oxygen 0.7 L/min.; (e) sample cooling with rate 1°/min. and (f) sample sintering time > 24 h.

Table 3. Lattice parameters and oxygen content of $YBa_2Cu_3O_{7-\delta}$ prepared at different conditions.

x	a (Å)	b (Å)	c (Å)	V (Å)^3	Oxygen content	Tc (K)
0.1	3.8412	3.8751	11.7004	174.16	6.84	120.5
0.2	3.8241	3.8777	11.706	173.58	6.69	86.5
0.3	3.8310	3.8838	11.6417	173.214	6.60	81

XRD for $Y_{x-1}Gd_xBa_2Cu_3O_{7-\delta}$ with x=0.1, 0.2 and 0.3 prepared by process1 showed single phase with orthorhombic structure as shown in figure 7. Comparable to XRD pattern of $YBa_2Cu_3O_7$ there is a small shift in 2θ with increasing x in agreement of reference [19]. Intensity represented by the peaks decreases with increasing the substitution except at x = 0.2. The lattice parameters a, b and c were found to vary unsystematically with increasing x, this results was similar to Ganepathi [20].

In general from Table 3 both compound volume cells are decreasing with increasing x, this is because of oxygen vacancies ordered Ba^{+3} and Gd^{+3} on Y^{+3} site in ordered arrangements [21].

Oxygen vacancies are thought to play a key role in the appearance or disappearance of superconductivity in $YBa_2Cu_3O_{7-\delta}$ compounds. Diffusion of oxygen through the superconducting material is suggested to be guidelines for annealing treatments intended to optimize the oxygen content and superconducting properties of $YBa_2Cu_3O_{7-\delta}$. Oxygen stoichiometry in the as - prepared sample was determined by Idometric titration method is listed in the Table 1. Decreasing sintering time causes reduction of oxygen content and this is properly due to incomplete oxidation of sample 1. We have noticed that the oxygen content increased continuously with increasing the sintering time with the flow of oxygen. But, however, it is found that increasing the sintering time with the flow of oxygen more than 24 h causes oxygen saturation such that the sample exceeds the recommended oxygen content > 7 as sample 3. Furthermore it was found that increasing of the oxygen flow more than 0.9 l/min, the oxygen content became more than 7 and we got non-superconducting samples.

Rapid cooling 1°C/min. yielded sample 4 with oxygen content > 7. It is obvious that decreasing δ causes semi saturation in the oxygen content in the structure which tends to decrease the amount of vacancies related to the site of O5, and the last cause to decrease T_C value or disappearance of superconductivity [22]. In fact, a slow cooling 0.5°C is considered the better way to produces a sample with acceptable oxygen content.

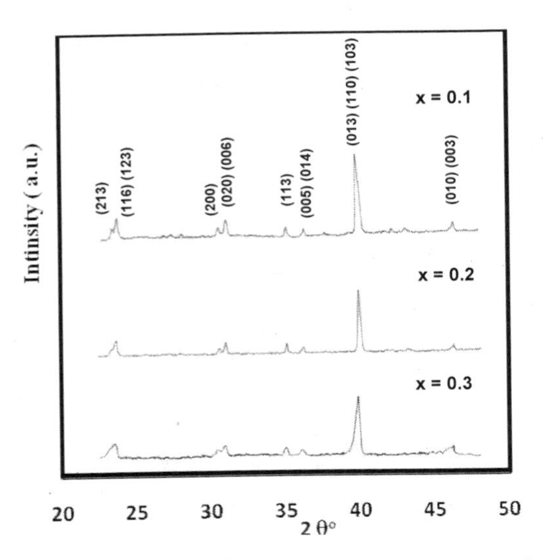

Figure 7. XRD pattern of $Y_{1-x}Gd_xBa_2Cu_3O_{7-\delta}$ compound with different x.

As for $Y_{x-1}Gd_xBa_2Cu_3O_{7-\delta}$ compound, it can be noted, from Table 4, the oxygen content and T_C both decreased gradually with increasing the partial substitution of Gd, this could be attributed to redistribution of charge in the lattice which affects the hole concentration in the CuO plane as well as create more distortions in the original lattice; and or high vacancy concentration on the bridging apical oxygen O4 sites [23].

Table 4. Lattice parameters, values of oxygen content and T_C for $Y_{x-1}Gd_xBa_2Cu_3O_{7-\delta}$ compound with different x.

x	a (Å)	b (Å)	c (Å)	V $(Å)^3$	Oxygen Content	T_C (K)
0.0	3.8314	3.8875	11.6744	173.885	7.00	86.0
0.1	3.8408	3.8892	11.6499	174.02	6.69	93.0
0.3	3.8354	3.8870	11.5595	172.33	6.68	80.5

The significant point which can be deduced from our data is that, all the samples by processes 1 revealed a superconductor behavior as prescribed in figure 8.

To explore why T_C decreased and the volume contracted with increasing Gd concentrations, our speculations are that:

(1) Gd^+ ion altruist additional oxygen atom causing an increasing of oxygen deficiency and Cu1-O4 band starts to be shorter and simultaneously T_C begins to decrease [24].

(2) Superconducting is dependent not only on the oxygen content but also depends on the arrangement of the oxygen ion [25] and also on the excess distortion resulted from partial substitution.

(3) Another researchers found that decreasing of T_C is connected with an increasing of the buckling of the CuO plane [26].

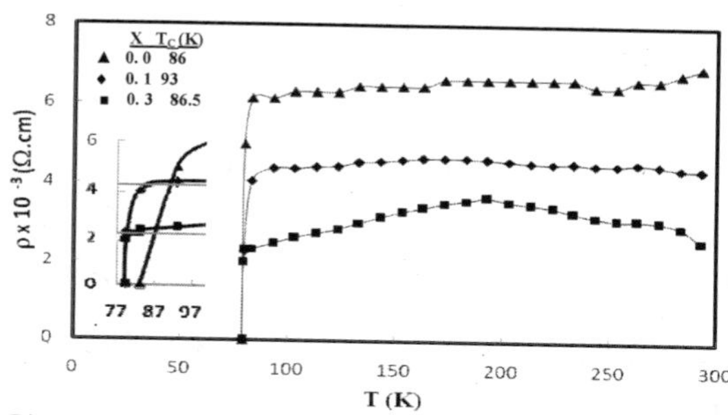

Figure 8. Temperature dependence of resistivity for $Y_{x-1}Gd_xBa_2Cu_3O_{7-\delta}$ prepared by process 1.

CONCLUSIONS

Optimum sintering conditions for our samples was sintering at 950°C for 29 h (5 h without oxygen and 24 h with 0.9 l/min. flow of oxygen). Albeit these optimum conditions showed superconductor behaviors for Gd concentration, but not for 0.2

Maximum value of T_C for the compound $Y_{1-x}Gd_xBa_2Cu_3O_{7-\delta}$ with (x= 0.0, 0.1 and 0.3) found to be 86.93 K and 80.5 K respectively.

It has been obtained from XRD that all HTSc samples are polycrystalline with orthorhombic phase.

The merit of the above results is that, the physical properties of $Y_{1-x}Gd_xBa_2Cu_3O_{7-\delta}$ compounds depend greatly not only on the elemental composition but also on the details of the preparation processes.

REFERENCES

[1]C.N.R. Rao, Chemistry of High Temperature Superconductors in Chemical and Structural Aspects of High Temperature Superconductors, Editor C.N.R. Rao, world scientific Co.Pte. Ltd. Singapore,1 (1988).

[2]A. Tavana and M. Akhavan, How Tc can go above 100 K in the YBCO Family. Eur. Phys. J. B (2009) DOI: 10.1140/epjb/e2009-00396-7.

[3]J. M. Tarascon, W. R. Mckinnon, L. H. Green G. W. Hull and E. M. Vogel, Oxygen and Rare Earth Doping of the 90K Superconducting Perovskite $YBa_2Cu_3O_{7-x}$, Phys. Rev. B., 36, 226-1 (1987).
[4]M. Georgiev and N. Balchev, Recent Developments of Home Research on High Tc Superconductivity and Colossal Magneto Resistance: An attempted survey. Bulg. J. Phys. 30, 60 (2003).
[5]G. Alecu. Crystal Structures of Some High- Temperature Superconductors, Romanian Reports in Physics, 56, 404 - 3 (2004) .
[6]Anderi Marouchkine, Room –Temperature Superconductivity, Cambridge International Science publishing United Kingdom, First published (2004).
[7]T. Iwata, M. Hikita and S. T Surumi, New Experimental Results Concerning the Difference between Various T_Cs of $LnBa_2Cu_3O_y$ (Ln=Lanthanide) as in Advances in Superconductivity, Proceeding of its International Symposium on Superconductivity (ISS88) edited by K. Ishiguro (Eds), Nagoya,197 (1988).
[8]F.Yu-Shan, S. Homg-Zhang, I.I Zhi-Hui, Y. De-Lin and H. Xing , Influence of the rare earth element substitution on oxygen adsorption-deposition properties of YBCO, Frontiers of Physics in China,3, 368(2006).
[9]A. Mohanta and D. Behera , Effect of Ga and Zn Doping in Coherent Transition of YBCO Superconductor, Indian J.of Pure and Applied Physics,47,676 (2009).
[10]J.E. Blendell ,C.K. Chiang, D.C.Cranmer, S.W. Freiman, E.R.Fuller, Jr., E. Drescher-Krasicka, W. L.Johonson, H.M. Ledbetter, L.H.Bennett, L.J. Swartzendruber, R.B.Marinenko, R.L. Myklebust, D.S. Bright, and D.E. Newbury, Processing –Property Relations for $Ba_2YCu_3O_{7-x}$ High T_C Superconductors, in Advanced Ceramic Materials, Special Issue, 2, 512- 3B(1987).
[11]G. W. Crabtree, J. W. Downey, B. K. Flandermeyer, J. D. Jorgensen, T. E. Klippert, D. S. Kupperman, W. K. Kwok, D. J. Lam, A. W. Mitchell, A. G. McKale, M. V. Nevitt, L. J. Nowicki, A. P. Paulikas, R. B. Poeppel, S. J. Rothman, J. L. Roubort, J. P. Singh, C. H. Sowers, A. Umezawa, B. W. Veal, and J. E. Baker, Fabrication, Mechanical Properties, Heat Capacity, Oxygen Diffusion, and the Effect of Alkali Earth Ion Substitution on High Tc Superconductors, Advanced Ceramic Materials, 2, 444-3B (1987).
[12]S. Kobayashi, T. Kandori, Y. Saito, S. Wada, Densification and Critical Current Density of $YBa_2Cu_3O_x$ Ceramics as in Advances in Superconductivity, Proceeding of its International Symposium on Superconductivity (ISS88) edited by K. Ishiguro (Eds), Nagoya, 259(1988).
[13]H.C.Ling, Effect of Cu and Sources on Powder Processing and Densification of Superconducting $YBa_2Cu_3O_x$ Ceramic as in Advances in Superconductivity, Proceeding of its International Symposium on Superconductivity (ISS88) edited by K. Ishiguro (Eds), Nagoya, 272(1988).
[14]B.A. Glowaki , R.J. Highmore, K.F.Peters,A.L.Greer and J.E. Evetts, A calorimetric Study of Oxygen Intercalation and Desorption in bulk Superconducting $Y_1Ba_2Cu_3O_{7-x}$, Supercond. Sci. Technol. 1,7 (1988).
[15]A.M.T. Bell, Calculated X-Ray Powder Diffraction Patterns and Theoretical Densities for Phases Encountered in Investigations of Y-Ba-Cu-O Superconductors, Supercond.Sci. Technol. 3,55-2,(1990).

[16]W.T. Elam, J.P. Kirkland, R.A. Neiser, E.F. Skelton, S. Sampath, and H. Herman, Plasma Sprayed High T_C Superconductors, Adv. Cer. Mat. 2/3B,411(1987).
[17]E. Pollertet J. Hejtmank, L. Matejkova, M. Nevriva, A.Triska,P.Vasek and D. Zemanova, Effect of annealing in oxygen on the superconducting properties of $YBa_2Cu_3O_{7-x}$, Physica C, 156, 533-4 (1988).
[18]B.Bender,L.Toth, J.R.Spann, S.Lawrence, J.Wallace, D.Lewis, Mosofsky, W.Fuller, E.Skelton, S.Wolf, S.Qadri, and D.Gubser, Processing and Properties of the High T_C Superconducting Oxide Ceramic $YBa_2Cu_3O_7$, Advanced Ceramic Materials, 2, 506- 3B, (1987).
[19]N. N.Wakim, Study the Effect of Gd Substitution for Y in High Temperature YBaCuO Superconductor, Ms. C. Thesis, Baghdad University College of Science, (1994).
[20]L. Ganapathi, L. Ganapathia, Ashok Kumara and J. Narayan, Characteristics of high-Tc $La_xY_{1-x}Ba_2Cu_3O_{7-\delta}$ Superconductors Physica C, 167, 669-(5-6), (1990).
[21] X. T. Xu, J. K. Liang, S. S. Xie,G. C. Che, X. Y. Shao, Z. G. Duan and C. G. Cui, Crystal Structure and Superconductivity of BaGdCuO System, Solid State Commu., 63, 649- 7, (1987).
[22] Y.Kubo and H. Igarashi, Theoretical Prediction of the Orthorhombic (II) Phase in the YBa2Cu3O7-δ System. J. Appl. Phys. 26,L1988 (1987).
23 J.L. Mac Manus- Driscoll, J.A.Alonso,P.C.Wang, T.H.Geballe and J.C.Bravman, Studies of Structural Disorder in $ReBa_2Cu_3O_{7-x}$ Thin Films as a Fucction of Rare Erth Ionic Radius and Film Deposition Conditions, Physica C, 232, 288(1994).
[24] J.X. Zhang, G.M. Lin, W.G. Zeng, K.F. Liang, Z.C.Lin, G.G. Siu, M.J. Stokes and W. Fung, An Elastic Relaxation of Oxygen Vacancies and High-T Superconductivity of $YBa_2Cu_3O_{7-6}$, Supercond. Sci. Technol. 3,113(1990).
[25] A.M. Neminsky and D.V Shovkun, Conductivity of $YBa_2Cu_3O_{6+x}$ Single Crystals Close to the Superconductor-Nonsuperconductor Transition, Physica C, 252,327-(3-4) (1995).
[26] B. Buchner. U. Calleib, H. D. Jostarndt. W. Schlabitz and D. Whohlleben, Correlation of Spectroscopic and Superconducting Properties of $REBa_2Cu_3O_{7-y}$ with the Rare Earth Ionic Radius, Solid State Comm., 73, 357- 5, (1990).

Magnetoelectric Multiferroics

FINITE-SIZE EFFECTS IN NANOSCALED MULTIFERROICS

J. J. Heremans (1), J. Zhong (2,3), R. Varghese (2,3), G. T. Yee (4), and S. Priya (2,3)
1) Department of Physics, Virginia Tech, Blacksburg, Virginia 24061, USA
2) Center for Energy Harvesting Materials and Systems, Virginia Tech, Blacksburg, Virginia 24061, USA
3) Department of Materials Science and Engineering & Department of Mechanical Engineering, Virginia Tech, Blacksburg, Virginia 24061, USA
4) Department of Chemistry, Virginia Tech, Blacksburg, Virginia 24061, USA

ABSTRACT

We present studies of single-phase multiferroic materials in nanoparticle or nanoplate form, where size effects modify the long range magnetic structure and hence influence the magnetic properties. Size effects as isolated here, are of particular relevance to understand the interplay between magnetic and ferroelectric properties in heterogeneous, multi-phase multiferroic structures. Bulk multiferroic $BiMnO_3$ shows ferromagnetic ordering below 105 K, but few characterization attempts have been performed on nanoscaled material. We have synthesized $BiMnO_3$ nanoplates and characterized the structure, composition, morphology and magnetic properties. Ferromagnetic behavior is not observed in the nanoplates, in contrast to the bulk, indicative of finite size effects. We have also synthesized $BiFeO_3$ nanoparticles (characterized by XRD, EDS and magnetometry), which exhibit ferromagnetism with spin-glass-like behavior, as a result of reduced dimensionality. Bulk perovskite $BiFeO_3$ instead shows antiferromagnetism. The difference in magnetic behavior for nanoparticles compared to bulk behavior, is, in both cases, attributable to finite-size effects.

INTRODUCTION

Magnetoelectric multiferroic single-phase materials are characterized by the presence of directly coupled ferroic orders, typically ferromagnetism and ferroelectricity. Attempts to identify single-phase systems exhibiting both ferromagnetic and ferroelectric orders hark to a study by Landau and Lifshitz analyzing the presence of magnetoelectric coupling on the basis of crystal symmetry, which led to theoretical and experimental work on Cr_2O_3 [1-4]. The magnetoelectric coupling in principle allows a technologically relevant control of magnetization by electric fields [5]. In particular, multiferroic materials offer potential applications in information storage, in spintronics and in sensors, where the magnetoelectric coupling can be used for multi-state functionalities [6]. However materials limitations exist on the types of single-phase structures in which the coupling can exist [7,8], and even then, in most single-phase materials the coupling is weak [9,10]. Several single-phase multiferroics have been reported in the literature, including $Ni_3B_7O_{13}I$, RMn_2O_5 (R: rare earths), $Ni_3V_2O_8$, $CoCr_2O_4$, $MnWO_4$, and $Pb(Fe_{2/3}W_{1/3})O_3$ [11-17], and the origin of the coupling between magnetic and ferroelectric degrees of freedom is under study. However, the search for high temperature single phase materials exhibiting a magnetoelectric effect of useful magnitude continues. Indeed in most single-phase materials studied so-far, the orders and their coupling occur much below room temperature, since Néel or Curie temperatures are low [18]. Several avenues exist however to circumvent the present limitations on single-phase materials. Bilayers in which individual layers are coupled via strain have shown much promise, and, as another avenue, here we present examples how the ferroic orders of materials can be altered by reduced dimensions, with as specific examples $BiMnO_3$ (BMO) and $BiFeO_3$ (BFO). Size effects can introduce changes in physical behavior and offer the opportunity to tailor the physical properties. For example, BFO is known to exhibit antiferromagnetism in bulk form, yet several studies have suggested the onset of ferromagnetic behavior in BFO thin films and nanoparticles [19, 20].

BISMUTH MANGANESE OXIDE NANOPLATES

Among the multiferroic single-phase materials, BMO is a promising candidate for applications, but experiments on BMO have focused on bulk or thin film materials, with few attempts at nanoscale dimensions. Perovskite-phase BMO is typically synthesized under pressures of 40 - 50 kbar in the temperature (T) range of 600 - 900°C which results in stabilization of the desired perosvskite phase [21]. In this work, we have attempted synthesis of BMO nanoplates by the co-precipitation method. Stoichiometric quantities of Bi_2O_3 and $MnCl_2$ were dissolved in HNO_3 and HCl and NaOH was subsequently added to form the hydroxide gel. The gel was continuously stirred for 4 - 6 hours with T maintained at 100°C. The resulting crystalline powder was subsequently filtered and dried, and was found to consist of two phases, which were characterized by various methods. X-ray diffraction (XRD), energy dispersive X-ray spectroscopy (EDS), X-ray photoelectron spectroscopy (XPS), SEM and TEM were employed to characterize the structure, composition and morphology of the phases. SQUID magnetometry was applied for characterization of the ensemble magnetic properties.

Figures 1(a) and (b) depict TEM and SEM micrographs of as-synthesized material. Although nanoscale plates (area marked as “1” in Fig. 1(a), and dominant in Fig. 1(b)) form the majority of the material, a concentration of needle-shaped particles (area marked as “2” in Fig. 1(a)) was identified along with nanoplates in the TEM micrograph. The XRD pattern shown in Fig. 2 did not detect the presence of a secondary phase and indicates single-phase BMO. The major fraction of the material is hence BMO, corresponding with the dominant plate shape. Peak broadening is visible in the XRD due to the reduced dimensions of the BMO nanoplates. TEM-EDS analysis (Table I-top) on the needle-like phase, in the area marked as “2” in Fig. 1(a), shows that the needle-phase has substantially lower Mn content than Bi content. On the other hand, EDS analysis on the BMO nanoplates (Table I-bottom), conducted on the area marked as “1” in Fig. 1(a), shows Mn content to be slightly off-stoichiometry, and higher than Bi. Thus, the minority needle shape is a Bi-rich phase, while the majority BMO phase has a plate shape, and dominates the XRD pattern. The majority phase also likely dominates the magnetic properties.

The presence of multiferroicity in BMO was proposed by Hill and Rabe [22] and was experimentally confirmed by Sugawara *et al.* [23,24] and Kimura *et al.* [25]. Bulk BMO exhibits a ferromagnetic transition around $T = 105$ K [26]. Measurement of ferroelectric polarization on the other hand has been hampered due to high electric leakage currents. We performed measurements of magnetization (M) at T = 10 K, 77 K and 298 K, as function of applied magnetic field (H). Magnetic hysteresis loops could not be identified from 298 K down to 10 K, precluding ferromagnetic behavior in the ensemble and hence also in the BMO nanoplates. Rather, paramagnetic behavior could be concluded in the sample, as shown in Fig. 3. Figure 3 depicts the T dependence of the susceptibility (χ, defined as $\chi = M(T)/H$) at fixed H under zero field cooling (ZFC) and field cooling (FC) conditions ($M(T)$ was measured under an applied H = 5000 Oe). The difference in $M(T)$ between ZFC and FC conditions never exceeds ~ 10 %, and hence irreversibility or spin-glass behavior are concluded to be absent. In short, no ferromagnetic transition was observed, indicating a loss of ferromagnetic order in these BMO nanoparticles.

In order to correlate the observed BMO slight non-stoichiometry to changes in valence state of the Mn ions, we performed a comparative XPS analysis. Figure 4 shows the XPS spectrum on four different materials: the BMO nanoplates, $MnCO_3$, Mn_2O_3 and MnO_2, with the latter three samples used as standard references corresponding to Mn valence states of +2, +3, and +4. If the XPS signal from the BMO sample coincides with one of the standards, the dominant Mn valence can be approximated. The results in Fig. 4 demonstrate that one peak position for the BMO nanoplates matches Mn_2O_3, indicating a Mn valence state in BMO of +3. However, XPS spectra of BMO samples also exhibit a signal in the binding energy range 650 eV - 645 eV. Hence other Mn valence states exist in the BMO nanoplates, as expected due to non-stoichiometry. This result also explains the presence of minor traces of needle shaped particles. The varying Mn valence can contribute to the deletion of the ferromagnetic transition, while another important contribution can be found in the nanoscale morphology.

BMO hosts three crystallographically distinct Mn^{3+} sites, and supports 6 different Mn-O-Mn superexchange pathways [26, 27]. The ferromagnetic order in bulk BMO originates in the Mn-O-Mn orbital configurations with ferromagnetic interactions outnumbering and dominating over the non-ferromagnetic configurations. A dominant collinear ferromagnetic ordering hence appears with an average Mn^{3+} magnetic moment of 3.2 μ_B [27], but the non-ferromagnetic interactions lead to partial frustration and a lower spin value than in the fully aligned case [26, 27]. Dynamic light scattering measurements in our samples show an average plate size of the BMO phase of ~120 nm in diameter. SEM and TEM micrographs (Fig. 1) imply a thickness much less than the diameter, hence the identification with a nanoplate morphology. In analogy to size effects mentioned for BFO, the low dimensionality of the BMO nanoplates can be invoked for the loss of ferromagnetism, resulting from alterations in the superexchange mechanism. We surmise that the low dimensionality of the nanoplates affects the length and the angle of Mn-O-Mn bonds, causing the delicate ferromagnetic order to disappear. The non-stoichiometric, Mn-rich composition of the BMO nanoplate phase may also play a role in suppressing ferromagnetism, since to compensate for the extra charge, changes in bond density of the various Mn-O-Mn orbital configurations may occur which affect the Mn-O-Mn superexchange interactions.

BISMUTH IRON OXIDE NANOPARTICLES

In bulk form the multiferroic perovskite BFO has a high ferroelectric Curie temperature T_C = 1143 K and a high antiferromagnetic Néel temperature T_N = 643 K [28]. BFO is a G-type antiferromagnet with the Fe^{3+} magnetic moments coupled ferromagnetically within the (111) planes and antiferromagnetically between neighboring planes [29,30]. The presence of Fe^{3+} ions with a magnetic moment of $\mu_{Fe} = 3.70\ \mu_B$ was confirmed by neutron diffraction [31,32]. Neutron powder diffraction showed that the structure could be described using modulated magnetic ordering models such as a circular cycloid, an elliptical cycloid and a spin density wave [33]. A cycloidal spatially modulated spin structure is hence present in bulk samples, responsible for a cancellation of the macroscopic magnetization [34]. A helical spin structure with a long wavelength of 62 nm, incommensurate with the lattice spacing [35,36], results from the fact that the axis of spin alignment precesses through the crystal. As a result of the complex cycloidal spin structure easily being perturbed by strain, it has been suggested experimentally [37-40] and predicted theoretically [34], that in thin films weak ferromagnetism is present. In this work the focus lies on exploring the magnetic behavior of the BFO nanoparticles, since few results are available on nanoparticles, while size effects can lead to a modification of magnetic order as thin films indicate.

The BFO nanoparticles here were synthesized via the sol-gel method. Bismuth nitrate ($Bi(NO_3)_3 \cdot 5H_2O$) and iron nitrate ($Fe(NO_3)_3 \cdot 9H_2O$) in stoichiometric proportions (1:1 molar ratio) were dissolved in 2-methoxyethanol ($C_3H_8O_2$). The solution was adjusted to a pH value of 4–5 by adding 2-methoxyethanol and HNO_3. This mixture was stirred for 30 minutes at room temperature to obtain the sol, which was then kept at 80°C for 96 hours to form the dried gel powder. The dried powder was calcined in a temperature range of 400–600°C for 1 - 3 h in air. The optimal calcination temperature was 450 °C for 2 hours. The structure and morphology of BFO nanoparticles were investigated by powder XRD, and SEM. Magnetic measurements were obtained by variable-temperature SQUID magnetometry.

A SEM micrograph of the as-synthesized nanoparticle ensemble is depicted in Fig. 5, revealing a spherical morphology with average particle size of 20 nm. The XRD pattern of the as-synthesized material, depicted in Fig. 6, indicates a single-phase perovskite structure. The (111) peak splitting indicates a rhombohedral structure for the nanoparticle ensemble, consistent with the reported structure of BFO ceramics [41].

Figure 7 shows the H dependence of the magnetization M for the BFO nanoparticles at T= 10 K, 77 K and 298 K. The magnetic hysteresis shows ferromagnetism from 10 K to room temperature. As mentioned above, in bulk BFO single crystals the magnetization measurements are expected to exhibit antiferromagnetism [42,43]. Yet, weak ferromagnetism is often reported in polycrystalline BFO ceramics and thin films [44]. Spontaneous magnetization in as-synthesized BFO samples has been ascribed to cumulative effects of mixed Fe^{2+}/Fe^{3+} valence formation, of suppression of the helical spin structure, of an increase in canting angle, and/or of iron-rich nanoclusters [37,38,40,41,45-47]. A spin-glass behavior has also been observed in the M (T) and M (H) measurements for BFO films annealed in air atmosphere, further enhanced by annealing the sample in an oxygen atmosphere and hence tentatively related to the formation of iron oxide nanoclusters or precipitates [40,47]. Yet in our samples no secondary phases are present according to the XRD and according to the EDS analysis contained in Table II. The BFO nanoparticle composition was found to be stoichiometric. Hence, we attribute the observed ferromagnetism to size effects resulting from the nanoparticle geometry influencing the cycloid spin structure. The wavelength of the fully developed incommensurate cycloid spin structure in bulk BFO is 62 nm [48], longer than the scale of our BFO nanoparticles. The cycloid spin structure characteristic of antiferromagnetism is interrupted by the particle surface in our BFO nanoparticles, an interruption that in thin films can induce ferromagnetic behavior [40,43].

Figure 8 contains the dependence on T of the magnetization M (T) under ZFC and FC conditions, with M (T) measured under applied H of 5000 Oe. At 5000 Oe the saturation magnetization is not reached, as can be observed in Fig. 7. Yet, the low-T value for the magnetization measured here, ~ 0.52 emu/g, compares well to the ~ 0.49 emu/g recently obtained for nanocrystalline films under FC conditions [49]. M (T) in Fig. 8 further shows irreversibility between FC and ZFC conditions over a wide range of T, a clear indication of spin-glass-like behavior (also observed in Ref. 49). For BFO, spin-glass-like behavior has been observed in both single-crystal and thin film samples (see *e.g.* Ref. 50). However, Fig. 8 contains evidence of spin-glass-like behavior in BFO nanoparticles. To provide context, the unique long-range spiral spin structure characteristic of bulk BFO will lead to spin-glass behavior differing from Ising systems. The spin-glass behavior may have long-range Coulumbic contributions to the electromagnons which are not pure spin waves [51]. Pinning of the incomplete spin spiral structure at the nanoparticle boundaries may also result in the experimental observation of spin-glass-like M (T). Further, spin-glass-like behavior in BFO has been ascribed to diffusion of domain walls, since the domain walls in BFO are known to influence both ferroelectric and ferromagnetic properties [52].

CONCLUSION

In both multiferroic materials $BiMnO_3$ and $BiFeO_3$, size effects modify the magnetic ordering, as demonstrated here by a synthesis of $BiMnO_3$ nanoplates and $BiFeO_3$ nanoparticles, followed by extensive compositional, structural and magnetic characterization. $BiMnO_3$ nanoscale plates were synthesized through co-precipitation. Magnetic measurements demonstrate a lack of magnetic ordering behavior which was explained on the basis of size effects and non-stoichiometry. $BiFeO_3$ nanoparticles were synthesized by a sol-gel method, and characterization shows the presence of $BiFeO_3$ at an average particle size of 20 nm (shorter than the cycloid length). Ferromagnetic properties and spin-glass-like behavior were observed in the $BiFeO_3$ nanoparticles. Spin-glass-like behavior can be ascribed to diffusion of domain walls, with possible contributions from pinning of the cycloid spin structure at the nanoparticle surface. Ferromagnetic behavior in the nanoparticles can be ascribed to an interruption or partial destruction of the long-wavelength cycloid spin structure characteristic of bulk $BiFeO_3$.

REFERENCES

[1] L. D. Landau and E. M. Lifshitz, *Electrodynamics of Continuous Media*, Pergamon Press, Oxford, 119 (1960).
[2] I. E. Dzyaloshinskii, Soviet Phys.- JETP **10**, 628 (1960).
[3] D. N. Astrov, Soviet Phys.- JETP **11**, 708 (1960).
[4] D. N. Astrov, Soviet Phys.- JETP **13**, 729 (1961).
[5] W. Prellier, M. P. Singh, and P. Murugavel, J. Phys. Condens. Matter **17**, R803 (2005).
[6] T. Zhao, A. Scholl, F. Zavaliche, K. Lee, M. Barry, A. Doran, M. P. Cruz, Y. H. Chu, C. Ederer, N. A. Spaldin, R. R. Das, D. M. Kim, S. H. Baek, C. B. Eom, and R. Ramesh, Nature Materials **5**, 823 (2006).
[7] N. A. Hill, J. Phys. Chem. B **104**, 6694 (2000).
[8] N. A. Hill, and A. Filippetti, J. Magn. Magn. Mater. **242**, 976 (2002).
[9] R. Ramesh, and N. A. Spalding, Nature Mater. **6**, 21 (2007).
[10] S. W. Cheong, and M. Mostovoy, Nature Mater. **6**, 13 (2007).
[11] G. A. Smolenskii, and V. A. Bokov, J. Appl. Phys. **35**, 915 (1964).
[12] W. Qu, X. Tan, R. W. McCallum, D. P. Cann, and E. Ustundag, J. Phys: Condens. Matter **18**, 8935 (2006).
[13] N. Hur, S. Park, P. A. Sharma, J. S. Ahn, S. Guha, and S. W. Cheong, Nature **429**, 392 (2004).
[14] G. Lawes, A. B. Harris, T. Kimura, N. Rogado, R. J. Cava, A. Aharony, O. Entin-Wohlman, T. Yildirim, M. Kenzelmann, C. Broholm, and A. P. Ramirez, Phys. Rev. Lett. **95**, 087205 (2005).
[15] Y. Yamasaki, S. Miyasaka, Y. Kaneko, J.-P. He, T. Arima, and Y. Tokura, Phys. Rev. Lett. **96**, 207204 (2006).
[16] K. Taniguchi, N. Abe, T. Takenobu, Y. Iwasa, and T. Arima, Phys. Rev. Lett. **97**, 097203 (2006).
[17] O. Heyer, N. Hollmann, I. Klassen, S. Jodlauk, L. Bohatý, P. Becker, J. A. Mydosh, T. Lorenz, and D. Khomskii, J. Phys.: Condens. Matter **18**, L471 (2006).
[18] M. Fiebig, J. Phys. D: Appl. Phys. **38**, R123 (2005).
[19] J. O. Cha, and J. S. Ahn, J. Korean Phys. Soc. **54**, 844 (2009).
[20] F. Gao, X. Y. Chen, K. B. Yin, S. Dong, Z. F. Ren, F. Yuan, T. Yu, Z. G. Zou, and J. M. Liu, Adv. Mater. **19**, 2889 (2007).
[21] Y. Yu Tomashpol'skii, E. V. Zubova, K. P. Burdina, and N. Yu. Venevtev, Sov. Phys. Crystallogr. **13**, 859 (1969).
[22] N. A. Hill, and K. M. Rabe, Phys. Rev. B **59**, 8759 (1999).
[23] F. Sugawara, S. Iida, Y. Syono, and S. Akimoto, Jpn. J. Phys. Soc. **20**, 1529 (1965).
[24] F. Sugawara, S. Iida, Y. Syono, and S. Akimoto, Jpn. J. Phys. Soc. **25**, 1553 (1968).
[25] T. Kimura, S. Kawamoto, I. Yamada, M. Azuma, M. Takano, and Y. Tokura, Phys. Rev. B **67**, 180401 (2003).
[26] A. M. Santos, A. K. Cheetham, T. Atou, Y. Syono, T. Yamaguchi, K. Ohoyama, H. Chiba, and C. N. R. Rao, Phys. Rev. B **66**, 064425 (2002).

[27] T. Atou, H. Chiba, K. Ohoyama, Y. Yamaguchi, and Y. Syono, J. Solid State Chem. **145**, 639 (1999).
[28] J. M. Moreau, C. Michel, R. Gerson *et al.* , J. Phys. Chem. Solids **32**, 1315 (1971).
[29] P. Fischer, M. Połomska, I. Sosnowska, and M. Szymański, J. Phys. C : Solid St. Phys. **13**, 1931 (1980).
[30] R. Przenioslo, A. Palewicz, M. Regulsk, I. Sosnowska, R. M. Ibberson, and K. S. Knight, J. Phys: Condens. Matter **18**, 2069 (2006).
[31] I. Sosnowska, R. Przenioslo, P. Fischer *et al.*, Acta Phys. Polonica A **86**, 629 (1994).
[32] I. Sosnowska, R. Przenioslo, P. Fischer, *et al.*, J. Magn. Magn. Mater. **160**, 384 (1996).
[33] R. Przenioslo, M. Regulski, and I. Sosnowska., Jpn. J. Phys. Soc. **75**(8), 084718 (2006).
[34] C. Ederer, and N. A. Spaldin, Phys. Rev. B. **71**, 060401 (2005).
[35] A. V. Zalessky, A. A. Frolov, A. K. Zvezdin *et al.*, J. Exp. Theor. Phys. **95**, 101 (2002).
[36] A. V. Zalessky, A. A. Frolov, T. A. Khimich *et al.* Europhys. Lett. **50**, 547 (2000).
[37] J. Wang, J. B. Neaton, H. Zheng, V. Nagarajan, S. B. Ogale, B. Liu, D. Viehland, V. Vaithyanathan, D. G. Schlom, U. V. Waghmare, N. A. Spaldin, K. M. Rabe, M. Wuttig, and R. Ramesh, Science **299**, 1719 (2003).
[38] H. Naganuma, and S. Okamura, J. Appl. Phys. **101,** 09M103 (2007).
[39] H. Bea, M. Bibes, S. Petit, J. Kreisel, and A. Barthelemy, Philos. Mag. Lett. **87** 165 (2007).
[40] P. K. Siwach, H. K. Singh, J. Singh, and O. N. Srivastava, Appl. Phys. Lett. **91**, 122503 (2007).
[41] S. T. Zhang, M. H. Lu, D. Wu, Y. F. Chen, and N. B. Ming, Appl. Phys. Lett. **87**, 262907 (2005).
[42] D. Lebeugle, D. Colson, A. Forget *et al.*, Phys Rev B **76**, 024116 (2007).
[43] F. M. Bai, J. L. Wang, M. Wuttig *et al.*, Appl. Phys. Lett. **86**, 032511 (2005).
[44] K. Takahashi, and M. Tonouchi, J Magn. Magn. Mater. **310**, 1174 (2007).
[45] J. B. Li, G. H. Rao, J. K. Liang *et al.*, Appl. Phys. Lett. **90**: 162513 (2007).
[46] W. Eerenstein, F. D. Morrison, J. Dho *et al.*, Science **307**, 1203a (2005).
[47] H. Béa, M. Bibes, S. Fusil *et al.*, Phys Rev B **74**, 020101 (2006).
[48] A. V. Zalesskii, A. K. Zvezdin, A. A. Frolov, and A. A. Bush, JETP Lett. **71**, 465. (2000).
[49] S. Vijayanand, M. B. Mahajan, H. S. Potdar, and P. A. Joy, Phys. Rev. B **80**, 064423 (2009).
[50] M. K. Singh, R. S. Katiyar, W. Prellier, and J. F. Scott, J. Phys.: Cond. Matter **21**, 042202 (2009).
[51] Yu. G. Chukalkin, and B. N. Goshchitskii, Phys. Status Solidi A **200**, R9 (2003).
[52] G. Catalan, H. Bea, S. Fusil, M. Bibes, P. Paruch, A. Barthelemy, and J. F. Scott, Phys. Rev. Lett. **100**, 027602 (2008).

FIGURES AND TABLES

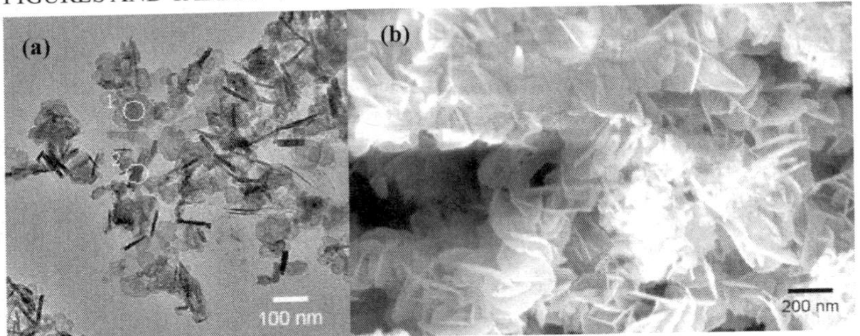

Figure 1. Micrographs of the BMO materials synthesis. (a): TEM and (b): SEM micrographs of the as-synthesized material, showing the $BiMnO_3$ nanoplates and the needle-like secondary phase.

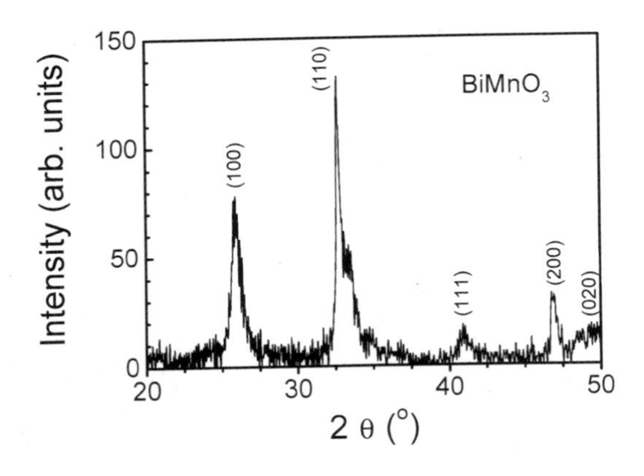

Figure 2. X-ray diffraction pattern (XRD) of the as-synthesized BMO material, showing a pattern dominated by the single-phase $BiMnO_3$ nanoplates.

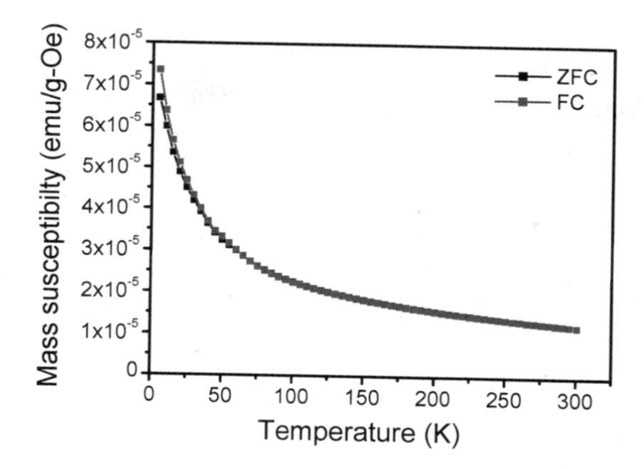

Figure 3. Temperature dependence of the mass susceptibility of the as-synthesized BMO material at an applied magnetic field of 5000 Oe. ZFC denotes zero-field cooled conditions, and FC denotes field-cooled conditions.

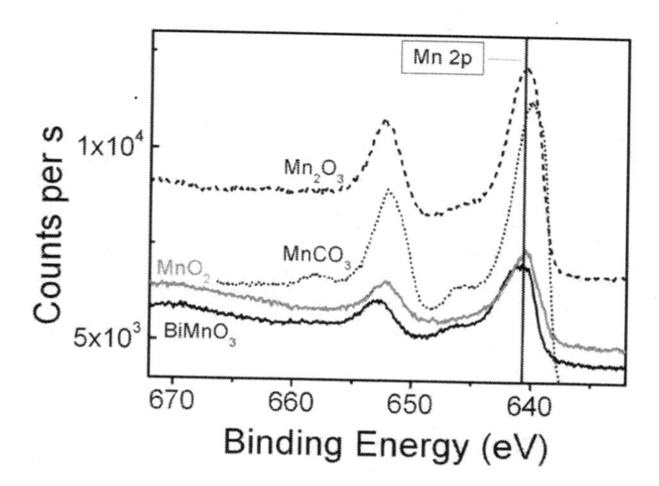

Figure 4. X-ray photoelectron spectroscopy (XPS) spectrum of the BMO nanoplates, and of standard samples of $MnCO_3$, Mn_2O_3 and MnO_2 used for comparison to identify the Mn valence.

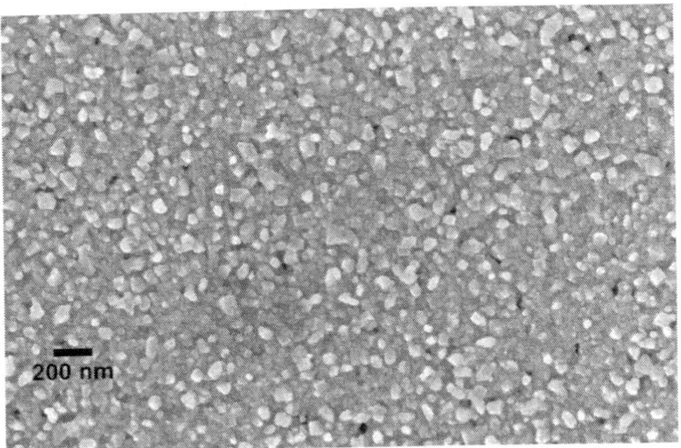

Figure 5. Scanning electron (SEM) micrograph of the as-synthesized BFO material, showing the $BiFeO_3$ nanoparticles.

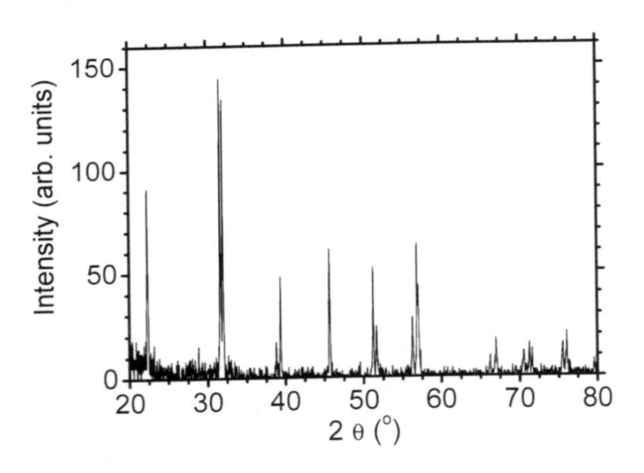

Figure 6. X-ray diffraction pattern (XRD) of the as-synthesized BFO material ($BiFeO_3$ nanoparticles), indicating a single-phase perovskite structure.

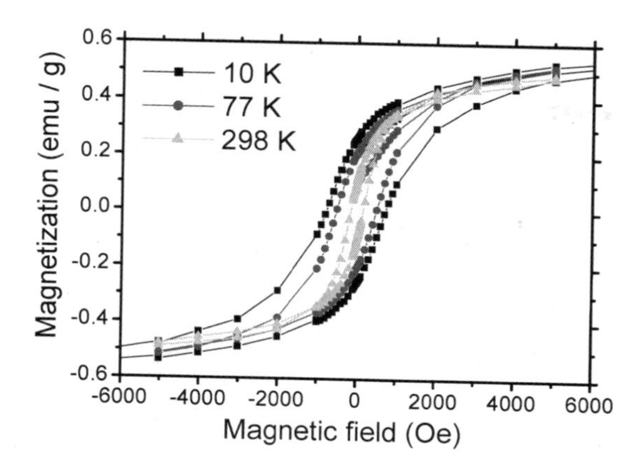

Figure 7. Magnetic field dependence of the magnetization (M) for the as-synthesized BFO material ($BiFeO_3$ nanoparticles) at T = 10 K, 77 K and 298 K. Ferromagnetic hysteresis is apparent from 10 K to room temperature.

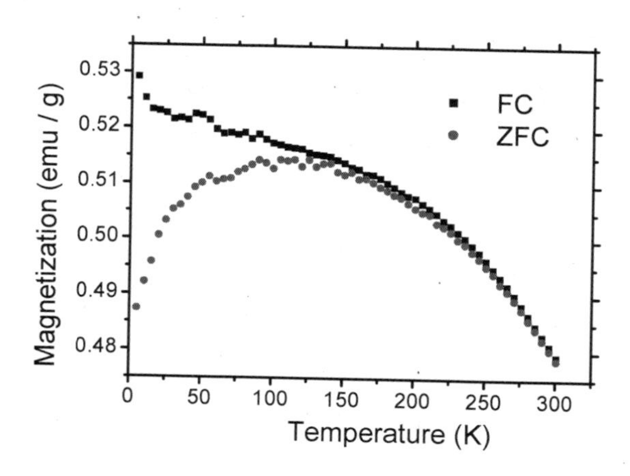

Figure 8. Temperature dependence of the magnetization (M) for the as-synthesized BFO material ($BiFeO_3$ nanoparticles) at an applied H = 5000 Oe (zero-field-cooled (ZFC) and field-cooled (FC) conditions).

Table I. (Top) EDS analysis of the needle-like phase in the as synthesized BMO matrix, conducted on the area marked as "2" in Fig. 1(a). (Bottom) EDS analysis of the nanoplate $BiMnO_3$ phase in the as-synthesized BMO matrix, conducted on the area marked as "1" in Fig. 1(a).

Element	Atomic %	Uncert. %	Correction	k-Factor
-------	-------	--------	-------	-------
Mn(K)	5.83	0.43	0.99	1.451
Bi(L)	94.16	4.43	0.75	6.708

Element	Atomic %	Uncert. %	Correction	k-Factor
-------	-------	--------	-------	-------
Mn(K)	58.28	2.97	0.99	1.451
Bi(L)	41.71	4.50	0.75	6.708

Table II. EDS analysis of the as-synthesized BFO matrix ($BiFeO_3$ nanoparticles).

Element	Atomic %	Uncert. %	Correction	k-Factor
-------	-------	--------	-------	-------
Fe(K)	50.28	2.97	0.99	1.451
Bi(L)	49.71	4.50	0.75	6.708

FUNCTIONALLY GRADED PIEZOMAGNETIC AND PIEZOELECTRIC BILAYERS FOR MAGNETIC FIELD SENSORS: MAGNETOELECTRIC INTERACTIONS AT LOW-FREQUENCIES AND AT BENDING MODES

S. K. Mandal, G. Sreenivasulu, V. M. Petrov, S. Bandekar, and G. Srinivasan
Physics Department, Oakland University
Rochester, MI 48309, USA

ABSTRACT

In a piezomagnetic-piezoelectric bilayer the interaction between the magnetic and electric subsystems occurs through mechanical forces. This study is on magneto-electric (ME) interaction in a bilayer of PZT and ferromagnetic layer in which the piezomagnetic coefficient *q* is graded. The grading is accomplished with the use of Ni (negative *q*) and Ni-Fe alloy or Permendur (positive *q*). At low-frequencies the ME coefficient shows a four-fold increase compared to homogeneous Ni-PZT bilayer. Bending resonance and consequent enhancement in ME coupling occurs at the lowest frequency, 800-1200 Hz, for a bilayer that is fixed at one end and free at the other end. The peak ME coefficient at resonance is in the range 12-15 V/cm Oe and is 10% higher for graded samples compared to Ni-PZT. The graded bilayers are promising candidates for use in ultrasensitive magnetic field sensors.

INTRODUCTION

The magneto-electric effect is the dielectric polarization of a material in an applied magnetic field *H*, or an induced magnetization in an electric field *E* [1]. The polarization *P* is related to *H* by $P = \alpha H$, where α is the ME-susceptibility. Most single phase compounds show weak ME interactions [2]. In a composite of piezomagnetic-piezoelectric phases the magnetoelectric coupling occurs through mechanical deformation and α is proportional to the product of the piezomagnetic coefficient *q* and the piezoelectric coefficient *d* [3]. Efforts so far have focused on ME interactions in bulk and layered composites of ferrites, manganites, or transition metals/alloys for the ferromagnetic phase, and lead zirconate titanate (PZT) or lead magnesium niobate-lead titanate (PMN-PT) for the piezoelectric phase [3].

This work is on ME interactions in composites consisting of functionally graded ferromagnetic and piezoelectric phases. Ferroics (ferromagnetic or ferroelectric) form an essential sub-group of functional materials whose physical properties are sensitive to changes in electric and magnetic fields. Recent works on functionally graded ferroics have resulted in the discovery of important phenomena including internal potentials, induced anisotropies, and spontaneous strain [4-8]. Since the piezoelectric (*d*) and piezomagnetic (*q*) coefficients are the critical parameters that determine the strength of ME coupling, grading one or both parameters will have substantial influence on α. The work was motivated by our recent theory that predicts a minimum of 50-60% stronger ME interaction in *q* and *d* graded ferrite-ferroelectrics as compared to homogeneous systems [9-11].

Here we discuss results of our studies of strain mediated ME coupling on bilayers of PZT and *q*-graded magnetic phase. Samples of homogeneous bilayer of PZT-Ni and *q*-graded samples of PZT-Ni and 60%Ni-40%Fe alloy and PZT-Ni-Permendur (P) were investigated. Nickel has negative *q*, whereas Ni-Fe alloy and Permendur, an alloy with 49% Fe, 49% Co and 2% V, both have positive *q*. Measurements of ME coefficients at low frequencies and at bending resonance reveal enhancement of ME coefficients compared to homogeneous bilayer of PZT-Ni. The grading of *q* perpendicular to the sample plane gives rise to a bending strain in a magnetic field that counteracts the flexural deformation due to asymmetry associated with bilayers and an increase in ME susceptibility. The strong ME coupling is of significance for use of graded composites as magnetic field sensors.

EXPERIMENT

Commercial samples of 99.9% Ni and Permendur were used. Alloys with 60%Ni and 40% Fe (Ni/Fe), however, were prepared by electrodeposition. Films of thickness 100 μm were deposited on Cu using the procedure described below. The Cu plate was first cleaned, one side was covered with lacquer and was immersed in a plating solution for the deposition of Ni/Fe. Details on the plating bath and process parameters are listed in Table I. Nickel sulfamate was used for stabilization of the plating solution and to achieve rapid deposition and low internal stress. Nickel chloride facilitates dissolution of the Ni anode. Boric acid acted as a buffer agent to stabilize the pH of the plating bath. Saccharin in the Ni-Fe bath acts as a strain reducing agent. The pH value of the solution was maintained to about 3.5 using sodium hydroxide and amidosulfonic acid. The surfactant sodium lauryl sulfate was added to the solution to avoid pinholes in the films. The thickness of the Ni-Fe layer was 100 μm for a deposition time of 6 hours. Following the deposition, the Cu plate was dissolved in ammonium hydroxide and the Ni/Fe film was annealed in N_2H_2 environment for 1 hour at a temperature of 400 °C.

Table I. Plating bath and process parameters for Ni-Fe electrodeposition

Nickel sulfamate tetrahydrate (g/l)	480
Iron (II) sulfate heptahydrate (g/l)	104.8
Nickel (II) chloride hexahydrate (g/l)	20
Boric acid (g/l)	20
Sodium lauryl sulfate (g/l)	1.5
O- Benzoic sulfimide sodium salt hydrate (Saccharin) (g/l)	3.0
pH value	3.5
Temperature (°C)	60
Cathode current density (A/cm^2)	0.03

Magnetosrtiction (λ) of Ni, Ni/Fe alloy and P were measured on 20 mm × 3 mm platelets using a strain gage and stain indicator. With the sample plane defined as *(1,2)*, in-plane λ was measured for two orientations of the bias magnetic field *H*: along the field direction, termed λ_{11} and perpendicular to *H*, termed λ_{12}. Figure 1 shows λ vs. *H* data for Ni. The parameter λ_{11} is negative, increases rapidly with *H* and shows saturation for $H > 300$ Oe. The magnetostriction perpendicular to *H*, $\lambda_{12} \approx \lambda_{11}/2$ and is positive. The piezomagnetic coefficient $q = q_{11}+q_{12} = d\lambda_{11}/dH + d\lambda_{12}/dH$ was estimated from data in Fig.1and its *H* dependence for Ni is shown in Fig.2. A large negative-*q* is evident for low bias fields and the magnitude of q decreases with increasing *H*. Figure 3 shows similar *q* vs. *H* data for Ni/Fe and P. It is clear from the data that both Ni/Fe and P have positive piezomagnetic coupling with Permendur showing a higher *q* than Ni/Fe.

Bilayer laminates were made by bonding the magnetic layers to PZT plates of dimensions 20 mm × 3 mm × 0.2 mm that were poled in an electric field. A 2 μm thick slow-dry epoxy was used for bonding. Following the application of epoxy between PZT and magnetic layers, the sample was to 60°C for 4-5 hrs.

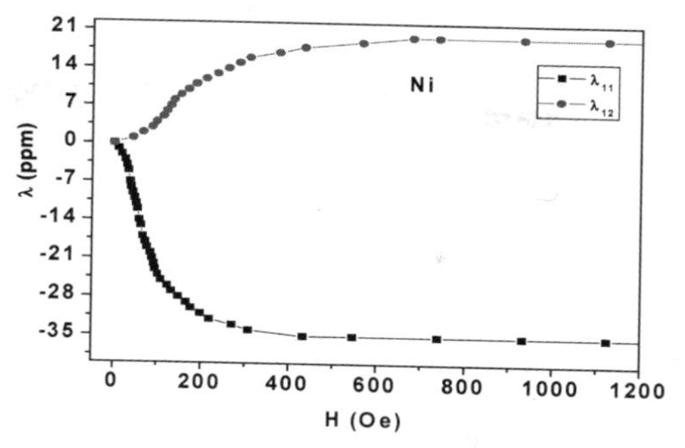

Figure 1. In-plane magnetostriction for Ni parallel (λ_{11}) and perpendicular (λ_{12}) to the bias magnetic field H.

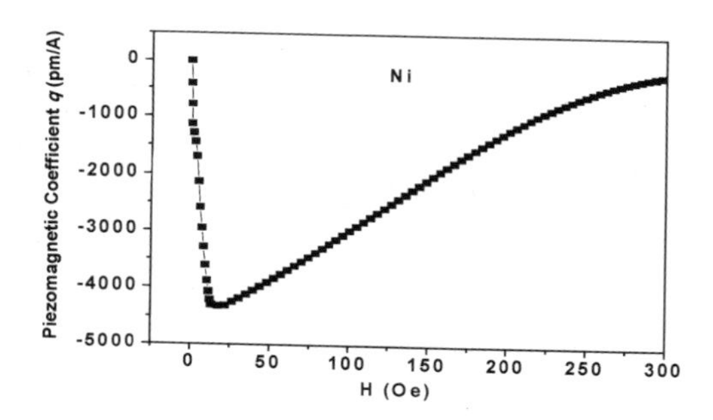

Figure 2. In-plane piezomagnetic coefficient $q = q_{11} + q_{12}$ as a function of H for Ni.

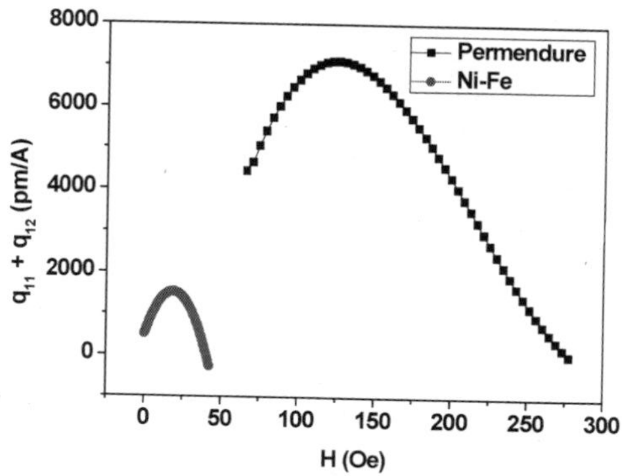

Figure 3. Similar data as in Fig.2 for Permendur (P) and Ni/Fe alloy.

For ME characterization, a 3-teriminal sample holder was used. An ac magnetic field field $\delta H = 1$ Oe and a bias magnetic field H were applied to parallel to the sample plane, along direction-*1* and the voltage δV across PZT was measured with a lock-in-amplifier. The ME voltage coefficient $\alpha_E = \delta E/\delta H = \delta V/t\ \delta H$, where t is the total thickness of PZT, was measured as a function of H and frequency f of the ac magnetic field.

RESULTS AND DISCUSSION

The influence of q-grading on ME coupling in bilayers with PZT is investigated for two cases: low-frequency ac magnetic fields and when the frequency of the ac field is tuned to resonance modes of the bilayer.

Low-Frequency ME coupling

First we consider low-frequency ME coupling in the bilayers. Data on the ME voltage coefficient (MEVC) were obtained as a function of H for $f = 30$ Hz. Figure 4 shows such data for a homogeneous bilayer of PZT and Ni layer of thickness 0.16 mm. A sharp increase in MEVC is observed in Fig.4 with increasing H; it reaches a maximum value of 220 mV/cm Oe and then decrease to near-zero value for $H > 500$ Oe. The H-dependence in Fig.4 essentially tracks the variation of magnitude of the piezomagnetic coefficient q with H that is shown in Fig.2. The peak value for MEVC is much smaller than reported values for a symmetric structure of Ni-PZT-Ni [11]. A bilayer, due to structural asymmetry, will have a flexural deformation in addition to the in-plane deformation due to magnetostriction. The out-of-plane bending deformation counteracts the in-plane strain, leading to weakening of the ME coupling.

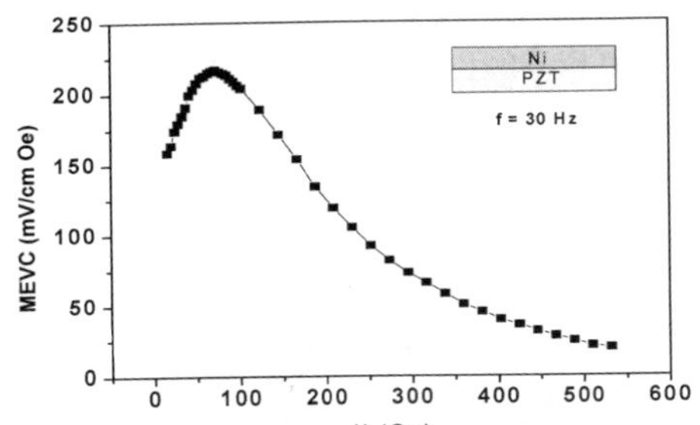

Figure 4. Low-frequency Magnetoelectric voltage coefficient (MEVC) as a function of bias magnetic field H for PZT-Ni bilayer.

Similar measurements were made on bilayers with graded q, i.e., by adding a layer of ferromagnetic phase with positive-q to PZT-Ni. Two such systems were investigated: Ni/Fe alloy of 0.15 mm in thickness and Permendur of similar thickness. Results for PZT-NI-Ni/Fe are considered next. Figure 5 shows MEVC vs H data for $f = 30$ Hz for the sample. The overall variation of MEVC with H in Fig.5 is similar to the data in Fig.4, but with the following departures. (i) A very sharp initial increase in MEVC with H, (ii) a peak in MEVC at a lower bias field compared to PZT-Ni and (iii) a peak MEVC of 780 mV/cm Oe which is a factor of 3.5 higher than for PZT-Ni.

Figure 6 shows the variation of MEVC with H for PZT-Ni-P. The addition of Permendur layer with positive q to PZT-Ni results in a factor of four increase in the peak MEVC compared to PZT-Ni.

It is clear from the data in Figs.3-5 that for positive grading of q by adding a layer of Ni/Fe or P, the MEVC is enhanced by a factor of 3 or more. Our studies on samples with

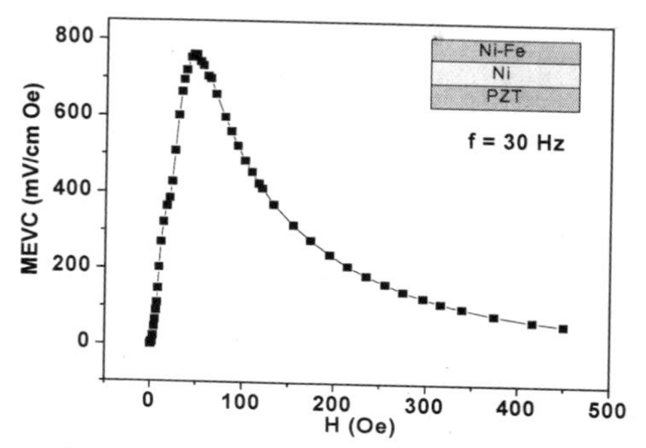

Figure 5. Data as in Fig.4 for a bilayer of PZT-Ni-Ni/Fe.

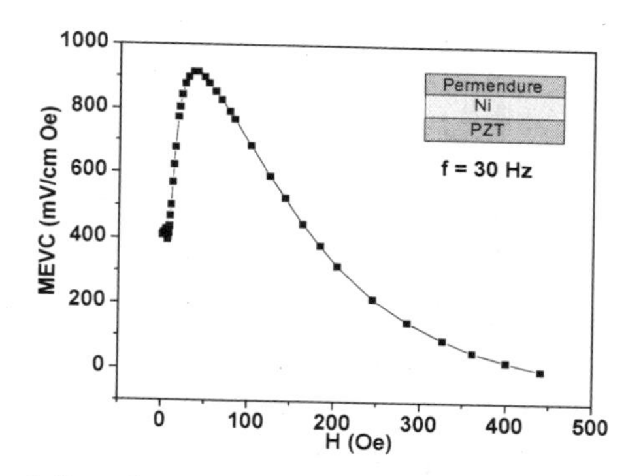

Figure 6. Low-frequency MEVC as a function of H for PZT-Ni-P.

negative grading of q, i.e., PZT-Ni/Fe-Ni and PZT-P-Ni, showed either an overall decrease or no change in MEVC. Similarly in our studies on PZT-ferrites, negative grading of q resulted in a peak value of α_E same as that for PZT-homogeneous ferrite [11]. Thus only when grading of q is positive, the grading induced bending moment counteracts the flexural deformation due to bilayer asymmetry, leading to an increase in α_E. One also observes a higher MEVC for PZT-Ni-P compared to PZT-Ni-Ni/Fe due to higher q-value for Permendur. Further, the peak values of MEVC in Figs.5 and 6 are comparable to values for a symmetric trilayer structure of Ni-PZT-Ni in which there are no flexural deformation [11].

Bending Resonance

Next we discuss the ME coupling when the frequency of the ac magnetic field is tuned to resonance modes for the bilayer. Since the ME coupling in the composites is mediated by the mechanical stress, one would expect orders of magnitude stronger coupling when the frequency of the ac field is tuned to acoustic mode frequencies in the sample than at non-resonance frequencies [14-18]. Several recent experiments and modeling efforts have dealt with ME interactions at such electromechanical resonance (EMR) in layered composites [2,3]. But a key drawback for this type of resonance ME effect is that the frequencies are quite high, on the order of hundreds of kHz, for nominal sample dimensions. The eddy current losses for the magnetostrictive phase can be quite high at such frequencies, in particular for transition metals and alloys and earth rare alloys such as Terfenol-D, resulting in an inefficient magnetoelectric energy conversion. In order to reduce the operating frequency, one must therefore increase the laminate size that is disadvantageous for any applications.

An alternative for achieving a strong ME coupling is the resonance enhancement at bending modes of the composite. The frequency of ac fields that must be applied to the bilayer for such bending oscillations is expected to be much lower compared to longitudinal acoustic modes. Recent investigations have showed a giant ME effect at bending modes in several layered structures [1,2,12].

We recently developed a model for resonance ME effects in layered magnetostrictive-piezoelectric samples at bending modes [13]. The frequency dependence of ME voltage coefficients were obtained using the simultaneous solution of electrostatic, magnetostatic and elastodynamics equations. The resonance behavior of ME effect in a bilayer was shown to be dependent boundary conditions. A giant ME interaction at the lowest frequency was predicted for a bilayer fixed at one end and free to vibrate at the other end.

Data on ME coupling at bending modes were obtained for homogeneous and q-graded bilayers. The samples were clamped at one end. A DC bias corresponding to maximum in MEVC in Figs.3-5 was applied to the sample. Then the ME voltage across PZT was measured as the frequency of the ac field was varied. Figure 7 shows such data for the PZT-Ni bilayer. With increasing *f*, MEVC remained constant for $f < 800$ Hz. The data in Fig.7 shows a resonance enhancement in MEVC for $f_r = 1$ kHz which corresponds to bending oscillations of the bilayer. The quality factor Q for the resosnce is 25. The MEVC at resonance is a factor of 60 higher than the maximum MEVC at 30 Hz (Fig.4).

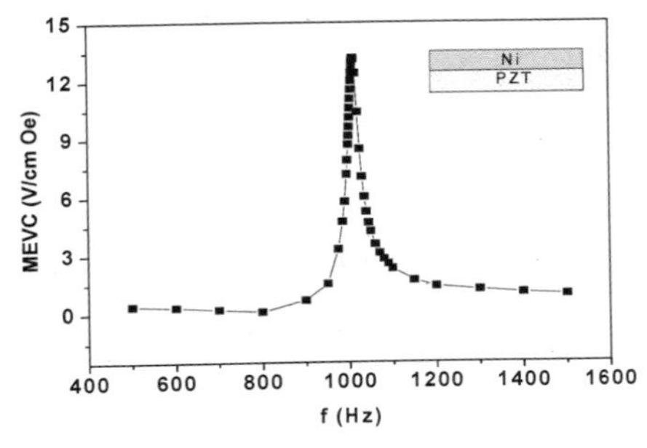

Figure 7. Frequency dependence of MEVC for PZT-Ni bilayer. The peak in MEVC occurs at bending resonance for the sample clamped at one end.

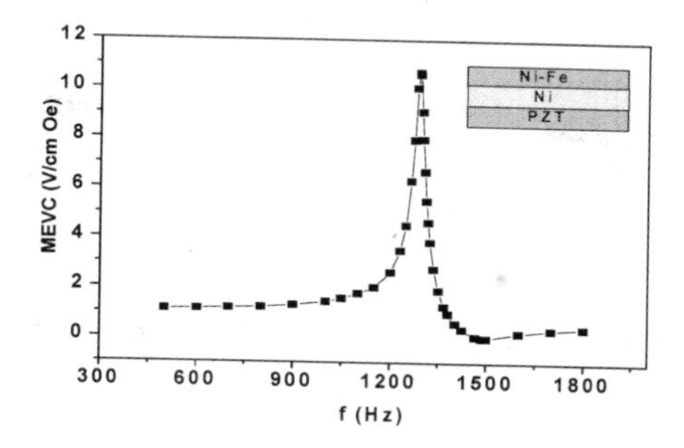

Figure 8. Data on MEVC vs. f for q-graded bilayer of PZT-Ni-Ni/Fe

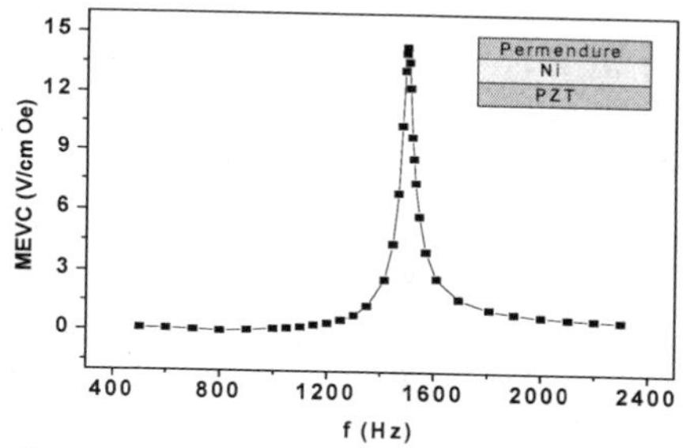

Figure 9. Frequency dependence of MEVC for a bilayer of PZT-Ni-P clamped at one end.

Results on measurements of MEVC variations with frequency of the ac magnetic field for the graded bilayers are shown in Figs. 8 and 9. Both the systems, PZT-Ni-Ni/Fe and PZT-Ni-P show a shift in f_r to higher frequency compared to PZT-Ni and is due to increase in the bilayer thickness for the graded samples. The grading results in an increase in the MEVC at resonance only for PZT-Ni-P. The resonance value of MEVC for PZT-Ni-Ni/Fe is in fact 20% smaller than for PZT-Ni. The MEVC at f_r is proportional to the amplitude of bending vibrations and it is clear from Figs.7-9 that q-grading does not lead to any substantial increase in the strength of ME interactions at bending modes.

CONCLUSION

In summary, studies on ME coupling in bilayers of PZT and ferromagnetic layers with grading of piezomagnetic coefficient show strong ME interactions due to grading induced bending moment. This bending moment counteract the flexural deformation due to asymmetry in a PZT-Ni bilayer and strengthen the ME coupling by a factor of 3-4 at low frequencies in PZT-Ni/Fe and PZT-Ni-P. But the enhancement of MEVC in q-graded samples at bending resonance is relatively small, on the order of 10% in PZT-Ni-P. The graded composites have the potential for use as ultrasensitive magnetic field sensors.

ACKNOWLEDGMENTS

The research was supported by grants from the National Science Foundation and the Defense Advanced Research Project Agency- HUMS program.

REFERENCES

[1] N. A. Spaldin and M. Fiebig, The renaissance of magnetoelectric multiferroics, Science **309**, 391 (2005).
[2] G. Srinivasan, Magnetoelectric composites, Ann. Rev. Mater. Res., **40**, 153 (2010).
[3] C. W. Nan, M. I. Bichurin, S. Dong, D. Viehland, and G. Srinivasan, Multiferroic magnetoelectric composites: Historical perspective, status and future directions, J. Appl. Phys., **103**, 031101 (2008).
[4] Z. G. Ben, S. P. Alpay and J. V. mantese, Fundamentals of graded ferroic materials and devices,Phys. Rev. B **67**, 184104 (2003).
[5] R. Nath, S. Zhong, S.P. Alpay, D. B. Huey, and M. W. Cole, "Enhanced piezoelectric response from barium strontium titanate multilayer films," Appl. Phys. Lett., **92**, 012916 (2008).
[6] J. V. Mantese, A. L. Micheli, N. W. Schubring, R. W. Hayes, G. Srinivasan, and S. P. Alpay, Magnetization-graded ferromagnets: The magnetic analogs of semiconductor junction elements, Appl. Phys. Lett., **87**, 082503 (2005).
[7] C. Sudakar, R. Naik, G. Lawes, J. V. Mantese, A. L. Micheli, G. Srinivasan, and S. P. Alpay, Internal magnetostatic potentials of magnetization-graded ferromagnetic materials, Appl. Phys. Lett., **90**, 062502 (2007).
[8] V. M. Petrov and G. Srinivasan, Enhancement of Magnetoelectric Coupling in Functionally Graded Ferroelectric and Ferromagnetic Bilayers, Phys. Rev. B **78**, 184421.(2008).
[9] V. M. Petrov and G. Srinivasan, Microwave magnetoelectric effects in bilayers of single crystal ferrite and functionally graded piezoelectric, J. Appl. Phys., **104**, 113910 (2008).
[10] S. K. Mandal, G. Sreenivasulu, V. M. Petrov and G. Srinivasan, Flexural deformation in a compositionally stepped ferrite and Magnetoelectric effects in a composite with piezoelectrics, Appl. Phys. Lett., 96, 192502 (2010).
[11] V. M. Laletin, N. Paddubnaya, G. Srinivasan, C. P. DeVreugd, and M. I. Bichurin, Frequency and field dependence of magnetoelectric interactions in layered ferromagnetic transition metal-piezoelectric lead zirconate titanate, Appl. Phys. Lett. **87**, 222507 (2005).
[12] J. Zhai, Z. Xing, S. Dong, J. Li and D. Viehland, Magnetoelectric laminate composites: An overview, J. Am. Ceram. Soc., **91**, 351 (2008).
[13] V. M. Petrov, G. Srinivasan, M. I. Bichurin, and T. A. Galkina, Theory of magnetoelectric effect for bending modes in magnetostrictive-piezoelectric bilayers, J. Appl. Phys. **105** 063911 (2009).

MAGNETIC AND ELECTRICAL PROPERTIES OF $0.7Bi_{0.95}Dy_{0.05}FeO_3$-$0.3Pb(Fe_{0.5}Nb_{0.5})O_3$ MULTIFERROIC

Agata Stoch[1], Jan Kulawik[1], Paweł Stoch[2,3], Jan Maurin[3,4], Piotr Zachariasz[3], Piotr Zielinski[5]

[1]Institute of Electron Technology, Kracow Division , ul. Zablocie 39, 30-701 Krakow, Poland;
[2]Faculty of Material Science and Ceramics, AGH - University of Science and Technology, al. Mickiewicza 30, 30-059 Krakow, Poland
[3]Institute of Atomic Energy - POLATOM, 05-400 Otwock-Swierk, Poland
[4]National Medicines Institute, ul. Chelmska 30/34, 00-725 Warszawa, Poland
[5]Institute of Nuclear Physics PAN, ul. Radzikowskiego 152, 31-342 Krakow, Poland

ABSTRACT

The $0.7Bi_{0.95}Dy_{0.05}FeO_3$-$0.3Pb(Fe_{0.5}Nb_{0.5})O_3$ multiferroic is a solid solution of the perovskite-type oxide ceramics, $Bi_{0.95}Dy_{0.05}FeO_3$ and $Pb(Fe_{0.5}Nb_{0.5})O_3$. The $0.7Bi_{0.95}Dy_{0.05}FeO_3$-$0.3Pb(Fe_{0.5}Nb_{0.5})O_3$ was synthesized using a standard solid state reaction technique. XRD, SEM + EDS, Mössbauer spectroscopy, impedance spectroscopy and magnetoelectric measurements were done. According to XRD analysis the investigated compound crystallizes in an R3c crystal structure with lattice parameters a = 5.632 Å and c = 13.891 Å. The stoichiometry of the material was confirmed by SEM + EDS observations and additionally confirmed by Mössbauer spectroscopy. Electrical properties were measured by impedance spectroscopy and grain interior and grain boundary resistivities, and activation energies were obtained. The magnetoelectric properties were measured. Antiferromagnetic, along with slight ferromagnetic, behavior was observed.

INTRODUCTION

Materials that exhibit ferromagnetic and ferroelectric ordering in the same phase are named multiferroics. Magnetoelectric materials are very attractive due to strong coupling between the magnetic and electric subsystems, which means that the magnetic field is able to induce electric polarization and vice versa. Magnetoelectrics are quite interesting because of their fundamental properties in modern physics but they also are very attractive for their practical applications.

The first single phase multiferroic perovskites were discovered in the early 1960s. However, very limited progress has been made during the last several decades[1-6]. The significant development of multiferroic materials started with the successful synthesis of multiferroic thin films[7].

Magnetoelectrics are widely used as transducers, actuators, detectors and other sensors, which are characterized by high dielectric permittivity and magnetic permeability. Strong coupling between electrical polarization and magnetization vectors are desirable for a new class of multifunctional materials, for example new mass storage devices.

Unfortunately, most multiferroic materials are characterized by a low magnetic ordering temperature, considerably below room temperature, and therefore the magnetoelectric effect is relatively small. Bismuth ferrite $BiFeO_3$ is a well known perovskite compound which simultaneously exhibits at ambient temperature ferroelectric (T_C = 1110 K) and antiferromagnetic (T_N = 610 K) ordering. The stereochemical activity of Bi lone-pair electrons induces ferroelectric polarization, while the partially filled 3d orbitals of the Fe^{3+} ions produce

G-type antiferromagnetic ordering. $BiFeO_3$ has a spatially modulated magnetic structure of a cycloidal type with a period of modulation of about 62 nm[4]. One way to suppress the spiral spin modulation is the chemical substitution of magnetically active atoms, especially rare earth trivalent ions, into the Bi sublattice.

Lead iron niobate $Pb(Fe_{0.5}Nb_{0.5})O_3$ is another type of magnetoelectric material where ferroelectric active Nb^{5+} cations are partially substituted by magnetic ions Fe^{3+} of a different valence state. As a consequence, $Pb(Fe_{0.5}Nb_{0.5})O_3$ exhibits antiferromagnetic ordering at temperatures below 143 K, as well as in ferroelectric material at above 385 K[5,6]. Additionally, $Pb(Fe_{0.5}Nb_{0.5})O_3$ is a perovskite relaxor, and is reported to have a very high dielectric constant, which is useful for multilayer ceramic capacitors[5,6].

Unfortunately, high current leakage, secondary oxide phases and other defects make $BiFeO_3$ difficult to obtain[2,3,4]. Therefore, a solid solution of Dy-doped $BiFeO_3$ and $Pb(Fe_{0.5}Nb_{0.5})O_3$ was synthesized. In this paper, we report the synthesis, crystal structure and magnetoelectric properties of $0.7Bi_{0.95}Dy_{0.05}FeO_3$-$0.3Pb(Fe_{0.5}Nb_{0.5})O_3$.

EXPERIMENTAL

A solid solution of the $0.7Bi_{0.95}Dy_{0.05}FeO_3$-$0.3Pb(Fe_{0.5}Nb_{0.5})O_3$ compound was obtained based on the following chemical reactions:

$$Fe_2O_3 + Nb_2O_5 \rightarrow 2FeNbO_4, \quad (1)$$

$$FeNbO_4 + 2PbO \rightarrow 2Pb(Fe_{0.5}Nb_{0.5})O_3, \quad (2)$$

$$0.95Bi_2O_3 + 0.05Dy_2O_3 + Fe_2O_3 \rightarrow 2Bi_{0.95}Dy_{0.05}FeO_3. \quad (3)$$

High purity Bi_2O_3, Dy_2O_3, Fe_2O_3 and $Pb(Fe_{0.5}Nb_{0.5})O_3$ oxide powders were weighted in stoichiometric proportions, mechanically activated for 7 h during a ball milling process and finally calcined at 1093 K for 4 h. The obtained sinter was granulated and then pressed into a disc shaped pellet. The pellet was heat-treated at 818 K for 2 h and next sintered at 1123 K for 10 h.

The crystallographic structure was checked by a Bruker-AXS D8 diffractometer. The microstructure was analyzed by scanning electron microscopy (SEM) with energy dispersive analysis (EDS) using Nova NanoSEM 200 equipment. ^{57}Fe Mössbauer spectroscopy measurements were performed on a Wisell 360H spectrometer with Oxford cryostat. Magnetic susceptibility was measured on a LakeShore 7225 susceptibilitimeter, and the magnetoelectric effect was probed by using a Stanford Research SR830m lock-in amplifier. The electrical properties of $0.7Bi_{0.95}Dy_{0.05}FeO_3$-$0.3Pb(Fe_{0.5}Nb_{0.5})O_3$ were characterized by a 7600 Precision QuadTech LRC Meter.

X-RAY STUDIES

The crystal structure and purity of the sinter were checked by X-ray powder diffraction (Philips powder diffractometer with Cu K_α radiation, $\lambda = 1.54056$ Å). A fitting program (FULLPROF software[8]), based on the Rietveld method, was used to analyse the diffraction pattern.

The simple ABO_3 perovskite crystallizes in a cubic Pm-3m crystal structure, although the ferroelectric materials distinctly show an off-centre shift of the B-site cation during polarisation ordering and therefore a small tetragonal distortion is observed. Additionally, the coexistence of ferroelectric and ferromagnetic orderings often leads to rotations of the oxygen octahedra resulting in rhombohedral distortion[9]. The crystallographic transitions are not distinct and therefore difficult to measure. A few probable crystallographic structures - Pm-3m, P4mm and R3c - were examined during the refinement procedure. The XRD pattern is presented in Fig. 1. The best result was obtained for the R3c structure. The lattice parameters are a = b = 5.632 Å and c = 13.891 Å. The rhombohedral angle is 59.94° which is close to 60° - the value for a cubic structure. The observed distortion is very small.

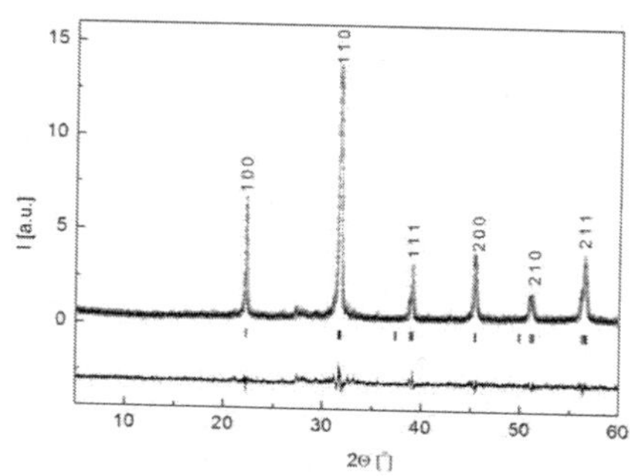

Figure 1. XRD pattern of $0.7Bi_{0.95}Dy_{0.05}FeO_3$-$0.3Pb(Fe_{0.5}Nb_{0.5})O_3$.

SEM AND EDS STUDIES

The surface of $0.7Bi_{0.95}Dy_{0.05}FeO_3$-$0.3Pb(Fe_{0.5}Nb_{0.5})O_3$ ceramic sample is presented in Fig.2. The material is a fine grained dense sinter with small crystallites of a few micrometers. Energy dispersive analysis was performed at various points corresponding to larger and smaller microcrystallites. An exemplary EDS spectrum is presented in Fig.3. In Table I the oxide content in the $0.7Bi_{0.95}Dy_{0.05}FeO_3$-$0.3Pb(Fe_{0.5}Nb_{0.5})O_3$ compound is presented. The EDS profile analysis is close to the theoretical oxide composition within the experimental error range and in this way the stoichiometry of the compound was confirmed. There are no unreacted starting oxides and oxides like Fe_2O_3, which can be precipitated due to strong evaporation of Bi_2O_3 during the synthesis process.

Figure 2. SEM micrograph of $0.7Bi_{0.95}Dy_{0.05}FeO_3$-$0.3Pb(Fe_{0.5}Nb_{0.5})O_3$.

Table I. Oxide composition of the $0.7Bi_{0.95}Dy_{0.05}FeO_3$-$0.3Pb(Fe_{0.5}Nb_{0.5})O_3$ compound [wt %] (theoretical values and EDS profile analysis).

oxides	theoretical [wt. %]	EDS analysis [wt. %]
Bi_2O_3	48.62	47.28
Dy_2O_3	1.99	2.42
PbO_2	22.55	24.05
Fe_2O_3	21.33	20.08
Nb_2O_5	5.51	6.17

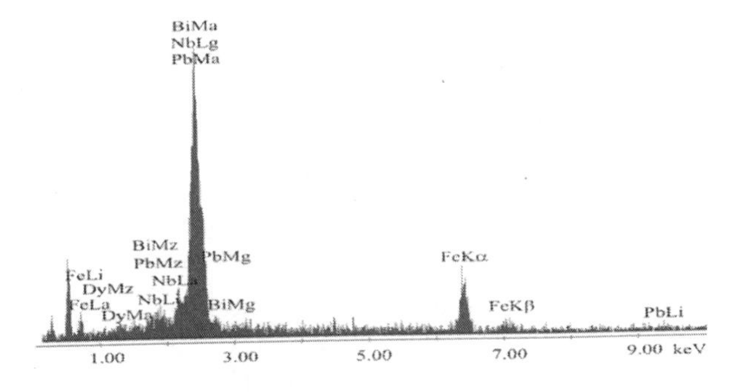

Figure 3. An exemplary EDS spectrum of $0.7Bi_{0.95}Dy_{0.05}FeO_3$-$0.3Pb(Fe_{0.5}Nb_{0.5})O_3$.

MÖSSBAUER SPECTROSCOPY MEASUREMENTS

The ^{57}Fe transmission Mössbauer spectrum of $0.7Bi_{0.95}Dy_{0.05}FeO_3$-$0.3Pb(Fe_{0.5}Nb_{0.5})O_3$ was measured at 11 K. The spectrum (Fig.4) was analyzed by the least squares method and the hyperfine interaction parameters such as isomer shift δ relative to metallic iron at 300 K, quadrupole split Δ and hyperfine magnetic field (μ_0H_{hf}) were obtained and summarized in Table II. The spectrum is composed of three Zeeman sextets and one small quadrupole doublet. The hyperfine interaction parameters such as Δ, δ, μ_0H_{hf} of the two first Zeemen sextets (subspectra No 1 and 2 – Table II, Fig. 4) are close to those previously obtained[10] for the $BiFeO_3$ phase. Therefore, we suggest these two subspectra are related to iron cations in a $BiFeO_3$ like phase. The hyperfine interaction parameters of the third subspectrum (component No 3 – Table II, Fig.4.) agree very well with parameters previously obtained for $Pb(Fe_{0.5}Nb_{0.5})O_3$ [11]. This allows us to suggest a local iron symmetry neighborhood close to iron in the $Pb(Fe_{0.5}Nb_{0.5})O_3$ phase. The subspectral area (parameter w – Table II) is proportional to the quantity of iron cations. The ratio of the sum of the subspectral areas of components 1 and 2 ($Bi_{0.95}Dy_{0.05}FeO_3$ – Table II) to the subspectral area of component 3 ($Pb(Fe_{0.5}Nb_{0.5})O_3$ – Table II) is about 0.20. The theoretical value of the ratio of iron content in $Bi_{0.95}Dy_{0.05}FeO_3$ to $Pb(Fe_{0.5}Nb_{0.5})O_3$ for the $0.7Bi_{0.95}Dy_{0.05}FeO_3$-$0.3Pb(Fe_{0.5}Nb_{0.5})O_3$ is 0.21. This could help to prove the stoichiometry of the material.

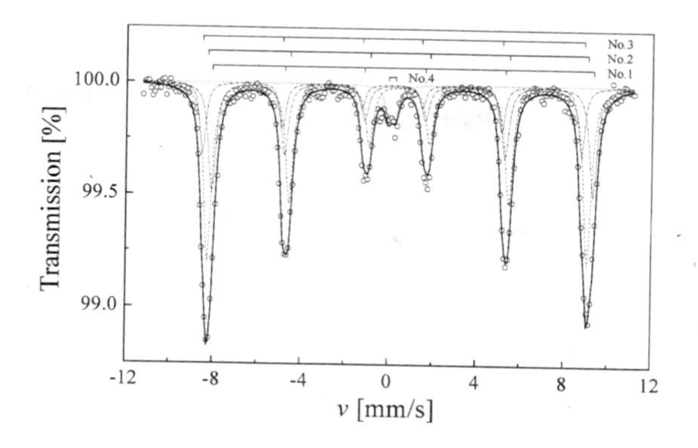

Figure 4. ^{57}Fe Mössbauer effect spectrum measured at 11 K.

A paramagnetic doublet (component No 4 – Table II, Fig. 4) probably characterizes the impurity phases $Bi_{25}FeO_{39}$ or $Bi_2Fe_4O_9$, which can be easily formed during the synthesis process of $BiFeO_3$. It was estimated that only 3.5% of Fe atoms were incorporated into the oxygen phases, therefore it is clear why the X-ray diffraction technique did not detect any additional phases. The values of hyperfine parameters δ and Δ are typical of Fe^{3+} in octahedral symmetry[10,11]. The derived magnetic hyperfine fields for both components are about 54 T, which are also values typical of ferric iron.

Table II. The ^{57}Fe hyperfine interaction parameters at 11 K (Γ - width line at half maximum, w - subspectral area, δ - isomer shift, μ_0H_{hf} - magnetic hyperfine field, Δ - quadrupole split).

No	Γ [mm/s]	w [%]	δ [mm/s]	μ_0H_{hf} [T]	Δ [mm/s]
1.	0.202(5)	29.97(1)	0.537(7)	53.93(7)	0.001(4)
2.	0.211(1)	50.53(4)	0.525(1)	53.73(4)	-0.058(1)
3.	0.164(3)	15.90(7)	0.231(6)	53.59(11)	0.365(12)
4.	0.200(9)	3.59(10)	0.209(5)	–	0.333(4)

ELECTRICAL PROPERTIES

Impedance spectroscopy is a very useful technique for analysing the electrical properties of a wide group of electroceramics. Impedance spectroscopy measurements were performed in a temperature range 450 – 700 K at frequencies from 10 Hz to 2 MHz. The recorded impedance spectroscopy spectra were analyzed by LEVM program with an equivalent electrical circuit proposed as a series-parallel connection of two resistors and constant phase elements (CPE)[11,12].

In Fig.5 the grain interior (R_{gi}) and grain boundary (R_{gb}) resistivities are presented in Arrhenius type plot. The conductivity of the material is a thermally activated process with activation energies E_{gb} = 1.02(4) eV and E_{gi} = 0.98(6) eV values typical of electroceramic materials[13].

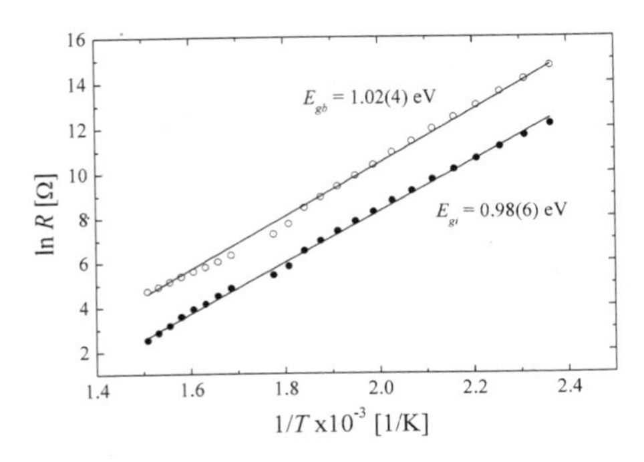

Figure 5. Arrhenius type plot of the grain interior R_{gi} and grain boundary R_{gb} resistivities.

The slopes of the plots are slightly different, which may be expected with a brick layer model without easy paths for systems with tetragonal symmetry[12]. The R_{gi} values are generally lower than the R_{gb} suggesting that electrical conductivity along the grains and across grain boundaries clearly dominates over conductivity along the grain boundaries[12].

The temperature dependence of the real part of the dielectric constant is presented in Fig.6. Two anomalies are clearly visible at 660 K and 760 K. The first small peak is very diffuse, and is related to the magnetic ordering temperature of the $BiFeO_3$ component. The second anomaly is much stronger and could be related to the ferroelectric ordering temperature for $0.7Bi_{0.95}Dy_{0.05}FeO_3$-$0.3Pb(Fe_{0.5}Nb_{0.5})O_3$. Both peaks are slightly frequency dependent, which is characteristic of relaxor behavior.

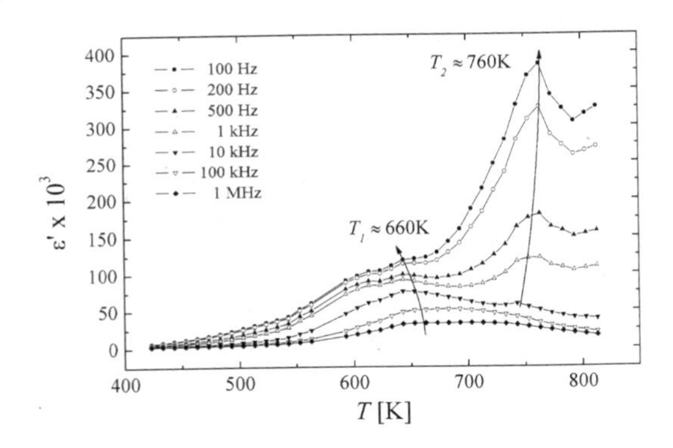

Figure 6. Dielectric constant ε' as a function of temperature for different frequencies. Two anomalies at 660 K and 760 K characterize magnetic and electric ordering temperatures.

MAGNETIC AND MAGNETOELECTRIC PROPERTIES

Magnetic hysteresis loops M(H) were recorded at 4.2 K and 277 K and are presented in Fig.7. Both magnetization curves M(H) reveal typical antiferromagnetic behavior with no magnetic saturation effect up to the maximal magnetic field of 56 kOe. A more detailed analysis of M(H) curves shows slightly ferromagnetic behavior – inset in Fig.7 for 4.2 K. The evaluated values of remanent magnetization M_R and coercive field H_C are $4.59 \cdot 10^{-3}$ emu/g and 35.7 Oe, respectively. Increasing temperature up to 277 K decreases M_R to $5.90 \cdot 10^{-4}$ emu/g and increases H_C to 370 Oe.

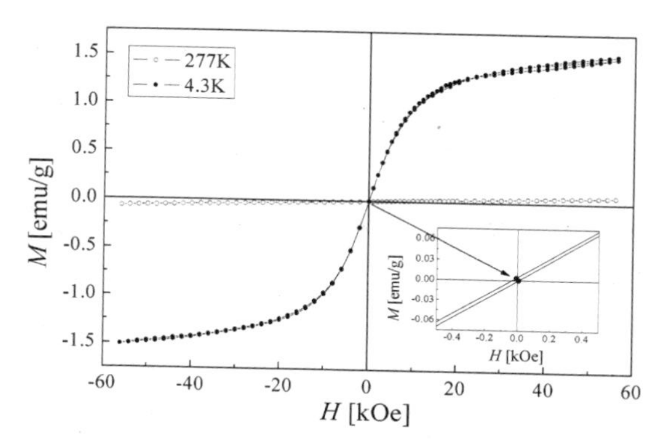

Figure 7. Magnetization curves M(H) taken at 4.2 K and 277 K.

In order to measure the magnetoelectric voltage coefficient, a disc shaped sinter of 0.15 cm height and 1.5 cm in diameter was placed into a static magnetic field H_{DC}, created by an electromagnet with additional modulation of magnetic field $H_{AC}(f)$ produced by Helmholtz coils. Both magnetic fields were oriented perpendicular to the surface of the sinter. An electric signal $\delta U(f)$ from the sample was detected by using a lock-in amplifier. The voltage δU was measured for different magnitudes of static magnetic field and frequencies of modulation field in the range 200 Hz – 20 kHz. A magnetoelectric voltage coefficient (ME) was derived using the formula

$$ME = \delta U/(\delta H \cdot d) \quad (4)$$

where d is the height of the investigated sample.

Fig.8 presents the dependence of the ME coefficient as a function of frequency at a static magnetic field H_{DC} of magnitude 350 Oe. At low values of frequency the ME coefficient rises rapidly, while on increasing the frequency a monotonic growth of ME is observed up to 20 kHz, where ME reaches saturation.

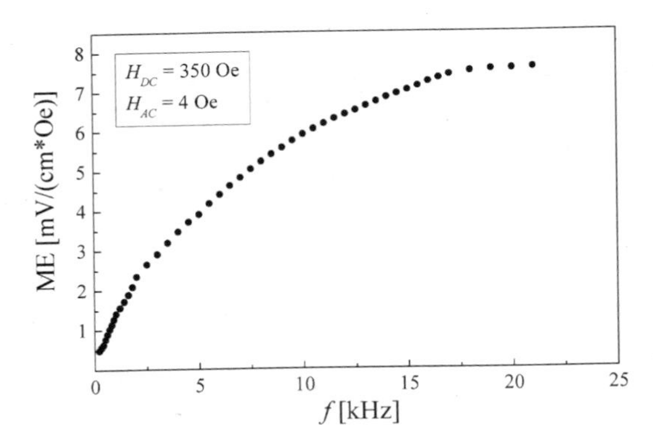

Figure 8. Magnetoelectric voltage coefficient (ME) vs. frequency for $0.7Bi_{0.95}Dy_{0.05}FeO_3$-$0.3Pb(Fe_{0.5}Nb_{0.5})O_3$.

Fig.9 shows the ME voltage coefficient vs. a static magnetic field for modulated frequency f = 5 kHz and H_{AC} = 4 Oe. Increasing the H_{DC} field causes an increase in the ME parameter up to about 2.3 kOe. Increasing the field further results in a decrease in the ME coefficient.

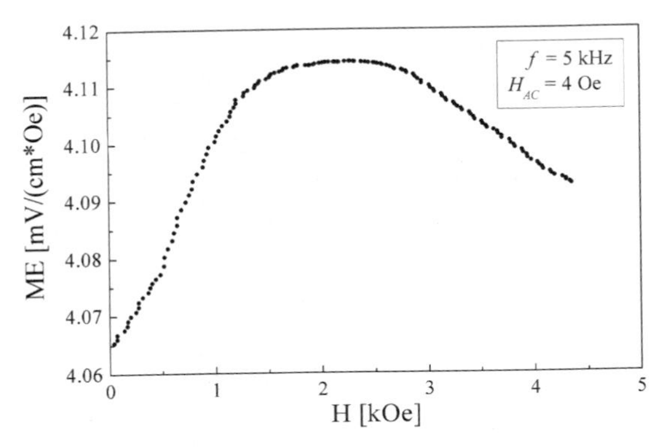

Figure 9. Frequency dependence of the magnetoelectric coefficient.

Fig.10 shows the AC magnetic field dependence of parameter α, which is the δU voltage divided by the width of the sample. The δU voltage increases almost linearly with an increasing ac field. Closer inspection suggests a slight quadratic dependence of the $\delta U(f)$.

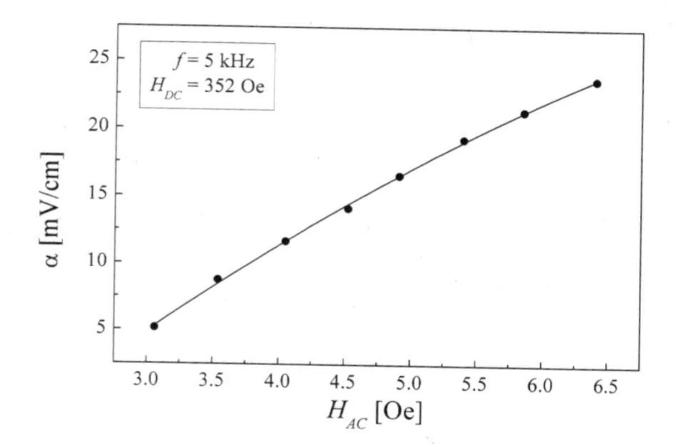

Figure 10. AC magnetic field dependence of the α parameter.

CONCLUSIONS

$0.7Bi_{0.95}Dy_{0.05}FeO_3$-$0.3Pb(Fe_{0.5}Nb_{0.5})O_3$ was synthesized using conventional a solid state reaction method. The purity of the obtained material was confirmed by X-ray and Mössbauer measurements. The X-ray studies showed that the measured compound crystallizes in an R3c crystal structure, which rhombohedrally distorted the ABO_3 perovskite structure. Mössbauer spectroscopy shows three Zeeman sextets characteristic of a magnetic state. One of them is related to the $Pb(Fe_{0.5}Nb_{0.5})O_3$ phase and the others to the $Bi_{0.95}Dy_{0.05}FeO_3$ phase. A small contribution of impurity phases at the level of 3.5% of iron atoms is observed.

The impedance spectroscopy results demonstrate that the conductivity of the $0.7Bi_{0.95}Dy_{0.05}FeO_3$-$0.3Pb(Fe_{0.5}Nb_{0.5})O_3$ material is a thermally activated process with an activation energy of about 1 eV. The temperature dependence of the real part of the dielectric constant shows two anomalies responsible for Neel and Curie ordering temperatures.

According to the magnetic results the investigated material is an antiferromagnet with no collinear alignment of magnetic spins. A very small ferromagnetic component is observed. Magnetoelectric investigations confirmed magnetoelectric coupling between electric and magnetic subnets with a strong dependence of the ME voltage coefficient on frequency. The above properties could be useful for developing new kinds of magnetic materials for sensors.

ACKNOWLEDGMENTS

The work was partially supported by the Polish Ministry of Science and Higher Education project No NN507347335.

REFERENCES

[1]J. F. Scott, "Data storage: Multiferroic memories", *Nat. Mater.* **6**, 256-257 (2007).

[2]H. Paik, H. Hwang, K. No, S. Kwon, D. P. Cann, "Room temperature multiferroic properties of single-phase $(Bi_{0.9}La_{0.1})FeO_3$–$Ba(Fe_{0.5}Nb_{0.5})O_3$ solid solution ceramics", *Appl. Phys. Lett* **90**, 042908 (2007).

[3]G. L.Yuan, S. W. Or, J. M. Liu, Z. G. Liu, "Structural transformation and ferroelectromagnetic behavior in single-phase BiNdFeO multiferroic ceramics", *Appl. Phys. Lett* **89**, 052905 (2006).
[4]S. T. Zhang, Y. Zhang, M. H. Lu, C. L. Du, Y. F. Chen, Z. G. Liu, Y. Y. Zhu, N. B. Ming, X. Q. Pan, "Substitution-induced phase transition and enhanced multiferroic properties of BiLaFeO ceramics", *Appl. Phys.s Lett.* **88**, 162901 (2006).
[5]Y. Yang, J. M. Liu, H. B. Huang, W. Q. Zuo, P. Bao, Z. G. Liu, "Magnetoelectric coupling in ferroelectromagnet $Pb(Fe_{1/2}Nb_{1/2})O_3$ single crystals", *Phys. Rev. B* **70**, 132101 (2004).
[6]J. Kulawik, D. Szwagierczak, "Dielectric properties of manganese and cobalt doped lead iron tantalate ceramics", *J. Eur. Ceram. Soc.* **27**, 2281-2286 (2007).
[7]J. Wang, J. B. Neaton, H. Zheng, V. Nagarajan, S. B. Ogale, B. Liu, D. Viehland, V. Vaithyanathan, D. G. Schlom, U. V. Waghmare, N. A. Spaldin, K. M. Rabe, M. Wuttig, R. Ramesh, " Epitaxial $BiFeO_3$ multiferroic thin film heterostructures", *Science* **299**, 1719 (2003).
[8]J. Rodríguez-Carvajal, "Commission on Powder Diffraction (IUCr)", *Newsletter* **26**, 12 (2001).
[9]K. F. Wang, J. M. Liu, Z. F. Ren, "Multiferroicity, The coupling between magnetic and polarization", *Advances in Physics*, **58**, 321 (2009).
[10]A. Stoch, P. Zachariasz, P. Stoch, J. Kulawik, J. Marin, "Hyperfine interactions and electrical properties of multiferroic $0.5Bi_{0.95}Dy_{0.05}FeO_3$-$0.5Pb(Fe_{0.5}Nb_{0.5})O_3$", *J. Phys.: Conf. Ser.* **217**, 012135, (2010).
[11] S. A. Ivanov, P. Nordblad, R. Tellgren, T. Ericsson, H. Rundlof, "Structural, magnetic and Mössbauer spectroscopic investigations of the magnetoelectric relaxor $Pb(Fe_{0.6}W_{0.2}Nb_{0.2})O_3$", *Solid State Sci.* **9**, 440 (2007).
[12] E. Barsoukov, J. R. Macdonald, *Impedance Spectroscopy* 2nd ed. Wiley-Interscience, Hoboken, (2005)
[13] E. J. Abram, D. C. Sinclair, A. R. West, "A strategy for analysis and modeling of impedance spectroscopy data of electroceramics: Doped lanthanum gallate", *J. Electroceram.* **10** 165 (2003).

MULTIFERROIC NANOFILM WITH BILAYER OF $Pb(Zr_{0.52}Ti_{0.48})O_3$ AND $CoFe_2O_4$ PREPARED BY ELECTROPHORETIC DEPOSITION

Dongxiang Zhou, Gang Jian, Yanan Zheng, Fei Shi
Department of Electronic Science and Technology
Huazhong University of Science and Technology
Wuhan Hubei, P. R. China

ABSTRACT

In this paper, the bilayer nano-films of $Pb(Zr_{0.52}Ti_{0.48})O_3/CoFe_2O_4$ (PZT/CFO) were prepared by electrophoretic deposition (EPD) from suspensions with superfine particles of PZT (30 nm) and CFO (40 nm). The PZT and CFO were synthesized via sol-gel and co-deposition method, respectively. Different solvents were tried. It is found that PZT in mixture solvent of acetic acid/acetylacetonel (1:1 volume ratio) and CFO in acetylacetone were the most stable suspensions (ζ= 50.2 mV and 38 mV). Effects of EPD process parameters such as suspension concentration, deposition time, and electrical field strength on the deposition weight were investigated. PZT layer was deposited onto $Pt/Ti/SiO_2/Si$ substrate, dried in air, and annealed at 750°C. CFO was deposited on annealed PZT layers. The PZT/CFO films were annealed at 700°C. The films exhibited excellent ferroelectricity (Ps = 30.2 $\mu C/cm^2$), ferromagnetism (Ms =160.7 emu/cm^3) and multiferroic properties (α_E = 5 mV/cmOe) simultaneously.

INTRODUCTION

Magnetoelectric (ME) multiferroic materials, which not only exhibit ferromagnetism and ferroelectricity simultaneously, but also show magnetoelectric coupling effect, draw much research attention for their potential application for storage units. The ME effect in composite materials was based on the well-known "product properties" theory.[1] 1-3 type $BaTiO_3/CoFe_2O_4$ composite films were fabricated via pulsed laser deposition (PLD) by Zheng et al.[2] Bilayer films or multilayer films (2-2 type structure) can overcome the shortcomings of large leakage current which exists in 1-3 and 0-3 type structure. Many material systems such as BaTiO3-CoFe2O4, $NiFe_2O_4$-$Pb(Zr_xTi_{1-x})O_3$, $BiFeO_3$-$Pb(Zr_xTi_{1-x})O_3$, and $Pb(Zr_xTi_{1-x})O_3$-$CoFe_2O_4$ have been studied.3-5 The preparation methods mainly focused on PLD, sol-gel, sputtering, and metal organic chemical vapor deposition (MOCVD).[6-10] These methods, however, have disadvantages such as difficulty in controlling chemical composition for sputtering, slow film growth rate for PLD, or uneven surface for sol-gel.

Electrophoretic deposition (EPD) is achieved via motion of charged particles in suspension towards an electrode under an applied electric field. This method is widely used in film preparation for its advantages such as fast and even film growth rate, precise controllability of composition, and mild reaction conditions. However less literature was reported to use EPD to prepare multiferroic films. Wu et al. used EPD to deposit the $BiFeO_3$ (BFO) ferromagnetic layer of BFO/PZT films.[4]

In this paper, 2-2 type $(Zr_{0.52}Ti_{0.48})O_3/CoFe_2O_4$ (PZT/CFO) bilayer films were deposited via EPD for the first time. In order to obtain stable suspensions, effects of different solvents and their volume fraction on the suspension stability were studied in detail. Deposition kinetic behaviors were investigated.

EXPERIMENTAL

Superfine particles of PZT (30 nm) and CFO (40 nm) were synthesized via sol-gel and co-deposition method, respectively. Solutions A and B were formed at room temperature by dissolving 0.1 mol $Pb(CH_3COO)_2$ and 0.052 mol $ZrO(NO_3)_2$ into 66.7 mL glycol ether and 0.048 mol titanate butoxide $Ti(C_4H_9O)_4$ into 95 mL mixture of glycol ether and glacial acetic acid, respectively. Then solution A was mixed with B with deionized water added slowly, and pH was adjusted to ~3.3 by glacial acetic acid to form PZT sol (~0.6 M). Sol to gel transition took place after 12 h in the air. The PZT gel was dried and calcinated at 700°C for 2 h to get PZT nanoparticles. A mixture solution of $CoCl_2$ and $FeCl_3$ with molar ratio of 1:2 was added into hot NaOH solution which acted as an oxidizer and pH adjuster to form CFO precursor. The CFO precursor was isolated, dried, and sintered at 800 °C for 2 h, and then CFO nano-particles were formed, with an average particle size of ~40 nm and regular spherical shape.

For preparing stable suspensions of PZT and CFO, sediment volume combined with zeta potential of dispersions was studied. Firstly, sediment volumes of suspensions formed by dispersed PZT and CFO particles in series solvents including water and organic solvents were studied. The most appropriate solvents for PZT and CFO were chosen, in which sediment volumes were the least. And then, modification of the chosen solvent systems was done, *e.g.*, mixing with other solvents with various volume fractions, sediment volumes or zeta potential were studied of suspensions formed in these solvent systems.

PZT and CFO films were alternating deposited onto substrates by EPD. Polished indium-tin oxide (ITO) conductive substrates (1.5×2.5 cm^2) were used as the anode and $Pt/Ti/SiO_2/Si$ as the cathode, keeping the distance between the two substrates fixed at 1.5 cm during deposition. A constant voltage power supplier (0~250 V) was utilized. A rectangular insulating tank was utilized for electrophoresis. Various deposition parameters such as suspension concentration, deposition time, and deposition voltage have been adopted in our experiment. Changes in electric current intensity ($\mu A/cm^2$) with time during deposition of PZT and CFO were also tested. The composite films were annealed at 700°C for ~30 min.

For ferroelectric and multiferroic measurement, Ag pastes were screen-printed on the top surface of the bilayer films, metallized, and formed ohmic contact via heat treatment at 550°C for ~5 min. Sediment volumes of these dispersions were measured after 3 weeks settling time at room temperature in 10 mL cylinders. The particle size and zeta potential of suspensions was examined by a light scattering commercial device (Zetasizer 3000 HAS, Malvern Instruments Ltd., UK), and the suspension concentration was diluted to 10 mgL−1 before zeta potential test. Conductivity of suspensions was measured by a conductivity meter. Ferroelectric hysteresis of the samples was recorded by a ferroelectric material parameters analyzer (ZT−I, CN). The multiferroic coefficient was measured at frequency of 10 kHz. Magnetic hysteresis loop at room temperature was measured with a vibrating sample magnetometer (VSM) (Model 4HF).

RESULTS AND DISCUSSIONS

Sediment volume is an effective reflection of stability of suspensions.[10,11] In a stable system, settling of the single particles is slow, and close sediment structure is formed, the sediment volume is small. While in the unstable and flocculated system, settling of particles is rapid, and the sediment has loose structure. To decide the most appropriate solvents for forming suspensions, a series of solvents were tested for both PZT and CFO, which was shown in Table I. Suspensions of PZT dispersed in water and ethanol had relatively higher stability, as sediment volumes were smaller than others. For preparation of CFO dispersion, acetylacetone was the best choice. Sediment volumes of CFO in acetylacetone were far smaller than other systems. For further improvement of stability of dispersion systems, measurements such as mixing other solvents with various fractions and adding dispersants were done.

Table 1. Sediment volumes of PZT and CFO dispersed in deferent solvents.

Solvents	Sediment volumes		Solvents	Sediment volumes	
	PZT (cm^3/0.3 g)	CFO (cm^3/0.2 g)		PZT (cm^3/0.3 g)	CFO (cm^3/0.2 g)
Water	0.14	0.51	Acetic acid	0.15	0.5
Methanol	0.2	0.45	Acetylacetone	0.1	0.2
Ethanol	0.23	0.51	2-methoxyethanol	0.15	0.46
n-Propanol	0.2	0.53	Glycol ether	—	0.5
Iso-propanol	0.22	0.9	Isopropyl ether	0.2	0.7
n-Butanol	0.22	0.52	Ethylene glycol	—	0.45
Turpentine	—	1.8	Ethyl acetate	—	1.1

Fig. 1 showed the values of zeta potential of suspensions of PZT in acetylacetone-acetic acid mixture solvent with different volume fraction of acetylacetone / (acetic acid + acetylacetone) = 0, 0.4, 0.5, 0.6, and 1. For all suspensions, the zeta potential was positive. This may be attributed to the adsorption of H+ to the surface of PZT powders. The maximal zeta potential (50.2 mV > 26 mV) was found at the volume fraction of 0.5. In other words, the optimal volumetric ratio of acetylacetone to acetic acid to form stable EPD suspension was 1:1. Thus we kept this volumetric ratio for preparation of PZT suspensions. Surfactant such as Tween 80 was also tested, however systems with surfactant added in did not exhibit better stability than PZT- acetic acid-acetylacetone (1:1) system.

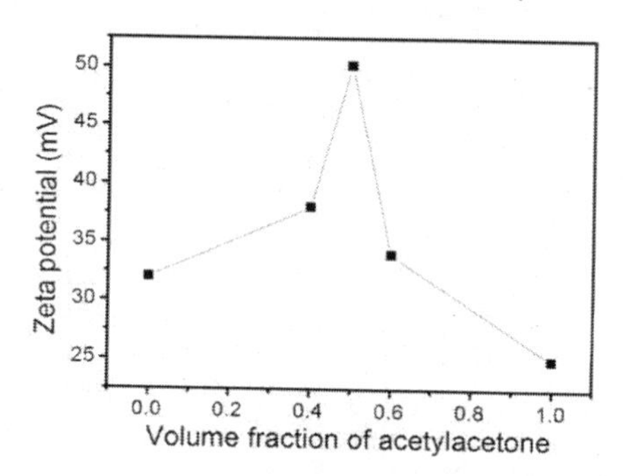

Figure 1. Zeta potential of PZT suspensions with different acetylacetone volume fraction in acetylacetone-acetic acid mixture solvent.

For further study of CFO-acetylacetone dispersion system, mixing with other solvents was also tested. Besides to ethanol organic solvents such as acetic acid and n-butanol were investigated.

However, none mixture systems had better stability than pure acetylacetone system, which was different from properties of PZT suspensions. The sediment volumes study showed that Tween 80 could not improve the stability of the dispersion systems either. In other words, the pure acetylacetone was the choice for CFO suspensions. The zeta potential values of CFO suspension in acetylacetone were measured as 38 mV. Thus acetylacetone was adopted for the preparation of CFO suspensions.

Fig. 2 depicted deposition weight ($mgcm^{-2}$) of PZT and CFO films varying with deposition time. The PZT deposition was conducted at concentrations of 10 and 20 gL^{-1}, and CFO deposition conducted on substrates and on deposited PZT layer. The voltage for both PZT and CFO was 20 V. The results were in accordance with the Hamaker equation,[12] *i.e.* deposition weight was proportional to deposition time, concentration and voltage within certain limits. The deposition weight of CFO was larger than that of PZT, which indicated deposition efficiency of CFO was larger. Deposition of CFO on PZT layer was a litter slower than on substrates directly, it attributed the resistivity of PZT layer. Fig. 2 also indicated that deposition conditions such as the concentrations and deposition time variations were in the linear scope for film growth.

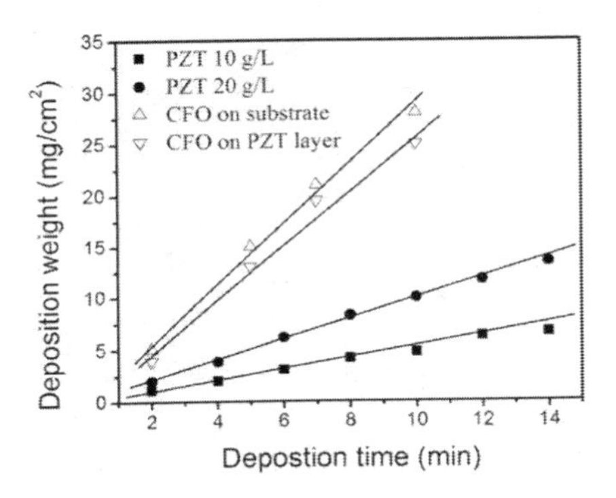

Figure 2. Deposition weight of PZT and CFO films as a function of deposition time.

Fig. 3 shows the effect of voltage on deposition weight of PZT films. Deposition time and concentration were constant (10 min, 30 gL^{-1}), voltage values were varied (20, 30, and 40 V). The results show that higher deposition weight can get at higher voltage. However, films deposited at 40 V have uneven surfaces with expansion scab scale on the edge of the films. Higher local electric field distribution was found on the edge of substrates, at which films'growth speed was faster. And higher voltage aggregated deposit of particles on the edge, so the different deposit speeds in the central and on the edge generated stress and strain in the films. The same phenomenon of voltage took place when CFO films were deposited. In a certain suspension concentration (20 gL^{-1}), films have cracking surface above deposited at 30 V even in short deposition duration (0.5 min).

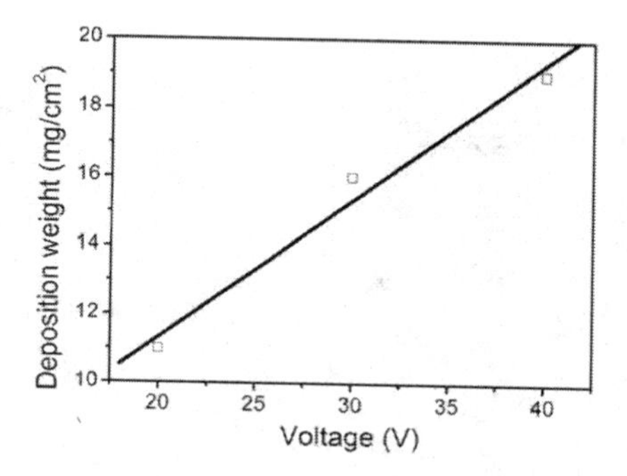

Figure 3. Deposition weight of the PZT film on substrates changing with voltage.

The microstructure characterizations are shown in Fig. 4 and Fig. 5. Fig. 4 presents XRD pattern of PZT/CFO bilayer film. The diffraction peaks correspond to PZT diffraction peaks and CFO spinel peaks. No intermediate phase is found. SEM images of bilayer samples are shown in Fig.5. The film shows evident nanostructure.

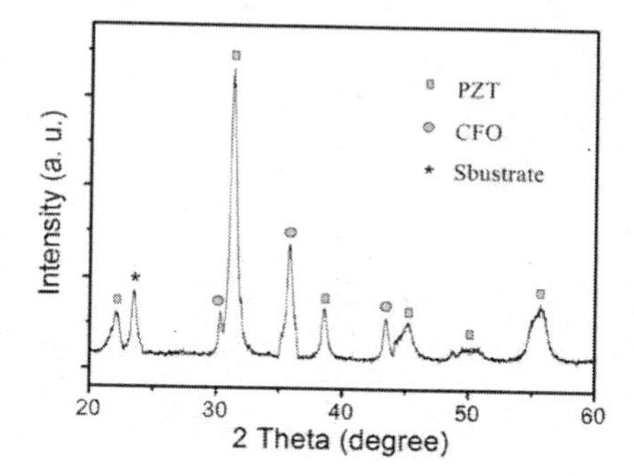

Figure 4. XRD patterns of PZT/CFO bilayer film

Figure 5. Surface image of PZT/CFO bilayer film

Fig. 6 reveals ferroelectric properties of the bilayer film and pure PZT film. Both films exhibit evidence of ferroelectricity. Asymmetry appears in ferroelectric hysteresis loops which may be attributed to the difference between top and bottom electrodes. Bilayer composite film's ferroelectricity is slightly worse than that of PZT single layer film. The saturation polarizations (P_s) of composite film and PZT film are 30.2 and 38.6 μCcm^{-2}, respectively, and their remnant polarizations (P_r) are 26.1 and 28.6 μCcm^{-2}. The coercive field (E_c) of the bilayer film is 290 kVcm−1, which is rather larger than that of PZT films (270 $kVcm^{-1}$). It may be attributed to the fact that movement of domains and domain walls of PZT is more difficult in bilayer films than in bulk materials and single layer PZT films due to the clamping effect from both of substrate and CFO layer.

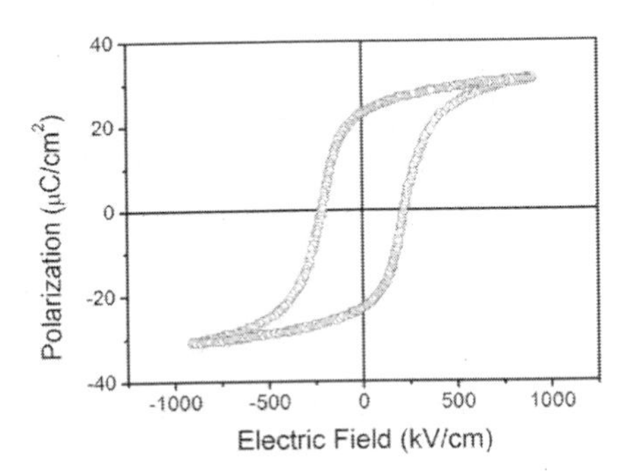

Figure 6. Electric hysteresis loops measured at room temperature of the PZT/CFO bilayer film.

In Fig. 7 is presented the magnetic field dependent magnetization of the composite film measured at room temperature by a VSM in plane. The loop has been saturated. The bilayer PZT/CFO composite film shows a typical ferromagnetic M-H loop with a saturation magnetization (M_s) of 160.7 $emucm^{-3}$ and a remnant magnetization (M_r) of 87.5 $emucm^{-3}$. The magnetic properties are comparable to results from Ortega et.al.13 The coercive field (H_c) in plane of the bilayer films (2510 Oe) is large enough to satisfy requirements as magnetic recording materials. The multiferroic coefficient α_E was also measured, the AC magnetic field frequency was 10 kHz, and the peak value 5 mV/come at magnetic field 4000 Oe.

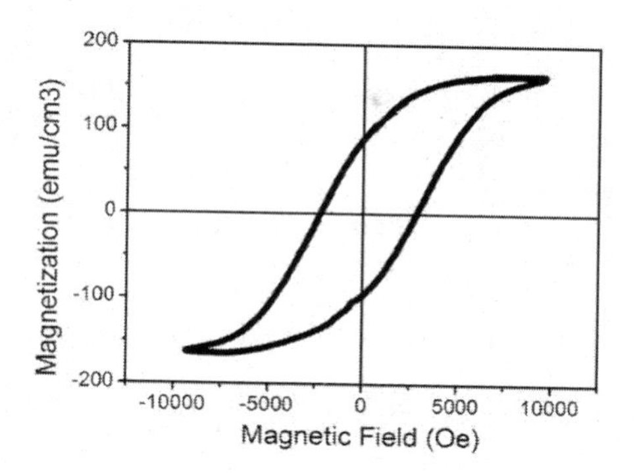

Figure 7. Magnetic hysteresis loop measured at room temperature of the PZT/CFO bilayer film.

CONCLUSIONS

In summary, EPD method was used to fabricate bilayer PZT/CFO composite film on $Pt/Ti/SiO_2/Si$ substrates by. Sediment volumes and zeta potential studies showed suspensions of PZT dispersed in ethanol-acetylacetone (1:1 in volume) and CFO dispersed in acetylacetone were the most stable dispersion systems, respectively. The zeta potential of the final PZT and CFO suspensions were 50.2 mV and 38 mV, respectively. The EPD parameters for each suspension such as deposition time, voltage, and suspension concentration allows the shaping of homogeneous and dense structure was studied. Microstructure, phases, ferroelectric, and magnetic properties of 2-2 type bilayer films have been studied. It is proved that no phase reaction happens. The obtained bilayer films exhibit sound ferroelectricity (P_s = 30.2 μCcm^{-2}) and ferromagnetism (M_s = 160.7 $emucm^{-3}$ (in plane)) simultaneously at room temperature. This deposition method provides a new approach to developing room temperature multiferroics.

ACKNOWLEDGMENTS

This work was supported by the National Natural Science Foundation of China (Grant No. 60871017/f010612) and Provincial Nature Science Foundation of Hubei in China.

REFERENCES

[1]J. V. Suchetelen, Product Properties: A New Application of Composite Materials, *Philips Res. Rep.*, 27, 28 (1972).

[2]H. Zheng, J. Wang, S. E. Lofland, Z. Ma, L. Mohaddes-Ardabili, T. Zhao, L. Salamanca-Riba, S. R. Shinde, S. B. Ogale, F. Bai, D. Viehland, Y. Jia, D. G. Schlom, M. Wuttig, A. Roytburd, and R.

Ramesh, Multiferroic $BaTiO_3$-$CoFe_2O_4$ Nanostructures, *Science*, 303, 661-663 (2004).
[3]Q. H. Jiang, Z. J. Shen, J. P. Zhou, Z. Shi, and Ce-Wen Nan, Magnetoelectric Composites of Nickel Ferrite and Lead Zirconnate Titanate Prepared by Spark Plasma Sintering, *J. Eur. Ceram. Soc.*, 27, 279-284 (2007).
[4]Yujie Wu, Jian-guo Wan, Chuanfu Huang, Yuyan Weng, Shifeng Zhao, Jun-ming Liu, and Guanghou Wang, Strong Magnetoelectric Coupling in Multiferroic $BiFeO_3$-$Pb(Zr_{0.52}Ti_{0.48})O_3$ Composite Films Derived from Electrophoretic Deposition, *Appl. Phys. Lett.*, 93, 192915 (2008).
[5]J. G. Wan, X. W. Wang, Y. J. Wu, M. Zeng, Y. Wang, H. Jiang, W. Q. Zhou, G. H. Wang, and J-M. Liu, Magnetoelectric $CoFe_2O_4$-$Pb(Zr,Ti)O_3$ Composite Thin Films Derived by a Sol-Gel Process, *Appl. Phys. Lett.*, 86, 122501 (2005).
[6]Jianhua Li, Igor Levin, Julia Slutsker, Virgil Provenzano, and Peter K. Schenck, Self-assembled Multiferroic Nanostructures in the $CoFe_2O_4$-$PbTiO_3$ System, *Appl. Phys. Lett.*, 87, 072909 (2005).
[7]Hong-cai He, Jing Ma, Jing Wang, and Ce-Wen Nan, Orientation-dependent Multiferroic Properties in $Pb(Zr_{0.52}Ti_{0.48})O_3$-$CoFe_2O_4$ Nanocomposite Thin Films Derived by a Sol-gel Processing, *J. Appl. Phys.*, 103, 034103 (2008).
[8]Ming Liu, Jing Lou, Shijian Zheng, Kui Du, and Nian X. Sun, A Modified Sol-gel Process for Multiferroic Nanocomposite Films, *J. Appl. Phys.*, 102, 083911 (2007).
[9]Kirby SD, Polking M, and van Dover RB, Epitaxial ($SrTiO_3$/NiO)n/MgO Multiferroic Heterostructure, *J. Vac. Sci. Technol. A.*, 25 [1], 37-41 (2007).
[10]S. Y. Yang, Q, Zhan, P. L. Yang, M. P. Cruz, Y. H. Chu, R. Ramesh, Y. R. Wu, J. Singh, W. Tian, and D. G. Schlom, Capacitance-Voltage Characteristics of $BiFeO_3$/$SrTiO_3$/GaN Heteroepitaxial Structures, *Appl. Phys. Lett.*, 91, 022909 (2007).
[11]Zhao J, Gong SP, Cheng CF, Zheng ZP, Liu H, and Zhou DX, Nanopowder and Fine-grained Ceramics for Multilayer PTCR Elements, *Key engineering materials*, 368-372 [Part I-II], 453-455 (2008).
[12]H. C. Hamaker, Formation of a Deposit by Electrophoresis, *Trans. Farad. Soc.*, 0, 279 (1940).
[13]N. Ortega, P. Bhattacharya, R. S. Katiyar, P. Dutta, A. Manivannan, M. S. Seehra, I. Takeuchi, and S. B. Majumder, Multiferroic Properties of Pb(Zr,Ti)O3/CoFe2O4 Composite Thin Films, *J. Appl. Phys.*, 100, 126105 (2006).

Multifunctional Oxides

SYNTHESIS AND CHARACTERIZATION OF TERNARY COBALT SPINEL OXIDES FOR PHOTOELECTROCHEMICAL WATER SPLITTING TO PRODUCE HYDROGEN

Sudhakar Shet,[1,2] Yanfa Yan,[1] Todd Deutsch,[1] Heli Wang,[1] Nuggehalli Ravindra,[2] John Turner,[1] and Mowafak Al-Jassim[1]

[1] National Renewable Energy Laboratory, Golden, CO, USA 80401
[2] New Jersey Institute of Technology, Newark, NJ 07102

ABSTRACT

We present an experimental investigation of ternary cobalt spinel oxides for photoelectrochemical water splitting to produce hydrogen. In this study, $Co_{1+\delta}X_{2-\delta}O_4$ (X = Al, Ga, In) thin films were deposited using RF magnetron reactive co-sputtering system. All the thin films were deposited on silver/stainless steel and on quartz substrate, because of the high temperature (800°C) oxide growth. We found that these thin films show excellent stability in solution and good visible light absorption. However, their performance as photoelectrochemical catalyst is limited by the poor transport properties induced by small polaron mobility.

INTRODUCTION

Photoelectrochemical (PEC) decomposition of water by visible light remains the most desirable hydrogen production method for post fossil-fuel energy employment.[1,2] Metal oxide photoelectrodes are of particular interest due to their low cost and relatively high stability in aqueous media. Despite the good catalytic activity of materials such as TiO_2,[3,4] oxides are generally limited by too large band gaps, which fail to absorb a significant fraction of visible light, resulting in poor solar to hydrogen conversion efficiencies under terrestrial conditions. After almost four decades of intensive research, no material has been found to simultaneously satisfy all the criteria required for widespread PEC application:[5–9] (i) low band gap (1.7–2.2 eV), (ii) low resistivity, (iii) low cost, (iv) corrosion resistant, (v) correct alignment of band edges with respect to the water redox potentials.

The majority of PEC oxide research has centered on trying to overcome issues associated with known photoactive materials (*e.g.* TiO_2, Fe_2O_3, WO_3) through doping or alloying. However, the incremental increases in efficiency presently achieved will not be enough to make PEC based hydrogen production commercially viable. We have recently been active in developing a unified approach of material design, synthesis and characterization to explore new classes of multiternary oxide semiconductor photoelectrodes. Similar to the way that the combination of several specific cations is required for multiternary copper oxides to exhibit high temperature superconductivity, it is our desire to combine multiple cations based on their individual chemical and physical properties (*e.g.* structural stability, catalytic activity, light absorption) to tune the material properties for enhanced oxide PEC response.

The potential of transition metal (Fe and Co) based oxide spinels has recently been highlighted *via* initial high-throughput experimental screening by Parkinson *et al.*[10,11] and our subsequent theoretical analysis.[12] Considering the magnitude of the band gaps alone, cobalt oxide (Co_3O_4) is too low (<1.7 eV) for direct PEC water decomposition. However, the spinel structure of Co_3O_4 contains two cation coordination environments, a four-fold tetrahedral site (A, 2^+) and a six-fold octahedral site (B, 3^+) giving the overall formula AB_2O_4. For Co_3O_4, the band edges are determined by a combination of the crystal field split Co 3d states on the octahedral and tetrahedral cobalt sites.[12–14] Isovalent substitution on the spinel cations sites can therefore, in principle, be used to influence both the band edge character and the magnitude of the electronic band gaps and optical absorption.

In this paper, we report experimental synthesis and PEC characterization to perform a comprehensive examination of how the pertinent chemical and physical properties of ternary cobalt spinels can be tailored towards those of an ideal PEC catalytic photoelectrode for solar driven hydrogen production. In particular, we explore each member of the CoX_2O_4 (X = Al, Ga, In) chemical series. Aluminum can be viewed as an ionic spectator in the $CoAl_2O_4$ lattice, which preserves overall charge neutrality, but contributes little to the density of states or optical absorption. However, the substitution of heavier group 13 cations helps reduce the electronic band gap and increase optical absorption through combination of the higher binding energy cation s states and reduced Co d–d crystal field splitting. Unfortunately, while we demonstrate that the band gaps can be tuned through a large range, the intrinsic charge transport properties result in poor PEC performance due to inefficient electron–hole extraction rates and high resistivity associated with small polaron carriers.

EXPERIMENTAL

$Co_{1+\delta}X_{2-\delta}O_4$ (X = Al, Ga, In) thin films were grown on Ag-coated stainless steel plates (Ag/SS) and quartz glasses for PEC measurement and optical characterization, respectively, using an RF magnetron reactive co-sputtering system. Here Ag/SS was required as the substrate, because of the high temperature oxide growth (800 °C). Two sputter guns were used for the Co–Al–O and Co–In– O syntheses, where the Co_3O_4 target was fixed and the secondary target was changed to Al and In_2O_3 for $CoAl_2O_4$ and $CoIn_2O_4$, respectively. The Co–Ga–O films were deposited using the different amounts of Ga_2O_3 powder on the Co_3O_4 target. The distance between the sputter guns and substrate was fixed at 11 cm. The substrates were rotated during deposition for enhanced uniformity and the working pressure was 1.22 Pa. The sputtering ambient environment was mixed Ar and O_2 (O_2:Ar = 5:1) and the RF power of each sputter gun was varied to obtain appropriate chemical stoichiometry. All of the samples were controlled to exhibit similar film thicknesses on the order of 500 nm, as measured by stylus profilometry. Structural characterization was performed by powder X-ray diffraction (XRD) measurements, using an X-ray diffractometer (XGEN-4000, SCINTAG, Inc.) operated with a Cu Kα radiation source at 45 kV and 37 mA.

PEC measurements were performed in a three-electrode cell with a flat quartz window to facilitate illumination of the photoelectrode surface.[15-31] The films (active area: 0.24 cm^2) were used as the working electrodes. A Pt sheet (area: 10 cm^2) and an Ag/AgCl electrode (with saturated KCl solution) were used as the counter and reference electrodes, respectively. A 0.5 M NaOH basic aqueous solution (pH ~13) was used as the electrolyte.[15-31] The PEC response was measured with a fiber-optic illuminator (150 W tungsten-halogen lamp) processed through a UV/IR cut-off filter (cut-off wavelengths: 350 and 750 nm). Light intensity with the UV/IR filter was 80 mW cm^{-2} as measured by a photodiode power meter[32-34]. The PEC response was then measured with respect to time under chopped light on/off illumination at constant applied potential bias. Temperature effects on the dark current, due to possible sample heating on illumination, were found to be negligible.

RESULTS AND DISCUSSION

The representative experimental optical data for each ternary cobalt oxide material, synthesized under a range of sputtering conditions, are shown in Figure 1. From both the measured optical absorption coefficient and a fit to the Tauc relation for direct transitions, tailoring of the optical bandgap from 1.5 to 2.25 eV can be observed. Taking into account that the separation between the H_2/H_2O reduction and O_2/H_2O oxidation potentials is 1.23 eV, and that due to the presence of unavoidable losses (e.g., component resistance, electron-hole recombination) an additional overpotential is required (raising the optimal voltage to 1.7 eV), this optical bandgap range appears quite promising for PEC water-splitting application.

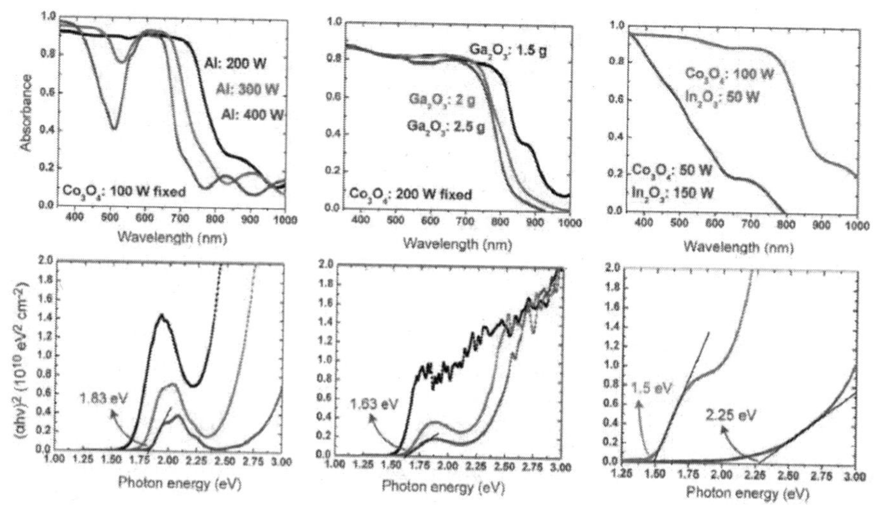

Figure 1. Measured optical absorption spectra (upper) and direct-bandgap fits (lower) of $Co_{1+}X_{2-}O_4$ compounds as a function of sputtering conditions: (left) X = Al, (centre) X = Ga and (right) X = In.

For $CoAl_2O_4$, the measured absorption features are consistent with previous reports.[35,36] The low-energy absorption peak for $CoAl_2O_4$ originates from the spin-allowed, parity-forbidden Co $^4A_2$ $^4T_1$ transitions, whereas the drop in absorption coefficient at higher energies results from a range of low-intensity spin-forbidden transitions. The shift to longer wavelengths with lower Al sputtering power can be understood as a transition toward the inverse Co_2AlO_4 spinel, which is known to possess a lower bandgap.[37] For $CoGa_2O_4$, the strong absorption profile is red-shifted to around 2.25 eV and begins to overlap with the low-energy absorption feature; this overlap is even more pronounced for $CoIn_2O_4$ at high Co_3O_4 sputtering power. At high In_2O_3 sputtering power, the Co–In–O samples exhibit the high levels of visible transmission expected from mixed-phase Co-substituted In_2O_3, i.e., the ternary composite is not fully formed. The time-dependent PEC response of each material was investigated under chopped-light illumination at a constant applied bias potential (*vs.* Ag/AgCl). The cathodic currents of $CoAl_2O_4$ and $CoGa_2O_4$ were found to increase on illumination, as in Figure 2.

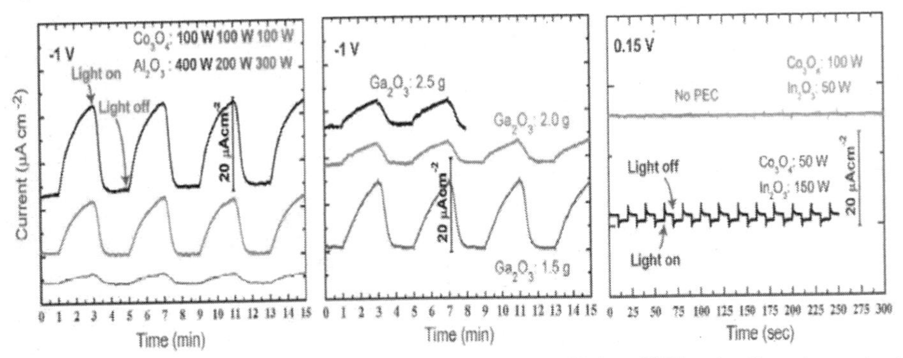

Figure 2. Time-dependent PEC response under light-on/light-off illumination at constant applied voltage for (left) Al, (center) Ga, and (right) In ternary cobalt oxides. For the Al and Ga spinels, illumination induces an increase in the background cathodic current (p-type response), whereas for In a small increase in the anodic current is observed (n-type response).

This p-type PEC response suggests the presence of intrinsic hole carriers, which prior calculations have identified as cation vacancies.[38] However, poor photocurrents on the order of 20 $\mu A\ cm^{-2}$ are observed in both cases (currents on the order of 10 $mA\ cm^{-2}$ will be required from commercially viable materials). It is worth noting that the PEC response of $CoAl_2O_4$ is better than that of $CoGa_2O_4$, which will be discussed in more detail below. Unfortunately, all synthesized $CoIn_2O_4$ films failed to exhibit any significant PEC response. Indeed, the weak PEC response varied from p-type to n-type with different samples, but no significant photocurrent was observed in either case.

In addition to the low generated photocurrents, the second discouraging trend emerging from Figure 2 is the long relaxation times between sample illuminations. For the majority of PEC materials, the recovery time on the removal of light is on the order of seconds or less; however, for these materials, it is on the order of minutes. This implies poor carrier transport kinetics, originating from confined electrical carriers (i.e., heavy hole effective masses). A more-detailed comparison between the relaxation times of $CoAl_2O_4$ and $CoGa_2O_4$ is shown in Figure 3. Ideally, the photocurrent decay for the chopped-light PEC system will closely follow a square wave; however, the response deviates greatly here.

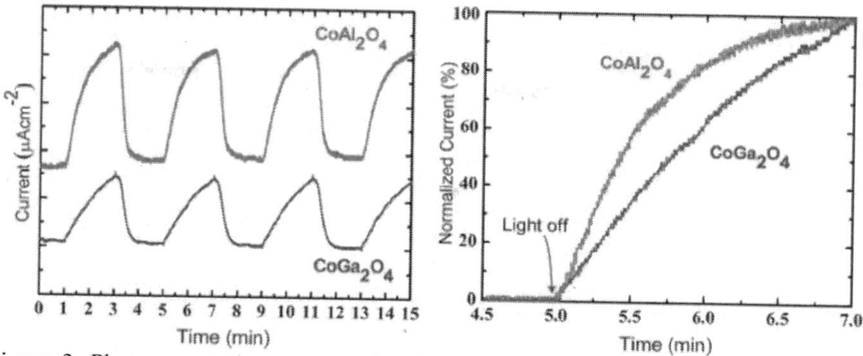

Figure 3. Photocurrent decay comparison between the $CoAl_2O_4$ and $CoGa_2O_4$ samples with the same bias voltage (−1 V).

Direct comparison of the normalized current of the Al and Ga ternaries in Figure 3 shows much faster decay in the former, indicating better carrier transport kinetics; hence, $CoAl_2O_4$ exhibits marginally improved PEC response. One encouraging outcome is that both materials exhibited no evidence of corrosion in solution for sustained PEC testing periods.

Based on the initial electronic structure analyses, it was anticipated that the Co–In spinel may possess both the lowest bandgap and highest n-type conductivity (through the presence of the In 5s conduction states). However, although the synthesized In-based ternary oxides did not exhibit any significant photocurrent on illumination, the n-type PEC response time was much shorter, indicating the beneficial influence of the delocalized In 5s orbitals in the lower conduction band. To explore the origin of the performance failure in more detail, we first measured the dark currents without illumination, as shown in Figure 4. Even for small applied potentials, the dark current is large, indicating that the film is not an intrinsic semiconductor; in fact, it is closer to a semimetallic state.

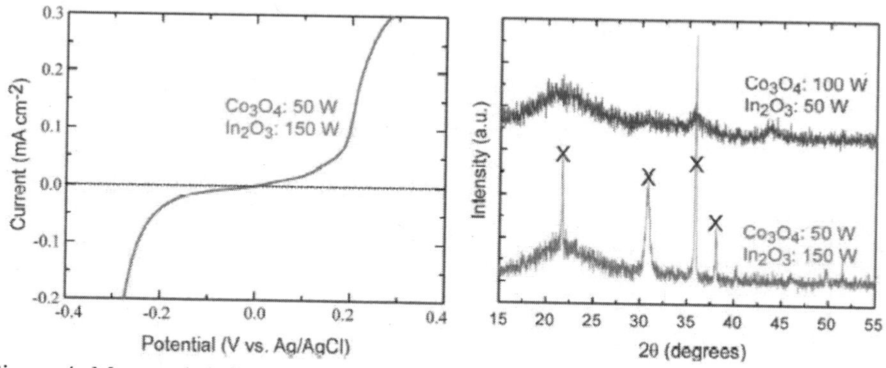

Figure 4. Measured dark current–voltage curve and powder XRD data for synthesized Co–In systems. The crosses in the XRD spectrum correspond to reflections associated with bixbyite In_2O_3.

The XRD curves are also shown in Figure 4. The measurements clearly show that the deposited films are not pure $CoIn_2O_4$, but that they undergo significant phase segregation into Co_3O_4 and In_2O_3 (this has been confirmed by TEM analysis). Taking into account that In_2O_3 itself exhibits degenerate electron conduction behavior as an n-type transparent conducting oxide[39-41], the presence of In_2O_3 in the film will contribute to both the inferior PEC response and high dark-current levels. Within the limitations of our co-sputtering system, we could not succeed in synthesizing a homogeneous $CoIn_2O_4$ film.

CONCLUSIONS

We have presented a experimental synthesis and characterization of a series of ternary cobalt spinel oxides. While these materials combine excellent stability in solution and good visible light absorption properties, their performance as photoelectrochemical catalysts is limited by the poor transport properties induced by small polaron mobility. This is an enhancement of the same effect that limits the performance of Fe_2O_3 based photocatalysts. While incremental improvements in charge carrier extraction efficiencies may be possible through the alteration and tuning of growth conditions and crystal morphology, it is unlikely that the limitations of late 3d transition metal oxides will be overcome to provide the performance required for commercial hydrogen production.

ACKNOWLEDGEMENTS

This work was supported by the U.S. Department of Energy through the UNLV Research Foundation under Contract # DE-AC36-99-GO10337.

REFERENCES

[1]M. Z. Jacobson, W. G. Colella and D. M. Golden, *Science*, **308**, p.1901, 2005.
[2]J. A. Turner, *Science*, **285**, p.687, 1999.
[3]A. Fujishima and K. Honda, *Nature*, **238**, p.37, 1972.
[4]A. Fujishima, K. Kohayakawa and K. Honda, *J. Electrochem. Soc.*, **122**, p.1487, 1975.
[5]B. D. Alexander, P. J. Kulesza, I. Rutkowska, R. Solarska and J. Augustynski, *J. Mater. Chem.*, **18**, p.2298, 2008.
[6]T. Bak, J. Nowotny, M. Rekas and C. C. Sorrell, *Int. J. Hydrogen Energy*, **27**, p.991, 2002.
[7]N. S. Lewis, *Nature*, **414**, p.589, 2001.
[8]O. Khaselev and J. A. Turner, *Science*, **280**, p.425, 1998.
[9]A. Kudo and Y. Miseki, *Chem. Soc. Rev.*, **38**, p.253, 2009.
[10]M. Woodhouse, G. S. Herman and B. A. Parkinson, *Chem. Mater.*, **17**, p.4318, 2005
[11]M. Woodhouse and B. A. Parkinson, *Chem. Mater.*, **20**, p.2495, 2008.
[12]A. Walsh, S.-H. Wei, Y. Yan, M. M. Al-Jassim, J. A. Turner, M. Woodhouse and B. A. Parkinson, *Phys. Rev. B*, **76**, p.165119, 2007.
[13]K. M. E. Miedzinska, B. R. Hollebone and J. G. Cook, *J. Phys. Chem. Solids*, **48**, p.649, 1987.
[14]Y. Jugnet and T. M. Duc, *J. Phys. Chem. Solids*, **40**, p.29, 1979.
[15]S. Shet, K. –S. Ahn, Y. Yan, T. Deutsch, K. M. Chrusrowski, J. Turner, M. Al-Jassim, and N. Ravindra, *J. Appl. Phys*. **103**, p.073504, 2008.

[16]S. Shet, K. –S. Ahn, T. Deutsch, H. Wang, N. Ravindra, Y. Yan, J. Turner, M. Al-Jassim, *J. Mater. Research* **25,** (2010) 69 Doi: 10.1557/JMR.2010.0017.
[17]K.–S. Ahn, S. Shet, T. Deutsch, C.-S. Jiang, Y. Yan, M. Al-Jassim, and J. Turner, *J. Power Sources* **176,** p.387, 2008.
[18]S. Shet, k.-S. Ahn, T. Deutsch, H. Wang, N. Ravindra, Y. Yan, J. Turner, M. Al-Jassim, *J. Power Sources* **195**, p.5801, 2010.
[19]H. Wang, T. Deutsch, S. Shet, K. Ahn, Y. Yan, M. Al-Jassim and J. Turner, *Solar Hydrogen and Nanotechnology IV, SPIE*, Nanoscience + Engineering, p.7408, 2009.
[20]S. Shet, K. Ahn, N. Ravindra, Y. Yan, T. Deutsch, J. Turner, M. Al-Jassim, *Proceedings of the Materials Science & Technology*, p.219, 2009.
[21]S. Shet, K. Ahn, N. Ravindra, Y. Yan, T. Deutsch, J. Turner, M. Al-Jassim, *Proceedings of the Materials Science & Technology*, p.277, 2009.
[22]K.–S. Ahn, Y. Yan, S. Shet, T. Deutsch, J. Turner, and M. Al-Jassim, *Appl. Phys. Lett.* **91,** p. 231909, 2007.
[23]K.-S. Ahn, Y. Yan, M.-S. Kang, J.-Y. Kim, S. Shet, H. Wang, J. Turner, and M. Al-Jassim, *Appl. Phys. Lett.* **95,** p.022116, 2009.
[24]S. Shet, K. –S. Ahn, H. Wang, N. Ravindra, Y. Yan, J. Turner, M. Al-Jassim, *J. Mater. Science,* (2010) DOI 10.1007/s10853-010-4561-x.
[25]Y. Yan, K. Ahn, S. Shet, T. Deutsch, M. Huda, S. Wei, J. Turner, M. Al-Jassim, Solar Hydrogen and Nanotechnology II. Edited by Guo, Jinghua. *Proceedings of the SPIE*, **6650**, p.66500H, 2007.
[26]S. Shet, K. Ahn, N. Ravindra, Y. Yan, T. Deutsch, J. Turner, M. Al-Jassim, Materials Science & Technology 2009, *Ceramic Transactions volume*, (2010) in press.
[27]K. Ahn, S. Shet, Y. Yan, J. Turner, M. Al-Jassim, N. M. Ravindra, *Proceedings of the Materials Science & Technology*, p.901, 2008.
[28]K.-S. Ahn, Y. Yan, S. Shet, K. Jones, T. Deutsch, J. Turner, M. Al-Jassim, *Appl. Phys. Lett.* **93,** p.163117, 2008.
[29]S. Shet, K. –S. Ahn, N. Ravindra, Y. Yan, J. Turner, M. Al-Jassim, *J. Materials* **62**, p.25, 2010.
[30]K. Ahn, S. Shet, T. Deutsch, Y. Yan, J. Turner, M. Al-Jassim, N. M. Ravindra, *Proceedings of the Materials Science & Technology*, p.952, 2008.
[31]S. Shet, K. Ahn, T. Deutsch, Y. Yan, J. Turner, M. Al-Jassim, N. Ravindra, *Proceedings of the Materials Science & Technology*, p.920, 2008.
[32]L. Chen, S. Shet, H. Tang, H. Wang, Y. Yan, J. Turner, and M. Al-Jassim, *J. Mater. Chem.*, 2010, **20**, 6962-6967, DOI: 10.1039/c0jm01228a.
[33]L. Chen, S. Shet, H. Tang, H. Wang, Y. Yan, J. Turner, and M. Al-Jassim, *J. Appl. Phys*, **108**, 043502 (2010); doi:10.1063/1.3475714
[34]S. Shet, K.-S. Ahn, Y. Yan, N. M. Ravindra, T. Deutsch, J. Turner, M. Al-Jassim, submitted to Journal of Thin Solid films.
[35]U. L. Stangar, B. Orel and M. Krajnc, *Journal of Sol-Gel Science and. Technology*, **26**, p.771, 2003.
[36]M. Zayat and D. Levy, *Chemistry of Materials*, **12**, p.2763, 2000.
[37]A. Walsh, S. -H. Wei, Y. Yan, M. M. Al-Jassim, J. A. Turner, M. Woodhouse and B. A. Parkinson, *Physical Review B*, **76**, p.165119, 2007.
[38]A. Walsh, Y. Yan, M. M. Al-Jassim and S.–H. Wei, *Journal of Physical Chemistry*,

C, **112**, p.12044, 2008.
[39] T. Minami, H. Nanto, S. Takata, *Japanese Journal of Applied Physics*, **24**, p.L605, 1985.
[40] K. H. Kim, K. C. Park, D. Y. Ma, *Journal of Applied Physics*, **81**, p.7764, 1997.
[41] K. J. Kim, Y. R. Park, *Applied Physics Letters*, **78**, p.75, 2001.

Author Index